THE GREAT RIFT VALLEY

BEING

THE NARRATIVE OF A JOURNEY TO

MOUNT KENYA AND LAKE BARINGO

WITH SOME ACCOUNT OF
THE GEOLOGY, NATURAL HISTORY, ANTHROPOLOGY
AND FUTURE PROSPECTS OF
BRITISH EAST AFRICA

J. W. GREGORY

Routledge
Taylor & Francis Group
LONDON AND NEW YORK

First published 1968 by Frank Cass & Co. Ltd.
First edition 1896

Published 2021 by Routledge
2 Park Square, Milton Park, Abingdon, Oxon OX14 4RN
605 Third Avenue, New York, NY 10017

Routledge is an imprint of the Taylor & Francis Group, an informa business

ISBN 13: 978-0-7146-1812-8 (hbk)

DOI: 10.4324/9781315827988

CASS LIBRARY OF AFRICAN STUDIES

TRAVELS AND NARRATIVES

No. 38

Editorial Adviser: JOHN RALPH WILLIS

A. Gregory, delt.

A CLUSTER OF INSECTS GROUPED TO RESEMBLE A FLOWER SPIKE.

For description see pp. 273–275.

THE

GREAT RIFT VALLEY

BEING THE NARRATIVE OF A JOURNEY TO MOUNT KENYA AND LAKE BARINGO

WITH SOME ACCOUNT OF THE GEOLOGY, NATURAL HISTORY, ANTHROPOLOGY, AND FUTURE PROSPECTS OF BRITISH EAST AFRICA

By J. W. GREGORY

D.SC. (LOND.)

F.G.S., F.R.G.S., F.Z.S.

OF THE BRITISH MUSEUM (NATURAL HISTORY)

'Or join the caravan in quest of scenes
New to the eye and changing every hour.'

JOHN ARMSTRONG, *The Art of Preserving Health*, 1744.

WITH MAPS AND ILLUSTRATIONS

LONDON

JOHN MURRAY, ALBEMARLE STREET

1896

To the Memory of

MY MOTHER

PREFACE

In the Presidential Address to the British Association at its last meeting, Mr. Thiselton-Dyer quoted a letter from Sir John Kirk, deploring the neglect of opportunities for scientific usefulness by British residents in West Africa. "Such chances," he said, "never will occur again, for roads are now being made and ways cut in the jungle and forest, and you have at hand all sorts of trees level on the ground ready for study. These bring down with them orchids, ferns, and climbers of many kinds, including rattan palms, etc. But, excellent as are the officers who devote their energy to thus opening up this country, there is not one man who knows a palm from a dragon-tree, so the chance is lost. Strange to say, the medical men of the Government service know less and care less for Natural History than the military men, who at least regret they have no training or study to enable them to take an intelligent interest in what they see around them."

Having felt the same regret on the other side of the continent, I had already divided this book into two parts—one devoted to a narrative of a journey in Eastern British East Africa, and the other to a general account of the natural history of the country visited. I have tried to make the latter sufficiently simple to be intelligible to most readers, and yet of use to travellers and residents in East Africa, by indicating the nature of the problems on which information is desired, and by suggesting lines of observation and research.

This part of the book is entitled "*Eastern* British East Africa," as attention is almost entirely confined to the part of the protectorate along, and to the east of, the Great Rift Valley. This area has been greatly neglected in comparison with the country to the west, which has been repeatedly described since it was first reached by Speke and Grant from the south, and by Baker from the north. The literature of the Victoria Nyanza basin (including Uganda) is voluminous, while that of the plateau region between it and the sea is scanty. References to most of the works upon it will be found in the text,[1] but it may be as well here to mention a few of the most generally accessible authorities. The principal accounts of exploration in this district are Krapf's *Travels in East Africa* (1860), Thomson's *Through Masai Land* (1885), and von Höhnel's *Discovery of Lakes Rudolf and Stephanie* (1892, English ed. 1894). Lugard's *Rise of our East African Empire* (1893), though dealing mainly with Nyasaland and Uganda, is indispensable to all students of this country, and gives the best account of its actual politics; while M'Dermott's *British East Africa*, and the Blue-books on the Uganda Railway, give its documentary politics. Ravenstein's twelve-sheet map is a complete compilation of all information available at the date of its publication (1889). The history of the growth of British influence in this region is told with masterly clearness and terseness in Keltie's *Partition of Africa* (1893 and 1895). For comparison of Kenya with Kilima Njaro, the other snow-clad peak of East Africa, reference may be made to Hans Meyer's *Across East African Glaciers* (English translation, 1891).

[1] See especially pp. 7-9, 214, 280, 316.

Of scientific literature on British East Africa there is unfortunately little to record. There is nothing which can compare with the magnificent series of works issued in description of German East Africa, such as Stuhlmann's *Mit Emin Pascha im Herz von Afrika*, Oscar Baumann's *Zur Nil-Quellen*, and the elaborate monographs in the volumes of *Pflanzenwelt Ost-Afrikas*, and *Die Thierwelt Ost-Afrikas*. The anthropological chapters in Stuhlmann and Baumann, and the botanical papers of Volkens, show that these authors unite the learning of the scientific specialist with the courage of the pioneer. The only work on Tropical Africa in English that can compare with these in scientific accuracy is *The Flora of Tropical Africa*, the last part of which was published in 1877; but this barely mentions British East Africa. The last 100 pages (vol. iii. pp. 425-525) include abundant references to Abyssinia, Somaliland, Monbuttu, Karagwe, Kilima Njaro, etc.; but of the 214 species therein described, British East Africa only contributed four.

Dr. H. R. Mill has remarked (*Knowledge*, Jan. 1896, p. 2) that in pioneer exploration England has led the way, but that in scientific geography we have always been beaten by our German rivals. The history of the exploration of Equatorial Africa is one to which Englishmen can look back with feelings of such just pride, that we may ungrudgingly admit the superiority of German scientific work in this region.

The Expedition of which the narrative is given in this volume was undertaken in 1892 and 1893. The delay in writing the account of it is due partly to my having returned —thanks to African fever—with lessened powers of work, and partly to arrears of work which had accumulated during my absence. Hence the book has had to be written in scraps at odd hours, generally at the end of days devoted to work on quite different subjects. This must be my excuse for the lack of uniformity and errors in style, of which I am sadly conscious. This delay explains that most of the book was written before the administration of the British East African Company had been superseded by a direct protectorate. The references to the Masai in the last chapter were printed before their recent massacre of a thousand men on the Uganda road showed that their capacity for mischief is not yet destroyed.

It may be explained, in reference to the spelling of native words and place-names, that the ordinary geographical rule has been followed as far as practicable—vowels being pronounced as in Italian, and consonants as in English. The vowels, therefore, are used as follows :—

a	as	*a*	in	father
ai	,,	*i*	,,	tight
ao	,,	*ow*	,,	how
e	,,	*a*	,,	fate
i	,,	*ee*	,,	fee
o	,,	*o*	,,	Tom
u	,,	*u*	,,	flute

An apparent discrepancy occurs in the spelling of personal specific names, these being sometimes spelt with small letters and sometimes with capitals (thus *Hemidactylus brooki* and *Senecio Johnstoni*). In deference to the wishes of my botanical colleagues, I have accepted capital letters in the case of plants.

In conclusion, there remains the pleasant task of expressing my sincere thanks for the helpful encouragement of friends at home, and for the unstinted and ever ready assistance of European residents in East Africa. I feel grateful to so many, that it is invidious to make selections ; but I will not deny myself the pleasure of thanking some because I cannot name all. To Sir William Flower, K.C.B., and Dr. Henry Wood-

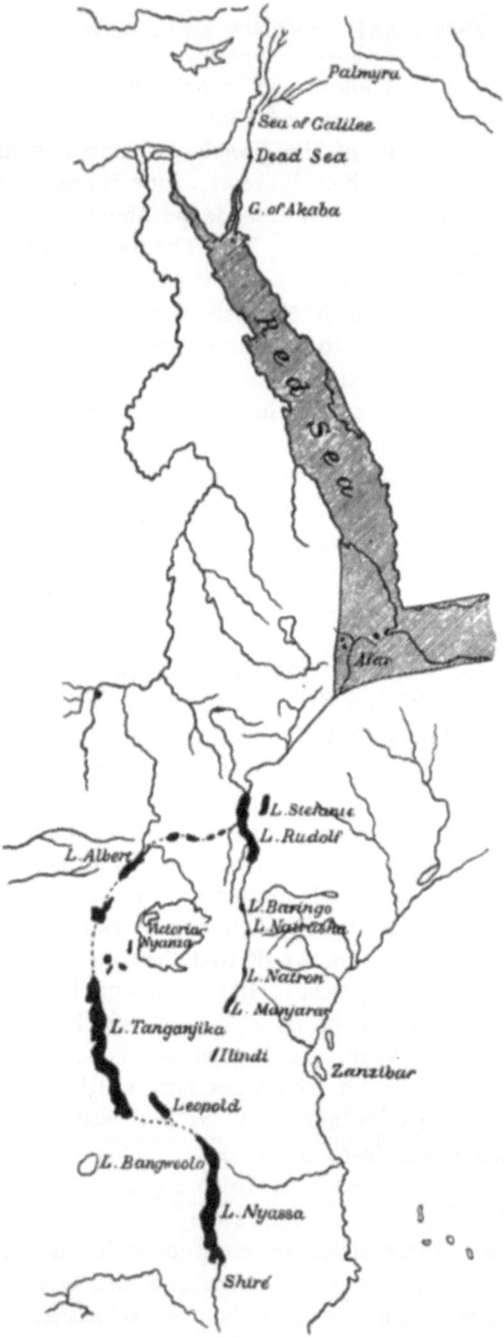

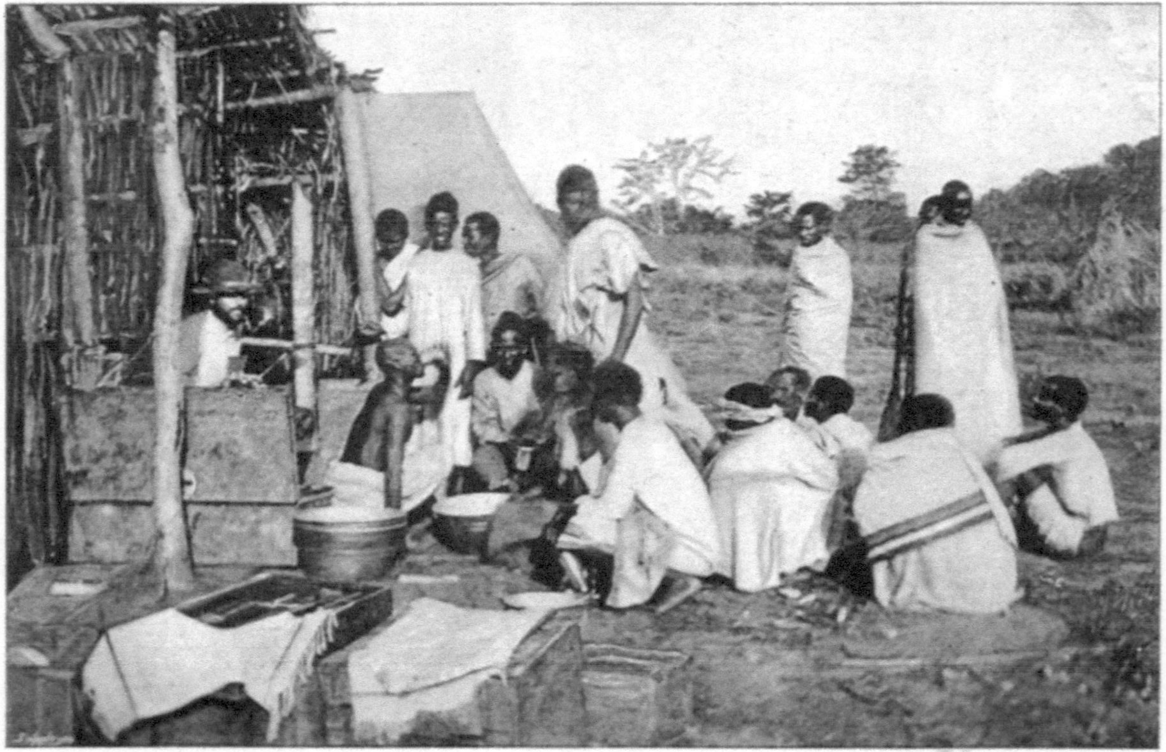

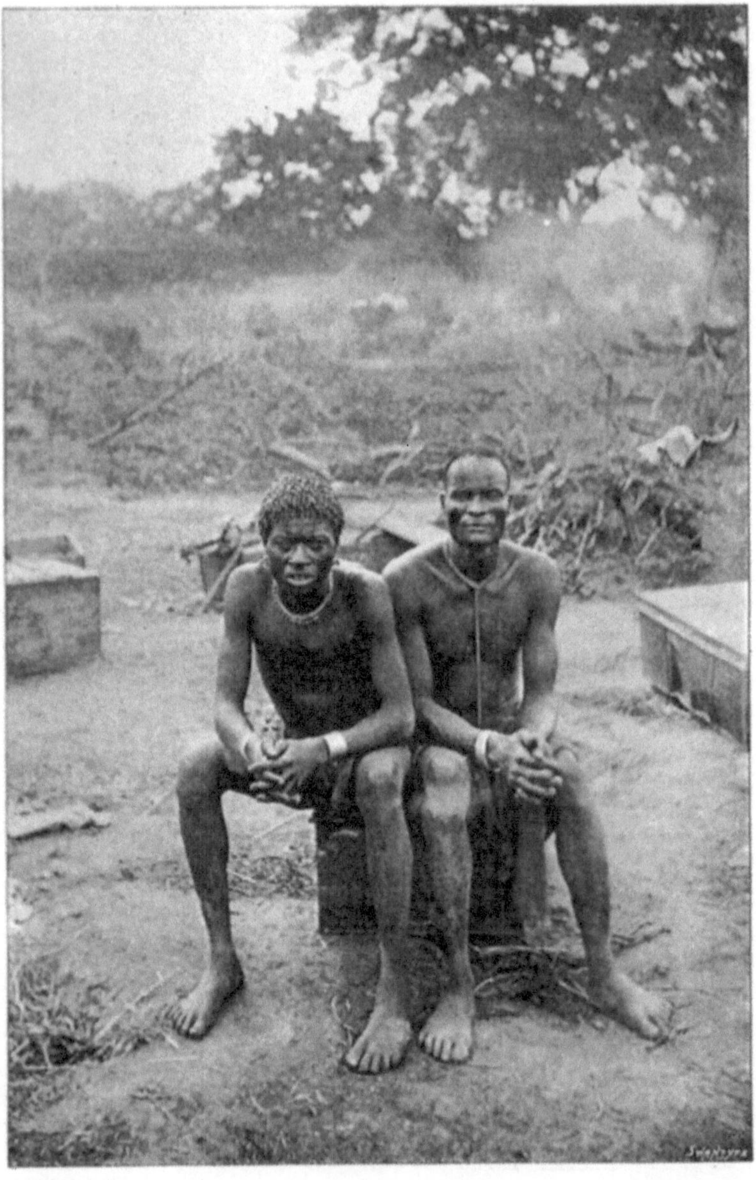

ward, F.R.S., I owe, amongst other things, the recommendation upon which the Trustees of the British Museum gave me the necessary leave of absence. To Sir Henry Tichborne, Captain W. H. Harris, and Mr. J. Benett-Stanford, I am grateful for many acts of kindness when with them in East Africa, and for a generous present of stores from those of the abortive expedition. I must express my best thanks to Mr. J. R. W. Piggott and Dr. Macdonald for hospitality during my two stays at Mombasa, and to Captain Rogers, Mr. C. W. Hobley, Mr. George Wilson, Mr. Ainsworth of Machakos, Mr. Watson of Kibwezi, Mr. Hall of Fort Smith, for help which removed many difficulties from my path. I must also thank those of my colleagues who have worked out the collections brought home, viz. Dr. A. Günther and Mr. G. A. Boulenger, Dr. A. G. Butler, Mr. E. A. Smith, Dr. R. Bowdler Sharpe, Professor F. Jeffrey Bell, and Mr. Oldfield Thomas, who have described respectively the fish and reptiles, the lepidoptera, mollusca, birds, land crustacea, and mammalia. New species of plants have been described by Miss A. L. Smith, Mr. Spencer Moore, Mr. E. G. Baker, Mr. A. B. Rendle, and Dr. Rudolf Schlechter, and lists have been contributed by Mr. J. Britten and Mr. A. Gepp. I must also thank Mr. G. R. M. Murray and Mr. C. E. Fagan for their interest in the Expedition, and Mr. C. J. Gahan for his help in connection with the questions concerning *Flata nigrocincta* (pp. 273-275), and Mr. E. G. Baker, who has read through the proof of Chapter XV.

I have also received much valuable assistance in the preparation of this book, which I gratefully acknowledge. Mr. J. Benett-Stanford has kindly lent me the negatives from which Plates Nos. II.-VII. and XVIII.-XX. have been prepared. Mr. Hallam Murray has taken great trouble in making from my rough sketches the admirable drawings of the Kedong and Höhnel Valleys, and the scenes by the snout of the Lewis Glacier (Nos. VIII. XIII. XV. and XVI.) He has also redrawn, and thus rendered available for process reproduction, the view of Kenya from the Kapte Plains, sketched by Mr. Ainsworth. For permission to use the photograph of the southern end of the Victoria Nyanza, I have to thank Mr. Gedge. For the *Frontispiece*, and Plates Nos. I. IX. X. XI. XIV. and XVII., I am indebted to my wife, who has taken great pains accurately to interpret my rough sketches; I also owe her much assistance in the revision of proofs and compilation of indexes. For permission to reprint Map No. II. and Figs. 3, 6, 11, and 13 from the *Geographical Journal*, and Fig. 4 from the *Quarterly Journal of the Geological Society*, I have to thank the Councils of the Geographical and Geological Societies. The conditions under which the book has been written are partly responsible for the defects in its style; that these are not more serious than they are I owe to Dr. W. R. Gowers, F.R.S., who has kindly read through nearly the whole of the manuscript, and given me the benefit of many most useful criticisms and suggestions. I must also thank Mr. Murray for his kind interest and advice.

The pleasure of looking back to these many acts of kindness is, however, lessened by the fact that the mortality in British East Africa has sadly shortened the list of those to whom I can express thanks. Bird Thompson and Rae of Witu, J. Bell Smith of Melindi, Edmonds of Borabini, the two brothers Dick of Mombasa, Charters of Kibwezi, and Purkiss of Fort Smith—a terribly high proportion of the few men who held British East Africa in 1892-93—have all since passed away. In such countries, as Carlyle said, "how much European heroism has to spend itself in obscure battle, to sink in mortal agony, before the jungles, the putrescences, and waste savageries can become arable, and the devils be in some measure chained there." Such sacrifices ought only to confirm British interest in the country and deepen our sense of responsibility. For we owe it to those who have perished, "worn-down swiftly in frightful travail, chaining the devils which are manifold," to make sure that their labours be not wasted nor their lives laid down in vain.

J. W. GREGORY.

3 Aubrey Road, Campden Hill, W.
January 1896.

CONTENTS

PART III

EASTERN BRITISH EAST AFRICA

LIST OF ILLUSTRATIONS

LIST OF ILLUSTRATIONS IN TEXT

MAPS (*at the End*)

[1] Not 1 : 250,000 as stated on map.

INTRODUCTION

In the early days of African exploration the interest of the geographical problems was so absorbing that but little attention was paid to those of other branches of natural science. This restricted range of interest was originally due to the fascination of the problems of the sources of the Nile and the course of the Congo, the exploration of the great lakes, and the discovery of the equatorial snow-capped mountains. It survived, however, owing to the once prevalent belief that Tropical Africa would never yield its fair share of help in the advancement of science. It was thought that all we had to expect from the exploration of this region was the record of new topographical facts and the removal of the blank spaces from our maps. Some of the problems its natural history presented to us were regarded as too complex to be solved with the available methods of inquiry. For example, the extent to which the tribes have intermarried and intermingled, have acquired new languages and lost all knowledge of their own, has so confused the race characteristics, that many authorities have sadly confessed it to be absolutely impossible to place African anthropology on a scientific basis. The evidence of this region on the remaining subjects was, on the other hand, regarded as too simple and monotonous to affect the development of scientific principles. Thus, when it was reported that from whatever side approached, in whatever direction traversed, the whole interior of the continent consisted of one vast expanse of gneiss and schist, geologists were ready, with Sir Roderick Murchison, to dismiss Africa south of the Sahara as a continent without a history.

So long has this view lasted, that even as late as 1891 we find Professor Henry Drummond saying: "Finally, the thing about the geology of Africa that strikes one as especially significant is, that throughout this vast area just opening up to science there is nothing new—no unknown force at work; no rock strange to the petrographer; no pause in denudation; no formation, texture, or structure to put the law of continuity to confusion."[1]

No doubt it is true that in Equatorial Africa the study of ethnography is attended with special difficulties, the flora of vast areas is poor in species, the rocks are monotonous in character, and the paleontological record is a blank. Nevertheless the old view has been abandoned. African exploration is now not undertaken only for the sake of obtaining fresh topographical details, of undergoing adventures with cannibals, and of "potting" big game. Instead of these we look to it to supply us with information on some of the doubtful chapters in the geological history of Europe, to give us insight into unknown or unfamiliar methods in the earth's great workshop, and to exert an important influence on the development of scientific principles.

Moreover, when studied from the modern point of view, the topographical details themselves acquire a fresh value. The investigation of the East African lake system has been the branch of exploration in which the widest general interest has been taken, ever since the native reports of the existence of the great inland seas were verified by Burton and Speke's discovery of Tanganyika (1857) and Livingstone's discovery of Nyasa (1859).

Probably the most interesting series of journeys in the records of African travel were those made to the lakes of the great eastern plateau, including as it does the many efforts to settle the long controversy as to the position of Tanganyika in the African river system: Speke's two journeys to the Nyanza; the circumnavigation of this lake by which Stanley proved its unity; Baker's visit to the Albert Nyanza; the expeditions in which Fischer discovered Lakes Natron and Naivasha, and Thomson demonstrated the isolation of Baringo; Teleki and von Höhnel's exploration of Basso Narok and

[1] H. Drummond, *Tropical Africa*, 4th edition (1891), p. 199.

THE SOUTHERN SHORES OF THE VICTORIA NYANZA. (After Gedge,)

THE WESTERN WALL OF TANGANYIKA. (After Giraud.)

Basso Ebor (Lakes Rudolf and Stephanie); and Baumann's recent march past Manyara and Eiassi.

The determination of the principal features in the topography of these lakes has, however, done more to stimulate than to satisfy our desire for knowledge about them. Mr. F. Galton,[1] in the discussion on Mr. Thomson's paper on his journey through Masailand, pointed out that the great depression or trough in which Naivasha and Baringo lie, is part of one "which begins with the Dead Sea, extends down the Red Sea, and ends at Tanganyika." This view has often been repeated, but it remained as a hazy speculation until Professor Suess of Vienna recently gave it scientific expression. To understand this let us examine a map of the lake system (Fig. 1) of East Africa. One of the first points shown by it is that the lakes are developed according to two absolutely different types. Some are rounded in shape, as the Nyanza, and others are long and narrow, as Tanganyika and the Nyasa. If we examine views of these lakes we find that the round lakes have low shelving shores, and that the long ones lie, like fiords, between high precipitous cliffs. The illustrations on the first plate show this contrast: the upper shows the southern shore of the Nyanza, and the lower, one of the bounding walls of Tanganyika. If we read descriptions of the lakes we learn the same facts. Thus Thomson[2] remarks that, "unlike most other African lakes, the Nyanza is not bounded by ranges of mountains. The ground descends gradually to its shores, and peacefully the water laps the muddy and marshy beach."

The map, moreover, shows us that these two types of lakes are not distributed haphazard, but on a definite plan. The long fiord-like lakes occur on two lines, which pass one on either side of the Nyanza and meet at Basso Narok (Lake Rudolf). Thence the line continues northward as a long strip of low land, dotted with lakes and old lake basins, and sinking in places below the level of the sea. This extends to the southern end of the Red Sea, which repeats the structure of these fiord-like lakes on a larger scale: it is long and narrow and, excluding some strips of coast deposits, has high precipitous shores. From its northern end the Gulf of Akaba leads to another valley with similar characteristics, and from

[1] *Proc. Roy. Geog. Soc.* new ser. vol. vi. (1884), p. 711. [2] *Ibid.* p. 707.

FIG. 1.—Map of East African Lake-Chain. (After Suess.)

this the Dead Sea and Jordan valley continue the same type of geographical structure, till it ends on the plains of northern Syria.

From the Lebanons, therefore, almost to the Cape there runs a valley, unique both on account of the persistence with which it maintains its trough-like form, throughout the whole of its course of 4000 miles, and also on account of the fact that scattered along its floor is a series of over thirty lakes, of which only one has an outlet to the sea.

This valley and its lake-chain are so different from anything else on the surface of the earth, that it is natural to ask whether different portions of it have been formed independently, or whether it was all formed at the same time and by the same process. The final answer to this question must be given by geology, but history affords us some useful hints. All along the line the natives have traditions of great changes in the structure of the country. The Arabs tell us that the Red Sea is simply water that did not dry up after Noah's deluge. The Somali say that when their ancestors crossed from Arabia to Africa there was a land connection between the two, across the straits of Bab el Mandeb. The natives of Ujiji, at the southern end of the line, have a folklore that goes back to the time when Lake Tanganyika was formed by the flooding of a fertile plain, rich in cattle and plantations. And at the northern end of the valley we have the accounts of the destruction of the towns of Sodom and Gomorrah.

There is geological evidence to show that great earth-movements have happened along this Rift Valley, as it may be termed, at a recent date, which makes it distinctly probable that these traditions are recollections of the geographical changes.

The structure of the Rift Valley has, therefore, very varied interests—geological and geographical, on account of its connection with the history of the eastern basin of the Mediterranean, and ethnographical, on account of its explanation of some of the best-known stories in our folklore. But it comes in contact with the problems of science on yet another side. Fig. 2 illustrates the structure of part of the surface of the moon, showing, in addition to the well-known "ring-systems" (usually called volcanoes), a series of long narrow clefts known as "rills." If all the air and water were removed from the earth, then the Rift Valley would present much the same aspect to an inhabitant of the moon as some of the larger of the lunar rills present to us. So the exploration of the Rift Valley has the additional attraction of offering the possibility of explaining the nature of some features in the surface of the moon.

We are now fairly acquainted with the rough outlines of the geography of most of the line of the great African "rill." Our knowledge, however, of the geological structure of its

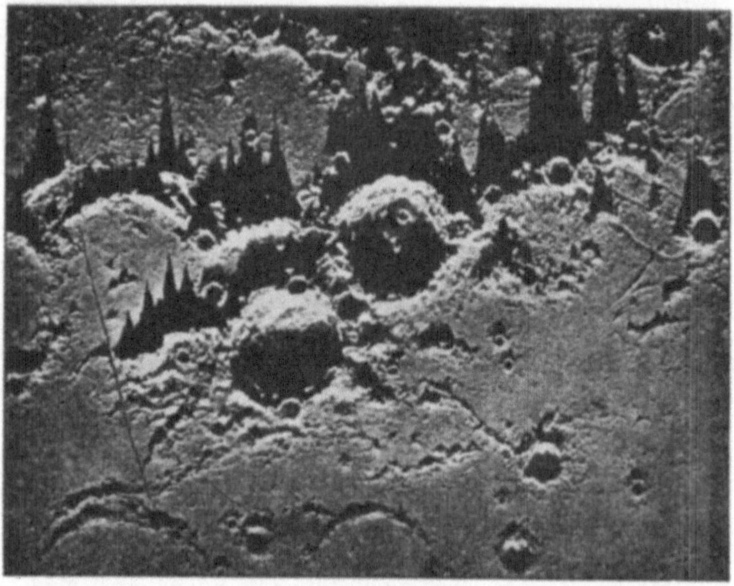

FIG. 2.—Rift Valleys of the Moon. Part of the Moon, including Mercator, etc. (From Nasmyth.)

African course is very imperfect, while one part of it in 1892 had not even been reached by Europeans. This was greatly to be regretted, because its study might be expected to yield much important evidence. The desire to obtain more precise geological information as to the structure of this Rift Valley was, therefore, the main reason for the journey described in this volume. Before proceeding to the narrative, however, it is advisable briefly to refer to the history of the exploration of the region of British East Africa, in which the route of the expedition lay.

The list of travellers who have worked in this area is a fairly long one, but the names of six men stand out pre-eminent,—Dr. Ludwig Krapf, Baron Carl Claus von der Decken, Dr. Gustav Fischer, Mr. Joseph Thomson, Count Samuel Teleki, and Lieut. Ludwig von Höhnel.

The Rift Valley was not first reached in British East Africa, but far to the south, toward the end of its two main branches. Of these, the western was entered by Burton and Speke on Tanganyika in 1857, and the eastern by Livingstone on the Nyasa in 1859. These pioneers were soon followed by others, and that district became familiar owing to the explorations and descriptions of many travellers and traders. Our knowledge of the equatorial regions of East Africa was, however, for long allowed to flag. Considerable attention had been previously attracted to them by the discovery there of two snow-capped mountains, Kilima Njaro and Kenya. The first was discovered by Rebmann in 1848; the second by Krapf in 1849. But little work of much scientific value was done till the journeys of Baron von der Decken in 1862-65. After his murder at Barderah on the Juba in 1865, scientific exploration stopped, and the next important additions to our knowledge of the region were based on native information. From time immemorial Arab and Suahili traders have sent caravans into the interior, and they discovered the main facts in its geography. Their information was collected by T. Wakefield of Mombasa and Clemens Denhardt of Lamu, and maps constructed therefrom. The map issued by the latter was based on thirty-three carefully recorded itineraries. It marked every important lake, river, and mountain in Masailand before a single European had set foot in that country. The outlines of the lakes and the courses of the rivers are often very incorrect, but the amount of information this map gave was enormous, and the facts are often remarkably accurate. In many points the structure of the country and the relations of the rivers are more fully and correctly represented than on the maps of later authors who have actually traversed the country. Thus in 1881 we find Lake Losuguta (Lake Hannington) inserted, whereas it was omitted from all other maps until that of von Höhnel in 1892. Settima is correctly mapped; and the separation of Lake Lorian from the Tana, proved by Chanler and Höhnel in 1893, was then shown, though it was not accepted by geographers.

The two most fundamental errors in Denhardt's map were the omission of the Rift Valley and the exaggeration of the size of Lake Baringo. These were serious, and most of the minor mistakes in the map result from them. They were corrected by Fischer and Thomson, with whom rests the credit of having first broken down the barrier of mystery and fear that for so long kept Europeans out of the Masai country. To Fischer belongs the honour of having first entered Masailand and demonstrated the occurrence of the Rift Valley in Equatorial Africa. He entered this valley in 1883, west of Mount Meru, near Kilima Njaro, and tracked it northward, past the still steaming volcano of Doenyo Ngai, and along the shores of Lake Natron, until he was stopped by the exhaustion of his food supply, on the steppes to the north of Lake Naivasha. His valuable ethnographical observations and collections, and his vocabulary of the language of the Masai, were the first reliable contributions to our knowledge of this tribe. His botanical and geological collections, moreover, gave important evidence as to the structure and natural history of this region.

Later in the same year Thomson continued the exploration of the Rift Valley still farther to the north. He determined the real size and position of Lake Baringo, discovered the temperate affinities of the flora of the high plateau of Laikipia, and was the first European to see Kenya from the west. He proved, moreover, the occurrence of a double series of volcanic rocks in the district, and gave a geological sketch map of the region.

In 1887-88 followed the most important of all the expeditions undertaken in British East Africa. Count Teleki, the famous Hungarian sportsman, then marched along the Rift Valley for three hundred miles farther to the north than had previously been reached. He discovered the two lakes, Basso Narok and Basso Ebor, which he named Lakes Rudolf and Stephanie. His discoveries were recorded by his accomplished companion Lieut. von Höhnel, on a map which is probably the best ever prepared by an African traveller. The map is so precise and instructive that, with the aid of the author's descriptions and sketches, and a small collection of rocks, it has enabled a very satisfactory account to be given of the geology of the area.

Numerous other expeditions have also added to our knowledge of the country. Piggott ascended the Tana (1889), and Lugard the Sabaki and the Athi (1890); Peters, with the German Emin Pasha Relief Expedition, fought his way across Laikipia (1889-90); F. J. Jackson and E. Gedge explored Mount Elgon (1890); Major Smith and James Martin opened new roads across the western edge of the Rift Valley to the Nyanza; Ainsworth mapped the region of Western Ukamba; the Railway Survey, under Captains Macdonald and Pringle, surveyed a line from Mombasa to the Nyanza (1892); and Hobley described the Taita Mountains and the Upper Tana (1894).

The materials thus collected by many different workers in Africa have been studied by Professor Eduard Suess of Vienna, who has shown that the facts reported have a greater significance than their discoverers knew. In a very remarkable memoir, entitled *Die Brücke des Ost-Afrika*, he has summarised our knowledge of the geology and structural geography of the whole line of country from the Nyasa on the south to Syria on the north, and proved the truth of the suggestion that the lakes along this line are due to a connected series of earth-movements. He has discussed the date of these movements, and indicated the important collateral problems upon which their study may be expected to throw light. With his usual insight into geographical problems, he has read more of the lessons of the country, from descriptions, than the travellers who wrote them, did from the country itself.

Suess's monograph pointed out that the part of the Rift Valley between the southern end of the Red Sea and the northern end of Basso Narok (Lake Rudolf) promises to be of especial interest. This had never been visited: new geographical as well as geological results were therefore promised by its exploration. I had always been keenly interested in the problems of African geology, and had watched the gradual opening of the Somali country, and longed to have a share in the work. When, therefore, in October 1892, the chance presented itself, I was of course eager to seize it. I was then asked to accompany, as naturalist, an expedition about to cross the Somali country to this one unexplored part of the Rift Valley. The expedition was ready to start before I had been asked to go. The members left London four days after I had received the invitation, and before my application for leave of absence could be submitted to the Trustees. At first I feared that the prolonged leave of absence required would be fatal to my chance of going. But, on the kind recommendation of Sir William Flower and Dr. H. Woodward, the Trustees gave me permission to accompany the expedition, on the understanding that the Museum should have its pick of the collections. A week later my hurried preparations were finished, and on 4th November 1892 I started overland for Brindisi, to join the others at Aden. Nine days later I arrived at Aden in the early hours of the morning. The *Malda* was due to leave at 3 A.M., so I transhipped at once. I then made my first acquaintance with the Somali; and this was ominous of my future relations with them, for I had to knock down two at the start. In their eagerness to secure me as passenger, rival crews tried to wrest the chronometers from my grasp. I had nursed these precious instruments throughout the journey with the tenderest care, I had never allowed any one else to carry them, and it was not to be expected that I should entrust them to the rough Aden boat-boys. As language was not strong enough to persuade them to keep their hands off the case, the instruments had to be protected in ways that the natives better understood.

A few hours later the *Malda* steamed to the south-east, towards the limestone cliffs of Cape Guardafui (Ras Alula). Having rounded these, the course was altered, and we went southward along the Somali coast towards the roadstead of Kismayu, where arrangements had been made for the expedition to land.

PART I

AN ABORTIVE EXPEDITION

" Kulekeza si kufuma "
(To aim is not to hit).

Suahili Proverb.

CHAPTER I

PREPARING FOR THE START

"In the reproof of chance lies the true proof of men."
Troilus and Cressida, i. 3.

THE voyage from Aden round the "Horn of Africa" and along the eastern coast was the pleasantest part of the journey towards "Lake Rudolf." The officers of the British India steamer, the *Malda*, did everything they could to make the passage comfortable both to us and to our noisy crowd of Somali. We had, moreover, the good fortune to number among our fellow-passengers Mr. J. R. W. Piggott, the present Administrator of British East Africa. He had himself led the way into the basin of the Upper Tana, and his wide experience of the country and its people was placed most generously at our service; but unfortunately most of his wise advice came too late. The plan of the expedition was altered, and it was decided to land at Lamu instead of at Kismayu, and ascend the Tana instead of the Juba. There were many advantages in this route, but some of us much objected to the change. It meant that, for the first two months at least, we should be in known country, instead of plunging at once into the unknown. Moreover, as far as we could learn, the climate was wholly unsuited to camels. The expedition had been planned on the basis of using these animals for transport, and if they failed us, we should fail. I felt disappointed, as I have a rooted distrust of change of plan at the last minute. The Tana valley is notoriously unhealthy, and we missed the march across the Borana country, west of the Juba. The steamer did not call at Kismayu, but kept southward until, on the morning of 21st November, we sighted the sandhills of Shella at the entrance to the harbour of Lamu. The *Malda* anchored a couple of miles from the shore to wait for the turn of the tide. The mails were sent ashore at once, and our chief went ashore with them to arrange for the disembarkation of the stores. He asked me to accompany him, and Mr. Piggott came with us and introduced us to Mr. Rogers, the superintendent of the district, his assistant, Mr. Macquarie, and Mr. Bird Thompson, the officer in command of the Indian troops at Witu, who was then staying at Lamu. The *Malda* entered the port in the evening, as the captain thought it prudent to wait till high tide before he threaded the channels between the reefs which constitute the harbour bar.

First thing next morning we set to work to unload the 300 tons of stores. The Somali worked well, with great energy and much noise; but their muscles proved weaker than their will or their voice. The Turks, on the other hand, were quiet and sluggish but immensely strong, and most of the heavy work had to be left to them. The Somali, however, made up in numbers and enthusiasm for their individual weakness, so between the two contingents of men the transhipment of the goods into dhows was finished early in the afternoon. We said farewell to our kind friends on board the *Malda*, and followed the men ashore. We landed close by the custom-house; the men paraded in the square in front of the old Portuguese fort (see Pl. II.), and then, under the guidance of one of Mr. Rogers's Askari, marched to a cocoa-nut grove about a mile to the west of the town. There camp was pitched around a bungalow in a hollow in the sandhills, near some brackish wells upon the shore. During the morning the *Juba*, a small steamer belonging to the Sultan of Zanzibar, called in on its way to Kismayu. So our chief seized the opportunity, and hastily made an agreement with one of the leading Hindu traders of Lamu for the supply of 110 camels and 40 donkeys. The trader's agent left at once in the *Juba* to purchase them. The price agreed on was a very high one. But the contract stipulated that they were all to be delivered in Lamu in a month's time, and it was verbally promised that they should be landed in three weeks.

As we could not move from the coast until the camels came, we settled in camp to prepare the equipment, drill the men, and

No. II. THE OLD FORT AT LAMU. *Page 14.*

(WITH SILK-COTTON TREE.)

repack the loads. In the intervals we made excursions through the plantations inland, or across the harbour to the island of Manda. During the heat of the day we rambled about the cool, narrow, tortuous streets and tunnel-like passages of the historic old town of Lamu.

The island consists of a series of sandhills overlying raised coral reefs. The dunes are covered with groves of cocoa-nut palms, with here and there a date-palm. In the hollows between the hills are massive, shady mango-trees and orchards of cashew-nut, surrounded by hedges of prickly Euphorbias, full of acrid juice. Beside the backwaters from the harbour stand clumps of screw-pine (*Pandanus*), and on the headlands wave the graceful feathery she-oaks (*Casuarina*), while the muddy shores are lined with dense thickets of mangrove. Whatever time I could spare from camp work was devoted to excursions in the woods and plantations, chasing shore crabs on the beach, dragging for algae in the estuary, and revelling in the delight of a first experience of Nature in the tropics. In addition to the other charms of the place, the climate seemed salubrious: the air was dry, and a cool bracing breeze blew daily from the sea. A week thus passed very pleasantly. Then our chief suddenly decided that some of us had better move to the mainland, so on 30th November Benett-Stanford and Tichborne, with most of the men, started in dhows to form a camp at the head of the creek of Mkonumbi. The chief and I remained a few days later to ship the rest of the stores and men. This was accomplished by the aid of the "chain gang," a very useful local institution, the services of which were lent to us by our kind friend Mr. Rogers. The sturdy Suahili easily carried off heavy boxes of ammunition which several of the Somali could barely lift. When the last of the loads and men were on the dhows, we returned to Lamu. We said good-bye to Mr. Rogers, to whose hospitality and help we were so much indebted, and then followed the others in the Company's launch to Mkonumbi. We found a comfortable camp had been pitched around some mango-trees, surrounded by a strong thorn zeriba or "boma." Here the organisation of the caravan and preparations for the march inland were busily pushed forward.

It may be advisable here to state the composition and objects of the expedition. It had been organised in order to explore in the less-known regions of British East Africa, by four English sportsmen, namely, by our chief, together with Sir Henry Tichborne, Mr. W. H. Harris, and Mr. J. Benett-Stanford, late of the Royal Dragoons. Dr. A. D. Mackinnon, who had seen much experience in British East Africa, and who had accompanied Mr. Jackson's caravan to Uganda, was the doctor ; a better man for the post could not have been found. Count Lovatelli (who subsequently distinguished himself by his courageous rescue of Mr. Tod in the fighting with the Somali at Kismayu) went with us by an arrangement between our chief and the Italian authorities ; and my services, as a naturalist, had been lent to the chief of the expedition by the Trustees of the British Museum. Sir Henry Tichborne was accompanied by his valet Gleave.

As a private expedition its scale was very extensive. It comprised 8 Europeans and over 300 natives, while a powerful baggage train had been arranged for. It had been hoped that the Egyptian Government would have lent 70 of its Soudanese troops to the expedition, but as the negotiations for this fell through, the chief defence of the caravan was entrusted to 150 Aden Somali. It was not thought advisable to rely entirely on these very excitable people, so ten stolid, imperturbable Turks had been engaged to act as crew of the Maxim gun. In case this jammed, we knew we could rely on them to stand their ground, while it was being again prepared for action. We were to pass near some of the so-called "sacred cities" of the Somali country, where it was thought possible that a "holy war" might be preached against us, and the Somali escort refuse to fight. So to prevent the expedition being dependent on any one set of men, an Abyssinian contingent was also engaged. As these were nearly all Christians, their nominal Christianity would, we hoped, ensure their constancy in case of any quarrel with the Somali. They were also engaged to act as porters as well as soldiers ; and further to supplement the camel transport, 80 Zanzibari had been enlisted by the British East Africa Company at Mombasa.

The goal of the expedition was Basso Narok or Lake Rudolf, which had previously been visited by one expedition only, that of Count Teleki in 1888-89. It was hoped to explore the western shore of the lake and follow up the two rivers that flow into its northern end. It was intended also to traverse the country of the Borana Galla and return by a long march across the head waters of the Juba to the coast opposite Aden. We thus hoped to run two lines across the biggest blank still left in the map of Africa. The caravan had been armed and equipped regardless of expense. We had with us a Maxim gun, 250 Sniders, 30,000 cartridges for the former, and 100,000 for the latter. We therefore felt on landing that, whether we succeeded in the accomplishment of all our plans or not, we were not likely to turn back without a long and determined struggle.

At this time our forces were divided. The Abyssinians were not ready when we left Aden; Harris and Mackinnon therefore stayed behind to take them on to Zanzibar in the French mail. Thence they were to cross to Mombasa, collect the Zanzibari, and march north to join us wherever our chief directed.

Thus at the beginning of December there were 6 Europeans, 150 Somali, 10 Turks, and all the stores in camp at Mkonumbi. Harris and Mackinnon were coming up from the south with 70 Abyssinians and 80 Zanzibari. A telegram was sent to them to land at Melindi on the south side of the Sabaki, and thence march overland to join the rest of us on the banks of the Tana. The camels were due to arrive by the 22nd December, and we continued the work of packing and overhauling stores, drilling the men, and preparing equipment, so as to be ready to march as soon as the transport animals should arrive. It was not long before the minor troubles of African travel began to worry us. Stores had been badly packed, and a great mistake had been made in buying the rice and dates in London instead of in Aden. We soon learnt to appreciate the wisdom of the Suahili proverb—

"Fulani amerudisha tende Manga"
(He has sent back dates to Arabia).

Our men would not take the rice, so 20 tons of it were sold, and the rest exchanged at a heavy loss. The cooking-pots proved useless and a source of danger to the men. The tents were unsuitable for the work; and two of them were missing, as well as other indispensable articles. Our men, moreover, proved incapable, lazy, and troublesome. We had to show them everything; all they seemed able to do on their own initiative was to loot in the neighbourhood and quarrel among themselves. Soon we had complaints from the chiefs of the neighbouring villages about the misdeeds of the Somali. When, a few days after we had landed, the rains began, we made the still more alarming discovery that the men had very frail constitutions. After every storm of rain, twenty or thirty of the Somali would be prostrated with fever. The condition of several of the men soon became so serious, that it was only owing to the unremitting care of Benett-Stanford that they were kept alive. Gleave was also taken ill, and caused us great anxiety. It was decided that, as thirty-three camels and several donkeys had been landed, and as a number of local Suahili porters had been hired, Sir Henry Tichborne and I should make a night march to Witu to try and persuade Dr. Rae to go to the rescue. I packed up my collections, my reserve luggage, and some manuscripts at which I had been at work during the voyage out, and left them to be returned to Aden. This, however, was the last I ever saw of them.

The night, however, was so cloudy that at first it appeared doubtful whether we could go, but at ten o'clock the moon broke through the clouds, and the order was given to the men to get ready. The Somali at first refused to go, and before they were persuaded to do so the clouds had again hidden the moon. In darkness, broken only by the fire-flies, on the night of the 14th December, we stumbled across the plain and through the swamps of Mkonumbi, in what we fondly imagined was the first march to "Lake Rudolf." After daybreak I hurried forward to Witu to arrange accommodation for the men, and as the late Dr. Rae kindly ordered the chain-gang to clean an empty hut, it was ready for them when they arrived, exhausted by the long night march.

The late Bird Thompson soon came in from the morning drill of the Indian troops, and received us with his usual kindness; Tichborne described the sorry plight of our men, and Dr. Rae, though very unwell, started off for Mkonumbi in the evening.

In the afternoon Thompson took me for a walk through the town, showed me the ruins of the Sultan's palace (Pl. III.) and some of the wells, and told me its history. Witu was

No. III. THE RUINS OF THE SULTAN'S PALACE AT WITU. Page 18.

(Destroyed during English Attack in 1890.)

at one time the "Cave of Adullam" of Eastern Equatorial Africa, and thither flocked the worst of the runaway slaves, the debtors, and the scum of the mongrel populations of the coast towns. Under that leading scoundrel, the "Sultan of Witu," they defied authority and tyrannised over the surrounding country. The suppression of this nest of thieves was absolutely essential to any peace or stability in the district. But Witu was a strong town; the approaches to it were unhealthy, and the Sultan could bring 1000 guns into the field. It might therefore have been long allowed to harbour its gang of ruffians and freebooters, but for the massacre of a party of nine Germans who visited the town in 1890, soon after the cession of the Witu district by Germany to England in exchange for Heligoland. Being thus responsible for the good order of the country, our government sent out a naval brigade, which easily defeated the attack made on it by the Witu army, commanded by my friend the Omari Mahdi. The town was shelled, and was set on fire by a war rocket, which knocked a hole through a great tree that stands beside the present entrance to the town. The rebels withdrew into the forests inland, where they have kept up a guerilla warfare ever since. Supplementary expeditions have driven them from their settlements at Pumwani and Jongeni, while Witu has been held by a garrison of Sepoys and connected with the coast by a road and telephone.

The latter had temporarily broken down, but communication by it was resumed the day after our arrival, and the late Mr. Bell Smith, the superintendent of the Melindi district, told us that Harris and Mackinnon had landed with 145 men, and would be at Ngatana on Christmas Eve. The plan at this time was for all the reserve food of the expedition and most of the heavy stores to be sent by sea in dhows to Kau, a town at the head of the estuary of the Ozi; they were there to be transhipped into canoes, which were to carry them through the Belezoni Canal (p. 31) to the Tana, and up this river to Ngatana. The whole expedition was to assemble at this point, and from it the real start was to be made. The Abyssinians and Zanzibari were to reach Ngatana by marching up the right or western bank of the Tana, and they were due there on 24th December. The baggage animals, the Somali, and the Turks were to proceed *via* Witu and the left bank of the river.

If the camels were landed by the date fixed in the contract, there was no reason why New Year's Day should not find us all in camp at Ngatana, ready to begin the march up the Tana valley by the end of the first week in January.

Unfortunately no reliable calculation had been made of the relation between weight of stores and means of transport at each point. All our food and loads were stacked at two different places, while our porters were foodless and idle at a third.

It was necessary that food should be taken to our Southern division, because the men had only sufficient to last till they reached the Tana at a station called Borabini, two days' march from the coast.

Tichborne and I accordingly started to Kau with some porters and donkeys laden with rice. On arriving there, however, we found that the dhow had not come from Lamu, and the canoes had not been collected to carry our goods up the Tana. With this not very pleasant news Tichborne returned to Witu, while I stayed at Kau to get canoes and take the food on to Harris.

As I could not start next day, I went down the estuary to the port of Kipini at the mouth of the river, to see if the dhow was there. There was no news of it, but the excursion was repaid by my shooting two hippopotami, seeing my first crocodiles, examining some coral reefs, and visiting some extensive cotton plantations and the tomb of the giant Fumo Liongwe, one of the best known characters in Suahili folklore.

The Somali left at Kau had a less pleasant day, and on my return demanded permission to go back at once to Witu, as the mosquitoes rendered their lives unbearable. These insects were certainly troublesome, though by sitting in the smoke of a fire one could keep them off; but the Somali said this remedy was as bad as the disease.

Sufficient canoes came to Kau next morning, and I prepared at once to start for the Tana. But just as they were being loaded, a letter arrived giving me a startling change of orders. The letter said that other arrangements had been made for the transport of food to Harris's party, and that I was to return at once to Witu. There I was to meet an advance guard that was being hastened up from a camp that had been formed at Fungozambo. I was to go on with them to Ngatana, at once and by forced marches, select the best site I could find for a camp beside the river, build a "boma," or thorn stockade, 50 yards square, and open up communications with Harris's force. My party would be lightly loaded, but it would be followed in two days by another under Count Lovatelli, who would bring me a tent, food, and medical stores.

At ten o'clock I was back in Witu, ready to march for Ngatana; but the men from Fungozambo were not there. I therefore went to sleep, to be ready, if necessary, for a night march. Thompson woke me about mid-day to complain of the behaviour of our men, who were seriously disturbing the peace of the town. A stormy interview between Tichborne and myself on one side, and the men on the other, seemed to quiet matters, and they promised to behave better.

At two o'clock next morning I was roused by a messenger from the coast, who brought word that more deaths had occurred and that our chief was ill. Dr. Rae was earnestly requested to return. The doctor was himself suffering from fever, but so urgent was the summons that he went. I was concerned about him, and accompanied him the first seven miles. On the way I made an annoying discovery. There were no preparations for firing the belt of forest at Pangani through which the road passes, and which is the haunt of the tsetse-fly (*Glossina morsitans*, Westw.) On the journey through it on our way up, I had found this dreaded fly. To expose our camels to an attack from it was to risk the whole safety of the caravan; I had therefore sent back a note to say that before the baggage animals entered this belt of forest, fires ought to be lighted in it and the animals driven through the smoke. This is the method commonly used to protect animals when passing through a fly-haunted district. I found out afterwards that the Galla in the Tana valley know all the fly-infested patches, and always adopt this precaution when driving their cattle through them.

As Dr. Rae seemed better, he went on with two men and I stopped there to prepare the fires, sending on a note urging that these should be lighted half an hour before the camels were allowed to enter the forest. The rest of the day was spent in building up stacks of wood and dead leaves; these were arranged as far as possible on the model of the "smudge-fires" used in Manitoba to protect the wheat from the early frosts of August, by raising a cloud of smoke above the fields.

For some time our men had been growing more and more discontented and disrespectful, and next day this broke out into open mutiny. To stop their depredations in the town, they had been ordered not to leave their hut without permission. Nevertheless, during the morning Thompson's Pokomo boy came rushing in to say that the Somali had gone into the town and that they were fighting the Sepoys. I seized my revolver and a stick, and ran as quickly as I could to the scene of action. The Somali had no chance against the Sikhs; one of them had been taken prisoner and the others were trying to rescue him. Failing to do so, they rushed back to their hut shouting "bundook" (guns).

I told the Sepoys to take their prisoner into the cantonments and then ran back after the Somali, arriving at their hut just as one of them was coming out rifle in hand, pushing a cartridge into the breech. I ran against him and bumped him back into the hut and mounted guard over the door. The Sepoys stood to arms and the Jemadar sent a messenger for Thompson, while an armed patrol of twenty men marched ostentatiously past the hut. The Somali declared that the Sepoys were killing their brother and that they must go and rescue him, but I refused to let them out. When Thompson arrived things had quieted down a little, and the Somali gave up their ammunition when ordered to do so.

Tichborne and I then had a "shauri" or conference with our men, who said that unless their brother was immediately released, they would return at once to the coast. They demanded that their ammunition should be given back to them, which was of course refused. Tichborne hurled at them all the Hindustani bad language that he had acquired during his various Indian shooting expeditions, and the men reiterated their demand for the release of their comrade and the return of their ammunition. Tichborne and the Somali kept this up till their throats were sore and we were tired of standing in the sun. The shauri ultimately broke up without any agreement being arrived at, so a Sepoy guard was kept on duty in an adjoining hut.

Later in the same day an explanation of our men's action was furnished by the arrival of the headman and some of the elders of the village of Fungozambo. They complained that they were being robbed and murdered by our men, over whom they said the European in command had no control; the Somali had looted the village, stolen all the "kuku" (fowls) and bed-cushions from the houses, and flogged the natives, who had now fled to the woods. We found out afterwards that these charges were substantially correct. For as soon as Benett-Stanford had left Fungozambo, the men informed Count Lovatelli that they were not going to obey him, as they had been told by the chief at the coast they were to take orders from no one but himself; so they were doing what was right in their own eyes. Thompson was very angry, and told us that he had ordered his men to parade at four o'clock next morning; he was then going to serve out eighty rounds of ammunition and march down to our camp. He was determined to put a stop to these lawless proceedings by whatever means he found necessary. He said, however, that if I could get to the camp before him and restore peace, so much the better for everybody. I had my dinner, mounted a donkey, and rode off for Fungozambo. About ten o'clock my donkey suddenly pricked up its ears and sprang from the path, and then, as I held it in, stood trembling in every limb. My boy said "Libah" (lion), so I fired several shots from my revolver into the air, the boy fired the Snider, and we stood waiting for something to happen. The donkey gradually recovered from its fright and we led it along the path, but it was so terrified that I could not ride it for some time. I heard afterwards from Thompson that the natives say there are lions at this place, and they will never pass it alone in the dark; it is probable that the donkey detected a lion, although neither I nor the boy heard anything.

A little after midnight we reached the camp. Benett-Stanford, Dr. Rae, and Gleave had arrived from the coast just before us, and our headman, Wasama, came in a few minutes later. They had been recalled to suppress the mutiny. We slanged the Somali vigorously. Wasama's language was moving. For some years he had been an interpreter on a man-of-war; his knowledge of our language was extensive and peculiar; his fo'castle English now rang through the camp like a recitation from a slang dictionary. For variety of epithet, power of expletive, and range of religions whose deities were invoked, I doubt whether his speech could have been strengthened by a committee of Colorado cowboys. Benett-Stanford let his corner of the camp know how the Royal Dragoons drill an awkward squad. Seeing the admirable effect of this powerful English, I tried a mild imitation of it on the group of men to which I directed my attention. Feeble though my efforts were, I fear my language on that occasion would have done more credit to a connection with Billingsgate than with the British Museum. The men were startled by our midnight arrival, and seeing that we meant business they became very respectful and promised implicit obedience. They swore that they had not in any way interfered with the villagers, but as some of them were wrapped in Suahili "mikeka" (sleeping-mats), and some were using the native wooden head-rests or "misamilo," we had no trouble in demonstrating the falseness of this statement.

Benett-Stanford and I had a long chat over the position of affairs, which was now rather serious. It was clear that if Harris reached Ngatana, as arranged, on Christmas Eve and found no food there, he would have trouble with his men. Everybody had warned us that if the Zanzibari did not get their rations they would desert. We knew he had left Melindi on the 17th December, and that he was determined to push on quickly. It was only six marches to Ngatana, so he could easily arrive there by the date appointed. This was now the morning of the 22nd; Ngatana was four days distant, but by hard marching it could be reached in three. So it was decided that I should get together a few camels and hurry on at once, and thus carry out the orders sent by our leader. Benett-Stanford kindly set about the preparation of the stores; while I, having dispossessed one of the thieves of a stolen sleeping-mat, snatched a couple of hours of welcome sleep.

At four o'clock Gleave woke me for breakfast: seven camels and twelve men under Sergeant Yussuf were ready. I was to leave most of these men at Witu and replace them by the most troublesome of the Somali there, as we were anxious that they should inflict no further annoyance on our kind hosts.

After a hurried meal we marched out of camp and stumbled along the track in the dark. We reached the forest shortly after daybreak, lighted the smudge-fires, and drove the camels through the smoke. The Somali could not quite understand this proceeding; they regarded it as a religious rite, and told me afterwards that they did not know that Christians worshipped fire. When I explained that it was to keep off the flies, they were delighted and said we English were "herribleh" (wise). We reached Witu early in the afternoon, but did not stay there, for as it is surrounded by forest we feared the dreaded tsetse-fly might occur there. So the camels were driven a mile to the south, and camp was pitched beside a deserted Galla village, on the hill from which the English guns shelled the town in 1890. The Somali were told to be ready to start at daybreak, and from the manner in which some of them received the order, we all went to bed expecting another row in the morning.

CHAPTER II

WITH THE ADVANCE GUARD TO THE TANA

"We eat our proper rations
In spite of inundations,
Malarial exhalations,
And casual starvations."
KIPLING.

WITU is the farthest outpost of civilisation in the Tana valley, and here we expected to begin regular "safari" or caravan life, and to say farewell to Europeans for the last time until we reached the Somali coast opposite Aden. My pleasure at getting away from the coast lands and starting for the "barra" was, however, lessened by my regret at parting from Bird Thompson.[1] He took a keener interest in the natives, he understood their point of view better, he knew more about their habits and beliefs, and was more popular amongst them, than any other man I met in Africa. To his advice I owe more than can easily be expressed. With his usual readiness to help us, he came over to camp to superintend our start. The Somali stood in awe of him, and so the camels were promptly loaded and we got away without a word of complaint.

We marched for four hours and then rested beside a small brackish lake, which must be the source of the Magogoni, the river which joins the Ozi at Kau. The acacias here were often pink, owing to the growth upon them of a species of *Loranthus*, a parasitic plant allied to the mistletoe.

At three in the afternoon we resumed our march, and left the woods of Hyphaene palms and entered the "barra" or open grassy plains beside the Tana. We crossed a dry channel

[1] He died off Teneriffe during the voyage home in February 1895.

R. M. Bird Thompson. Dr. Rae.

No. IV. THE GARRISON OF WITU.

Page 26.

through which the river flowed a couple of generations ago, and camped upon the plain beside a water-hole in this channel. A lion came roaring round the camp at night, but it was too dark to hope for a shot; and knowing that roaring lions do not fight, it did not disturb my rest. A more serious annoyance was the loss of a young wild cat or Serval, which I had purchased in Witu and was trying to tame. It was a most ferocious little brute, only twice the size of a kitten of the same age, but with temper enough for fifty. It bit and scratched like a fury. This did not hurt me, for my leggings were thick; but the men's legs were bare, and they did not like the animal. They declared it would grow into a tiger, and begged me to kill it. As I would not do so, I believe they let it escape on purpose.

At daybreak next morning we resumed our march across the plain. There was a good deal of game, but we could not stop to shoot. I almost tumbled on to three black buffalo, which fled before I could snatch a rifle from the men. I gave chase at their request, but fortunately did not come up with the animals. Life is too good to be thrown away by shooting at buffalo with a Snider. During the heat of midday we rested for a couple of hours by some Borassus or Palmyra palms (*Borassus flabelliformis*). Our rest was cut short by a threatening storm, which caught us on the march; the rain was heavier than I had ever experienced, and fell on us like a waterspout. To face it was impossible, so we crouched behind the camels to get what little shelter these gave. When the rain was at its worst we could not see five yards along the line. The storm travelled from north to south, down the valley, extending for about two miles in width, and turning the whole of the belt it traversed into a swamp. The men said the camels could not go through it and we must camp where we were. However, we were due at Ngatana that night, so I could not consent to this delay, and we pushed ahead through the swamp and slush.

As we approached the Tana the grass became higher and denser, and we had hard work in forcing a way through it. At length we reached the banks of the river and struck a native track, which we followed through the woods until we emerged into a clearing opposite the village of Vuju. It was dark when we arrived there, so it was hopeless to try and find Harris's party that day; but we had reached the trysting-place in time.

The ground was so swampy that we could not light a fire or cook any food. As the expedition only provided three tents for eight Europeans, I had had to start without one. It poured with rain all night, so we all, men, camels, and donkeys, huddled together as closely as we could. I fixed my ground sheet like an umbrella on a pole above me, to keep off the rain, but it was not much of a success. Even the camels seemed to despise it, for they twice deliberately kicked it over.

Thus supperless and sleepless we spent Christmas Eve. The rain stopped for a while towards morning, but a dense mist hung over the country and buried everything in its cold malarial pall. At ten o'clock the mist lifted sufficiently to show us the opposite bank of the river, where we descried two natives hiding in the reeds. We shouted "Yambo" to them, and told them that a friend of Bwana Kilemba (*i.e.* Bird Thompson) wanted to see their chief.

They went away with the message, and returning some hours later took me across the river in a dug-out canoe, to have a conference with a large party of Wa-pokomo on the opposite bank. They were very much afraid of the Somali, so I went alone. We could not talk, but I smiled and nodded, and this seemed to allay their suspicions. They allowed me then to send back for an interpreter, and when he came I asked where were the two white men and their large party of Zanzibari? The chief told me that they had not come anywhere near this district; so he gave me a couple of guides to accompany a Somali whom I sent off at once with a letter to Harris and Mackinnon, in the hope that he would meet them on the road. The Wa-pokomo again landed me on the other side, and after giving me some unripe bananas, went away. It soon began to rain again, but we had got some dry wood from the natives and were able to light a fire. I had my Christmas dinner of Indian corn, tinned beef, and baked bananas. The rain increased in strength until it put out our fire, and this night was even less pleasant than the last, for anxiety about the non-arrival of Harris's force was added to the discomfort of the wet and cold.

Next day I sent some of the men with the camels and donkeys back to Witu and crossed with the remainder to the other side of the river, so as to be able to keep a better watch for Harris's force. But the men did not start without a row. When I told Jarma that he was to be one of those to return with the camels, he refused to do so. He said that he had been told by the chief to obey the orders of no one but himself, and that he had only told him to come to Ngatana. He was therefore going to stay where he was, until he had orders from the chief to go elsewhere. Before he knew what I was doing I had seized his rifle and ammunition. I had him arrested and tied to a tree. The other Somali asked what I was going to do with him. "Flog him, of course," I replied. At once there was a hubbub in camp. The men threw down their guns and danced about in the wild, noisy, excitable way to which I had now become accustomed. "Kowadku, labadku, sadehadku" (By the first, by the second, by the third), they shouted. "You cannot flog a Somali. You may flog a Pokomo; you may flog a Zanzibari; you may flog an Habeshi (Abyssinian); but you cannot flog a free Somali." They were simply told to wait for a few minutes and see. There were some Wa-pokomo at the other end of the village, watching us and wondering what the excitement was all about. I called one of them and told him he was to flog the Somali. This was adding insult to injury, for the Somali hold the Wa-pokomo as their slaves. There was therefore another outburst of "Kowadku, labadku, etc.," but it made no impression. I was resolved that the man should be flogged. It was not till Jarma, seeing I was in earnest, began to whine and promise to do anything I told him, if I would only let him off, and when I reflected that striking a Somali meant starting a blood feud, that I decided it was useless proceeding to extremities, as the man was thoroughly cowed. I had him untied, returned him his rifle but not his ammunition, and he started back with the others without a murmur. Our headman subsequently told me that this Jarma was a hopelessly bad character, and that he had tried in Aden to persuade our chief not to take him, as he would be sure to give us trouble. However, I had no more bother with him after this.

The camels and the other men returned, leaving me with

only eight men; they were so exhausted by exposure that it would not have been fair to set them to work at once on the erection of a boma. So we waited, keeping watch on one side of the river for the reinforcements and stores which it had been promised should follow at once, and on the other side for the Southern division of the expedition.

On the 27th my messenger returned with a letter from Harris, stating that his men were all at Borabini, a mission-station some miles down the river, where they were blocked by floods. He was in urgent need of food for his men, as the stores sent by our chief had been wrongly addressed and had failed to reach him. It was obvious from this that they were not coming up the river at present. This placed me in rather a difficulty. We were being watched by the spies of the rebel chief, Fumo Omari, who was anxious to capture our cartridges, sixty thousand of which were being sent up the Tana in canoes. It appeared to me highly probable that if this bait were offered to the rebels they would strike for it, and I could not hope to resist their attack with my present force; I therefore engaged a mau, loaded it with sacks of rice for Harris's men, and rushed down the river to meet the ammunition.

A long day's canoe journey took me to the German Lutheran mission-station at Ngao, then under the charge of Herr Beking, who told me a pathetic story of its past history.

After much trouble the missionaries had succeeded in gaining leave to settle there. A house was then erected on a spur on the left bank of the river, to be destroyed in one night by a flood which simply swept away the whole site of the station. A better locality was chosen on the right bank, and the missionaries set to work to rebuild the house. Then Frau Beking died. The station was finished; and shortly afterwards the Witu rebellion broke out, when the missionaries had to fly to the coast to save their lives. The buildings were burnt down and the plantations devastated. With splendid patience the station had been rebuilt for the third time, but Herr Beking said he did not know how long it would last. He told me the people were then all anxious to turn Christians, but from purely political motives. They said they were now European subjects and they must learn the European religion.

The missionaries were striving to take advantage of this opening, and were teaching the natives blacksmith's work, carpentry, reading, writing, and better methods of agriculture.

After leaving Ngao I reached Borabini or Golbanti, where I found Dr. Mackinnon staying in the mission-house as the guest of the Rev. Mr. Edmonds. Harris had left that morning for Witu, while the flotilla of maus, bringing up the ammunition, arrived in the evening. At my request the cartridges were landed and stored in the mission-house, and the canoes sent back for goods which would be a less tempting bait to the rebels. The mission-station was a strong stone building surrounded by a powerful stockade; the house was provided with an iron roof so that it could not be set on fire, the windows were protected by iron shutters, and the walls were loopholed. A stand of arms also showed that the Methodist Missionary Society does not belong to the peace-at-any-price party. These precautions, however, are necessary, for the founder of the station and his wife were both massacred by Masai.[1] The mission-station is now so well fortified that our 61,000 cartridges were quite safe, the more so as the garrison was reinforced by 150 of our own men.

I spent a most interesting evening with Dr. Mackinnon and Mr. Edmonds, neither of whom had I previously seen, and next day continued the journey down the Tana. The country became one vast swamp, and sank below the level of the river. The water was pouring through the banks by numerous gaps, natural and artificial, and was thus irrigating the rice-fields on either side. Some of these had been already sufficiently flooded; so the water had been cut off, and the ground was green with the young rice shoots. At two o'clock we reached Charra, a group of huts beside a grove of cocoa-nut palms, the farthest point from the sea at which they occur in this district. Here we left the Tana and entered the Belezoni Canal. This is a narrow ditch connecting the Tana and the Ozi. The mouth of the former river is closed by a bar, so that dhows cannot enter it except with great difficulty and at certain states of the tide. The Tana runs actually parallel to the sea, which is at a distance of only a few hundred yards; yet the river flows on for some miles before it succeeds in bursting through the sand-

[1] This incident is referred to in Rider Haggard's *Allan Quatermain.*

dunes which separate it from the ocean. Its mouth, moreover, is continually being silted up by the tide, and reopened on new lines by freshets. The sandbanks, therefore, constantly shift their position in the struggle between the waters of the river and of the sea.

All this renders the mouth of the Tana very difficult of navigation. The Ozi, on the other hand, has a fine broad open estuary, and the water over the bar at its mouth is usually deep enough to be crossed by the shallow draught dhows that do most of the coasting trade. In order that the Ozi might become the port of the Tana, the rivers were connected by a canal; this was cut by forced Pokomo labour under the direction of Sultan Hamad, the great-grandfather of the present chief of the Witu rebels. The channel was originally 10 feet wide and 6 feet deep, but it is now much less.

Owing to the high level of the Tana a powerful torrent poured along the canal from that river to the Ozi; the two canoemen had only to push aside the stems of the water plants and the current carried us along. As we swept through the rushes we raised crowds of mosquitoes, which had their revenge for being disturbed in their afternoon nap. So we were not sorry when, after a passage of three-quarters of an hour, the canoe shot out into the quiet waters of the Ozi. At six o'clock I reached Kau, where my old friend Suleiman the Akida, or headman of the town, urged me to stay the night. We had a chat, but, as usual, neither of us understood what the other was saying. He showed me a paper which he said was a testimonial or "chit" given him by the superintendent of the district. It was, however, a letter from Mr. Rogers to one of the members of the Borabini Mission; but as it referred to matters then of ancient history, I did not destroy the Arab's faith in the value of his treasure.

The Akida provided me with a man to ferry me across the Kirimanda, a river that had to be crossed on the way, and also with two Arab soldiers or "Kiroboto" as an escort against lions. One of them was armed with a picturesque sword, and the other with a rusty muzzle-loader. They scornfully refused to carry my mackintosh or satchel, so that they were useless; and as the gunner would occasionally point his gun at me, I was more afraid of him than of all the lions in the district.

Tichborne, Harris, Lovatelli and Gleave were all in Witu.

From their accounts the whole expedition seemed to have drifted into confusion and disorder. The next morning news came up from Lamu that a dhow, which was bringing forty-six of our camels from the Somali coast, had been wrecked on a coral reef; this seemed for a moment the deathblow to the expedition. On thinking the matter over, however, it was quite clear that the story was untrue, for no dhow on the coast could carry so many camels. A general consultation was held, and it was decided that a determined effort should be made to get all the stores and men to Ngatana as quickly as possible.

I accordingly started back to Borabini to hurry the porters up to Witu, and to arrange for the flotilla of maus to continue carrying the goods to Ngatana. A native guide was engaged at the preposterously high price of eight rupees, from whom I learnt that among the Suahili the only essential in a guide is ability to ask the way. As we met no one, this power was not very useful, and we lost our way. We spent from eleven till half-past three wading through a dismal swamp with the water always up to our waists, and frequently to our shoulders. We left the swamp less than three hundred yards from where we had entered it, and then plunged into a jungle of grass and rushes, 12 feet in height, in which the guide again lost himself, and I took the lead. We found the Galla village of Dibbe at sunset, but the natives would not enter the Pokomo district in the dark, so we tried to find our way by ourselves. We lost the path, found an unrecorded lake known as Somite, and finally, by accident, stumbled on another Galla village.

I was now dead tired, and, though the guide begged me not to stop the night in this village, declined to go any farther. A Galla lent me some skins in which I wrapped myself. The Galla, however, have rather a bad character, so I slept with one eye watching my sullen host. He could not give me anything to eat, so I started off at dawn and soon reached the river. Some Pokomo fishermen ferried me across, and pointed out the smoke of the fires of Ngao. I arrived there just in time for breakfast, and was very glad to get it, as my one frugal meal of the day before had been my only food for thirty-six hours.

I had intended to go on to Borabini at once, but Herr Würtz, a colleague of Herr Beking's, who had just returned from Lamu, offered to show me his manuscript Ki-Pokomo dictionary, and to tell me some of the legends of the people about the geographical changes of the Tana valley. This was a temptation I could not resist. I spent a most interesting day with these hard-working and intelligent missionaries, and then went on to Borabini to see Mackinnon.

Mr. Edmonds, the missionary in charge of the station, kindly sent for a Galla elder from the adjoining village, and he gave me the Galla version of the stories, which the German missionaries had told me according to the Pokomo traditions. From this man's information it is clear that great changes in the course of the Tana have happened during the last two generations.

That night we had a scare. We were called up after midnight by the statement that an attack was impending. The Galla flocked into the stockade, the Pokomo fled to the swamp, and our men stood to arms. The rumour was very precise. Twelve shots had been heard up the river. "Was it at Ngao?" we asked. "Very likely." "About an hour ago?" "Probably." Then it was the German Mission welcoming the New Year, we concluded, and so went back to bed again. It turned out that this was the cause of all the excitement. But that innocent little celebration gave several hundred people a miserable night in the rain.

The next morning I started up stream again for Ngatana. The river, however, was in full flood, and progress was very slow. A series of heavy rain-storms fell upon us and further delayed our journey. We reached Dsundsa, a large Pokomo village on the right bank of the river, just after sunset, but thick clouds made it so dark that it was impossible to proceed. The elder of the village gave me a hut, where I sheltered while my garments were being dried by the fire. I had been sitting half the day in a pool of water, which the storms had left in our canoe. By three in the morning the rain had ceased, the clouds had dispersed, and the moonlight was sufficient to enable us to continue our voyage. We therefore started up stream again, only to be caught, an hour later, in another deluge of rain.

Shortly after daybreak we lost some time by an incident that nearly became an accident. One of the canoemen went ashore to try to spear a bird. The bowman happened to drop his paddle into the river, and while trying to recover it he lost his hold of the bank. The rapid current at once swung the canoe out into the stream, and swept us broadside on towards some "snags" in a rapid. Fortunately a galvanised iron washing basin with which I had been baling was ready to hand, and using this as a paddle I managed to steer the canoe down the rapids, and then to the bank in the quiet water below them.

At the next village I went ashore for some food. A native had killed a crocodile the day before, and it was now being cooked. The man who killed it was the hero of the village; he looked very happy, for the feat conferred on him the rights of manhood, and he could now be married. I made the usual sign of hunger, and the kind-hearted natives pressed on me the best they could offer. Crocodile intestine cut open and baked in the ashes of a wood fire is esteemed the greatest delicacy in the Pokomo bill of fare, and a supply of this was at once offered me. I pointed to some old Indian corn cobs beside the fire, and signed to them that I should prefer some of these. The natives appeared a little hurt at my low tastes, and pressed the crocodile's guts so warmly upon me that I felt bound to accept them, and grin with gratitude, if not with pleasure.

I gave the people some strings of beads, and returned to the mau with my treasure. My canoemen both smacked their lips at the prospect of a share in such a feast, and though I chewed the cartilage till my jaws ached, the men had a far larger share of the spoil than they had expected.

Late the same afternoon (2nd of January 1893) we reached our camp at Vuju in Ngatana, where I found that all was well. There was no news from Witu, so I settled down to rest and wait, hoping that the tax on my patience would not be a long one. We had now spent six dreary weeks on the low coast-lands, and I was yearning for the mountains and high plateaux of the interior.

CHAPTER III

COLLAPSE AND RETURN

> "Wot makes the soldier's 'eart to penk, wot makes 'im to perspire?
> It isn't standin' up to charge, nor lyin' down to fire.
> But it's everlastin' waitin' on a everlastin' road
> For the commissariat camel, an' 'is commissariat load."
>
> KIPLING.

A STARTLED cry of "Webbi! asan! dig!" (River! red! blood!) roused the camp at dawn next morning. I knew the first word meant river, and the last meant blood. I seized my revolver and cartridges, crawled through the low doorway of the hut, and ran across the village clearing to the river bank. I quite thought that Fumo Omari's men were already crossing the river, and that our sentry's cry of "Blood" meant that he was wounded. The mist hid the opposite shore, and in vain I scanned the river for any sign of a foe. The man then pointed to the water, and said it had turned to blood. During the night an extraordinary change had occurred in it; instead of the usual muddy brown its colour was now a dark blood-red. The floods in the upper part of the river must have washed into it some material coloured by red oxide of iron, and effected this startling change in its hue. "This is a bad river; we shall never go up it; this is a sign," said my old cook. I said it was only the result of "rain-wash," and joked about it, to try to prevent the men attaching any importance to the incident. But nothing I could say would shake its significance to them.

The cook, Hirsi, was by far the oldest man in camp, and as such his opinion was always received with respect. My assertion that rivers often changed their colour, and that it was as natural for them to do so as for mosquitoes to bite, was received with incredulity, and with the remark "Hirsi knows."

The water was now useless either for washing or for cooking, and we had to draw our supply from the swamp beside the village. This event made a deep impression on the men, who were gloomy and depressed for the rest of the day. Many times afterwards they reminded me of the incident, and warned me of the folly of trying to ascend a river that occasionally changed to blood. The last request made to me by the Somali, when I parted from them at Mombasa, was to admit that after all Hirsi had been right when he interpreted the "fal" or omen of the stream.

The day was not a lucky one, and it was only the first of a series which furnished the most unpleasant experiences I can remember. For things at the coast, in spite of the splendid efforts of Harris and Benett-Stanford, had gone amiss from the first. The Somali were out of hand, and the transport had completely collapsed. Up at Ngatana we knew nothing of this, and day by day we expected the reinforcements, medicines, and stores which the chief had promised should leave Fungozambo the day after we did.

We soon learnt by bitter experience the playful little ways of the camel. We were waiting for the "commissariat camel and his commissariat load," and we had several unpleasant illustrations of that animal's viciousness while doing so. One of the camels, which we had to keep with us, as it was too ill to march back to Witu, stupidly fell into the Tana. By two hours' hard work we cut a slope down to the river and hauled it ashore. As soon as it reached the top of the bank it shook itself, and then, as if to demonstrate the truth of Kipling's accusation,

> "And when we saves 'is bloomin' life, he chaws our bloomin' arm,"

it calmly and deliberately walked up to a Somali and proceeded to munch his shoulder. I was fortunately standing by with a spade, and at once opened a flank attack, which made the camel leave the man, but not before his shoulder had been seriously injured.

The drawbacks of our life here were increased by illness. My legs gave me a good deal of trouble; they had been sun-blistered during the canoe voyage down the river, and the over-exertion of the march from Witu to Borabini had produced a crop of ulcers. For more than a week I could not walk. Fever broke out in camp, and my two best men both had it in a serious form. Twice a day scouts went out to look for Count Lovatelli's party, which I expected to bring the medicine and stores we now urgently needed. Day followed day without any news, and we became very anxious. At length the welcome tidings came in of the approach of the party, which was nearly three weeks overdue.

Wasama marched into camp with ten camels, sixty porters, and fourteen Somali, but no medicines. He told me that there had been a good deal of illness and many deaths at Mkonumbi, and some at Fungozambo; that thirty-five Somali and the few survivors of the Turkish contingent had been sent back invalided, and that both our leader and Count Lovatelli had gone away. He added that the former had told one of the Somali that he was only taking the invalids to Mombasa in order to send them to Aden, and that he would then come back to us. Only twenty more camels had been landed, and five of these were useless. Wasama was very disheartened about the whole expedition, and very bitter in his complaints about mismanagement. We sent the men back the same day to bring up another batch of stores, and then set to work strengthening the camp. This, however, was interrupted by the outbreak of an epidemic of malarial fever combined with gastritis.

A day or two afterwards a canoe came up from Mackinnon, stating that he had just received a telegram from a friend in Mombasa to say that our leader had passed through that town on his way into the interior, and to ask if I knew anything about the matter.

As my services had been lent to our chief and not to the expedition, and I had been instructed before leaving London that in case of any disagreement or breaking-up of the expedition I was to keep with him, I felt very awkwardly situated. The last communication I had had from him was the letter at Kau, ordering me to take the advance guard at Ngatana. At

No. V. THE ZANZIBARI CAMP AT NGATANA. *Page* 38.

first I thought it hardly credible that he had gone off into the interior on an altogether different errand, without one word of explanation to Mackinnon or myself.

One of the many inconveniences of not knowing what contracts our late chief had signed happened next day. Greatly to our surprise a dhow was seen coming up the river; the Suahili in charge brought me a letter from Macquarie, stating that he had found the boat on the beach at Lamu. By an accident he had discovered that it had been bought and paid for by the expedition, and so sent it on. He asked me to count the oars, sails, etc., and then keep it till the others came up to decide what was to be done with this useless encumbrance. I thanked the man, and told him, as well as I could, to tie our boat to a tree. He refused, as it was his boat, and he was going up stream. Pointing to the paper, "Chombo changu" (My dhow), I said. "Hapana! chombo changu" (No! my dhow), he indignantly repeated, as he sprang into the boat and pushed off before I could follow. He merrily waved his hand and proceeded up stream. Two miles farther up the river the boat had to come close in to our bank in order to pass a snag. Having served out extra cartridges to twenty Somali, we marched thither and waited, hiding in the rushes. The boat came up, the crew suspecting nothing. A sudden rush and the dhow was ours, the Suahili and his men were prisoners. We pulled down the sail, and the current quickly carried us back to camp, the Suahili owner raging and cursing like a madman. Neither oaths nor entreaties had any effect on us, for the simple reason that not one of us understood a word that the poor man was saying. He gave it up at last, and sat in the bottom of the boat, an amusing picture of rage and bewilderment, alternating with despair. I was myself beginning, however, to feel very unhappy about this business. The man's indignation seemed so genuine that I was sure there was a hitch somewhere. Anyhow my orders were clear: I was to take possession of the boat that would be brought up the Tana by the bearer of the letter. This I had done, and I intended to stick to it till some solution of the mystery was forthcoming. On our arrival at Vuju we soon discovered the mistake, for there we had a man who knew both Somali and Kisuahili. This dhow had passed ours on the way, and as the men despaired of ever getting our clumsy boat to Ngatana, they had sent the letter on by the owner of the captured vessel, who was going up the river elephant-shooting, for which he had a permit from the British East Africa Company. I apologised, and gave him some coffee. We laughed over the matter; he gave me a fine pair of horns of the waterbuck, and then went on his way. It was lucky we caught him by surprise, for his party was well armed, and would probably have resisted capture. If his men had had time to get their weapons loaded, accidents might easily have happened.

On the 14th seventy more porters came up under Omari, our head Zanzibari; most of them returned the same day for another batch of stores. So many of the camp garrison were down with fever and dysentery that I had to retain a number of the newcomers in order to keep the camp work going. Most of my small stock of medicines had long since been exhausted, and the condition of the men was very serious. One great difficulty in treating the sick was that, owing to their fanatical prejudices, they could not be persuaded to take proper nourishment. Half of the available food was barred by the condition of their stomachs, the other half by the articles of their creed.

The Somali, as Islamites of the Islamites, would not take any meat that had not been killed by one of themselves. As the Abyssinians were Christians I thought I should be spared this trouble with them. One evening, therefore, I sent them my soup, though I could ill afford to spare it. As the sick men were no better when I went round to see them a couple of hours afterwards I made inquiries, and found that they had thrown the soup into the Tana; they explained that though it had been prepared by Christians, it had been made by such very different Christians from themselves that they could not drink it.

The result was the men sickened, weakened, and died. We had started for Ngatana with a stock of medicine calculated to last twelve men for four days. It had to last for a month, and during part of this time to supply ninety men, of whom on an average fifty were ill.

On the 17th I was taken ill myself; I tried to go my rounds as usual, but had to be carried back to the tent. Next

No. VI. THE DOCTOR "AT HOME. *Page* 41.

(A Scene in Camp at Ngatana.)

morning I was still weaker, and it was with the greatest difficulty I could write a brief note to Witu asking that Mackinnon should be sent up, as I could do no more. In the afternoon I was still weaker, and completely lost the power of focusing my sight. Wasama came to me for medicine. There were still a few tabloids left in some envelopes fastened together by a paper clip. But my eyes were so useless that I could not read the names on any of them; I could only find the quinine by licking a tabloid out of each envelope until I came to it. But other medicines could not be distinguished by taste, and the men accordingly had to go without. Several times they came and begged for medicine, but I was powerless to help them.

Next day the men were worse, but intensely to my relief Harris, Tichborne, and Benett-Stanford arrived from Witu. They found the whole camp in charge of a Somali corporal, as the headman and all the sergeants were ill. A message was at once despatched to Borabini by canoe, to tell Mackinnon to come up immediately. Two more Abyssinians, however, died before he arrived, in spite of Benett-Stanford's unremitting care and the fact that proper medicines were now available. Under Mackinnon's able treatment the men soon improved, but several of them were for some time in a very critical state of health. Mackinnon said he was called up at night more times during his first week at Ngatana, than he had been during his two years with Jackson's great caravan in Uganda.

After the arrival of the others I had a very easy time. They made me drop all camp work, and nursed me with the greatest tenderness and care. They did a good deal of shooting, and thus kept the camp supplied with fresh meat. Harris was especially successful in this, and when he went out rarely returned without a waterbuck or topi. The expedition stores furnished us with all sorts of European dainties. Under these conditions I soon recovered strength, though I suffered from a complete loss of memory. I could remember what had happened during the past few weeks, but everything else was in a fog. I could not recollect the simplest scientific terms, and any effort to read led to such violent headaches that I resolved on perfect mental rest.

My old friend, the son of the Pokomo chief, taught me

how to "mau," and together we had some delightful canoeing trips on the river, punting up stream in the morning by a pole against the bank, and paddling down again in the evening.

About this time our strength was further diminished, as Sir Henry Tichborne was recalled to England to fulfil his functions as High Sheriff of Hampshire. Benett-Stanford accompanied him to Mkonumbi, and returned with the news that 600 men from the island of Siyu had joined the rebels of Jongeni, and that the allies were marching to attack us. We strengthened our outposts, and prepared camp for defence. For some time after this we lived in a succession of alarms. The outposts would fire at one another, or mistake a herd of antelope for the enemy and give the signal. The Somali on these occasions would rush about the camp in a wild fury of excitement. I always felt very glad that my duty then was with the Zanzibari, who used to take up their stations in a very quiet business-like way. The promised attack never came off, though the natives were watching us closely, and several times guns were fired at night some distance from our camp, and certainly not by any of our own men.

After Stanford's return future plans were discussed. The remnant of the Abyssinian contingent was in a state of utter collapse; many of the men could not even carry their own food for a week, and some of them would not have lived through a hard day's march. It was decided, therefore, to send them back, to reduce our kit and repack it in suitable form, and then go up the Tana and possibly strike north for Basso Narok. As soon, however, as we began work on this plan, the hopeless nature of the task became apparent. There were tons of stuff that was absolutely useless, and indispensable articles were wanting. Our geographical equipment, in particular, was hopelessly inadequate. I had been told in London that no African expedition ever started with such a perfect scientific outfit, but it had neither barometer, thermometer, nor theodolite. It was also stated that a good library had been provided, but the only books we had on the district were the few scientific memoirs that I had taken with me and a copy of Höhnel's map. We were preparing to further explore the regions visited by Peters, Thomson, Piggott, Dundas, James, Fischer, and others,

No. VII. THE ABYSSINIAN CAMP AT NGATANA. *Page* 42.

(The mule is wrapped in calico as a protection against flies.)

but we had none of the maps or reports of any of these expeditions. Our camp equipment was less imperfect, but it was miserably unfitted for the work. The tents were rotten, and one of the three had been blown to pieces. After a week's work we were all in despair, though Harris and Stanford were still resolved to go on.

On 8th February a letter came from Witu, saying that the Administrator officially warned us against attempting to go near the Juba, as hostilities had broken out there and severe fighting had taken place between the Somali and a naval brigade. The day previously it had been decided that all the Abyssinians must be sent back, and that one of us must go with them to Aden, to arrange for their payment and transhipment back to Massowah. This decision, which was a necessary one, coupled with Mr. Piggott's warning, was the last blow that settled the fate of the expedition.

It was resolved that we should all return to Mombasa. There we could discuss plans, and either arrange for a short shooting trip to Kilima Njaro, or abandon the expedition altogether. I said that I, at any rate, should make another attempt to get inland. Harris and Benett-Stanford generously offered to set me free at once, and to let me take whatever men and stores I wanted. The offer was tempting, for it would have saved a couple of months' time. But in spite of the tons of material in the camp, it was impossible to equip a caravan of even thirty porters properly out of it. So I declined this kind offer, and thought it best to return to the coast with the others, and there make an entirely fresh start.

We all gave up the expedition very reluctantly, but it was the only thing that could be done. When the trouble came and our leader left us, Harris and Benett-Stanford did all that men could do to carry the expedition on. But our caravan fell, and all their efforts could not raise it. They worked with judgment, pluck, and perseverance, and did not give in till the absolutely hopeless nature of the task was apparent to every one of us.

Mackinnon's African experience was an extensive one. His verdict was that the only thing that could be done was to return to Mombasa, dismiss all the Abyssinians and all the

Somali but about six, engage another lot of the latter, and practically start the whole thing anew.

On the 12th of February Harris and I crossed the Tana with a few Somali and all the Zanzibari, to commence the march back to Mombasa; Mackinnon and Benett-Stanford left next day with the camels and Somali to return to Lamu *via* Witu.

In spite of the regrets with which I said farewell to Benett-Stanford and Mackinnon, I was very glad to leave that camp. I had spent seven weary weeks in it. The memory of the first four was so nauseous that I never left a place with such pleasure as I did that fever-haunted camp in the swamps and forest of Vuju. Both Harris and I were glad, moreover, to escape from the crowd of Somali, and to have to do with Zanzibari instead. We found these so much more useful, reliable, and obedient.

We resolved to cross the deserts to the south of the Tana and to strike a tributary of the Sabaki marked on the maps as the Ndeo. We met some Galla, and tried to persuade them to go as guides. They refused, for they said there was no such river. This information I proved to be correct during my return march from the later expedition. The Ndeo is lost in the deserts long before it approaches the Sabaki; it was, therefore, very lucky that we gave up the attempt to reach it.

We did not do so, however, without another try, and I went to Borabini to engage guides at the Galla settlement there. I started overland, but was taken ill and had to hire a canoe. I reached Borabini prostrated by another attack of malarial dysentery. The two missionaries were away, but fortunately for me Mr. W. W. A. Fitzgerald, a specialist on tropical agriculture, on the staff of the British East Africa Company, had arrived there just before me. He nursed me through a rather serious illness, delaying his departure for Witu till Harris could come down. Thanks to their careful nursing I shook off the attack. We rested at Borabini, until I was strong enough to sit on a donkey, when we continued our march to the coast. We reached this at Marareni, a village occupied during only part of the year by the collectors of the "orchilla-weed," which abounds there. Thence we marched along the coast to Mambrui, where we were kindly received by the Lewali or Arab Governor. We made a branch excursion to the Magarini Hills, where the British East Africa Company had some extensive plantations. The work was carried on, under the supervision of Mr. Weaver, by freed slaves and some of Emin Pasha's old Soudanese soldiers, who had been brought to the coast by Stanley.

The next march took us to Melindi, where we had the pleasure of meeting the late Mr. Bell Smith. Harris continued overland to Mombasa with the Zanzibari, but as several of the Somali were too ill to march, and we were anxious that they should get to Mombasa in time for the German steamer, I went south with them on a dhow.

We ought to have done the passage in a day. But we made a late start, the winds were contrary, and at nightfall we had only reached Takaaungu, about half way. The captain would not sail in the dark, so we ran into the harbour and anchored. I insisted on being landed and sleeping on shore, for the stench of the dhow, and the quarrels of the Somali with the other passengers and the crew, were not conducive to slumber. We started again at daybreak, and drifted slowly southward. About eleven o'clock the breeze freshened, and we had to take down the awning. After this we had to sit huddled up on the poop in the full glare of the midday sun, which made the deck so hot that it could not be touched without discomfort.

We arrived off Mombasa late in the afternoon, and stood in at once through the narrow channel that leads to the harbour. To the south rose a dazzlingly white cliff of raised coral rock, worn into hollows and caves by the waves that splashed its face with clouds of spray; to the north stretched a wide sheet of surf over the Leven Reefs, on which Vasco da Gama tells us that the native pilots tried to wreck his ships. We passed the new hospital and the famous old fort which guards the town, and threaded our way through a crowd of shipping, which included a British cruiser, a German merchant vessel, and native craft of all sorts and sizes. There were "batili" from Muscat, vessels with a long projecting prow like the "counter" of a racing yacht, massive "bagalas" from Bombay, with square sterns and high poops, reminding one of old Spanish galleons, and graceful light "mtepes" from Lamu, with lines not unlike those of an English racing cutter. With shortened sail and wind abeam, we worked slowly up the harbour against the falling tide, and anchored beyond the quay. I went ashore at once, expecting to arrange for the transference of the Somali to the German steamer which was to leave the same night. I therefore told the men to stop on board until my return. As there was no news from Benett-Stanford, the men could not be sent on, so I hired a house for them in the town and went back to the dhow. Here I found that, with characteristic disobedience, all the Somali, except my cook Hirsi, had gone ashore the moment I was out of sight. I had all the baggage transferred to a Customs shed or "go-down," and while I was watching its removal the Somali came up for their "posho" or food money. The usual rate is sixteen pice[1] for a Somali and eight for a Zanzibari. I gave Hirsi twenty pice as a reward for his obedience, and refused to give the others more than seven as a punishment, for I said they were worth less than Zanzibari. They regarded the insult as worse than the injury, and in a fit of passion declared that rather than accept seven pice they would take nothing. I took away their rifles, belts, and ammunition, and told them to wait for Harris, for they would get nothing more out of me. They stormed and raged for their sixteen pice posho—of course without avail.

While this was going on, the last of the luggage was brought ashore and stowed in the go-down. When it was all safely landed, and the Customs were responsible for the safety of the luggage, and the Mombasa police for the future behaviour of the men, I was free.

The clerk gave me a receipt for the baggage, and I took my place on Mr. Piggott's trolly. The Somali redoubled their protests, and in an appropriate sense of clamorous hurly-burly, my connection with the great Lake Rudolf Expedition came to an end.

The trolly soon carried me out of the hubbub, and I could quietly think over plans for another try. I had learnt one lesson from my late leader, who had at least taught me that

[1] A pice is a sixty-fourth of a rupee, so sixteen pice are worth about fourpence.

the best way to conduct an African expedition is not to throw it up as soon as difficulties occur, and to leave one's own responsibilities to others. So I looked back to the past with gratitude for the one lesson, and to the future with hope. For I felt that if the Zanzibari proverb be true—

> " Kupotea njia, njiko kujue njia "
> (To lose the way, that is the way to know the way)—

then, having lost the way under the guidance of another, I had learnt the first lesson in the art of finding the way for myself.

PART II

TO BARINGO AND MOUNT KENYA

"Let it be virtuous to be obstinate."
Coriolanus, v. 3.

CHAPTER IV

AT MOMBASA—A SECOND START

"Kongowea ya mvumo, maangavu maji male
Haitoi lililomo. Gongwa isingenyemele
Msiotambua ndumo na utambaji wa kale."

Verses of Muyaka on Mombasa, collected by Rev. W. E. Taylor.

(Mombasa roars like the wind through the fan-palm,
Surf breaks o'er its reefs in calm as at spring tide;
It ever is sending forth, yet ne'er has it fail'd,
Think not 'twill be peaceful as you hear not its seething,
For you know not its war-cries, nor its story of old.)

THE town of Mombasa is situated on an island, which, owing to the healthiness of its climate, the fertility of the neighbouring country, the convenience of its harbours, and the strength of its position, has exercised an influence on East African history out of all proportion to its size. The town stands on the edge of a cliff on the northern shore; to the east stretches a plateau of coral rock, healthy and dry, and open to the bracing breezes from the sea. To the west the country is richly wooded and fertile, but in places swampy and malarious, for the red sands of which it is composed yield a soil very different from that formed from the coral rock. To the north and south of the island are the harbours of Mombasa and Kilindini, which, with the exception only of Delagoa Bay, are admittedly the finest on the eastern coast of Africa. A shallow creek, which can be forded at low tide, separates the island from a wide tract of rich food-bearing country, which now, as at the visit of Ibn Batuta in the fourteenth century, forms the granary for the inhabitants of the island. With all these advantages, it is not surprising that

Mombasa has long been recognised as occupying one of the most important situations on the coast of Tropical Africa, and that, when it is first mentioned in history, it was already an important town.

Of the foundation and early history of Mombasa we know nothing. It is said that there are some old manuscripts at Lamu, in the possession of the great Mazrui clan, which could tell the story of the Arab settlements on this coast. These precious documents, however, are so jealously guarded that no European has been allowed to see them. Until their evidence is forthcoming, the history of Mombasa begins with the visit to it in 1331 of the famous Arabian traveller, Muhammad Ibn 'Abd Allah, or, as he is generally known, in order to distinguish him from the six other authors of the same name, Sheik Ibn Batuta.[1] He spent one night in the town during the course of his 75,000 miles of wanderings. His account of the place seems to be as accurate as most of his records are. His experience of the people was very different from that of his successor Vasco da Gama, who called in April 1498 on his first voyage to India. The Arabian found the people "generally religious, chaste, and honest." Da Gama describes them as hostile and treacherous. According to his own account, the king sent on board a present of fowls and fruit, and begged him to bring his ships into the harbour: pilots were lent for this purpose, who had been really ordered to run the ships on the reefs. The Portuguese records do not agree among themselves, and it appears probable that no such treachery was intended. Da Gama went north to Melindi without personally landing on the island, though the two men whom he had sent ashore had been kindly treated by the natives.

From this time onward the history of Mombasa fully justifies its native name of "Mvita" or "battle." Two years later, after Da Gama's return to Lisbon, the town had to pay the penalty of having aroused the Portuguese suspicions. A fleet was sent to India to annex and proselytise; its commander, Cabral, was ordered "to begin with preaching, and if that failed, to proceed to the sharp determination of the sword."

[1] The standard edition of his voyages is *Voyages d'Ibn Batoutah*, 4 vols., Soc. Asiat. Paris, 1853-59. For most of the later history I am indebted to Burton's *Zanzibar*.

Cabral apparently preferred the latter method of persuasion. In 1500 he avenged the supposed treachery to Da Gama by looting Mombasa. Three years later another expedition passed along the coast, and made the town pay tribute ; and in August 1505 a fleet of twenty ships, under the command of Francisco Almeyda, attacked and destroyed it. In 1508 the Portuguese formally took possession of the island, and appointed a resident governor. But they had by this time annexed more of the world than they could conveniently manage ; to help in doing this they proceeded to classify their possessions. They divided Arabia and Ethiopia into three provinces, in the first of which they placed Mombasa. These proceedings resulted in a *pax Lusitanica*, the value of which we may estimate from the fact that as the Mombasa people were said to have treated the natives of Melindi and Zanzibar badly, it was necessary in 1528 to send a powerful force under Don Zuna da Cunha to raze the city to the ground. This, however, was not so easily effected ; the town was so well defended that it was not taken till after a siege of four months. The next opportunity for mischief that presented itself was on the occasion of a visit from a Turkish fleet under Ali Bey in 1586. A generation had by this time grown up which knew not Da Cunha, and so could not restrain its innate passion for intrigue. The natives placed the island under the suzerainty of the Porte, but neither of the high contracting parties gained much by this arrangement. The Portuguese Viceroy of India sent Alfonso de Melo Bombeyro with a fleet of eighteen ships to punish this act of rebellion, and Mombasa was promptly reduced to ashes. The town soon grew up again, but only to be looted by a tribe of barbarians named Zimbas, who came from the south. Their occupation of the island was, however, a very short one. The Portuguese seem then to have grown tired of this constant series of evictions, and built the fort which is now the most picturesque building in East Africa. This secured peace on the island until 1630, when the fortress was captured by stratagem. The members of the garrison were shot by arrows, although they had surrendered on condition that their lives should be spared. A force from India was at once despatched to recapture the citadel and punish the apostate mission boy, who planned both the revolt and the massacre. For three months the rebel leader, Yusuf bin Ahmed, or, to give him his baptismal name, Don Jeronymo Chingoulia, held his own; then he dismantled the fortress and fled to Arabia, in a ship that he had captured from his foes.

In 1660 trouble came from a fresh quarter. The Iman of Oman, more generally known as the Sultan of Muscat, laid siege to Mombasa and, after a struggle of five years, captured the fort. The town, however, held its own, and it was not till 1698 that the Portuguese were for the last time expelled from the island. The victory had been gained by the united forces of the Iman and of the Mazrui, the leading clan of Arab settlers on the coast. The latter gained most of the fruits of success; for though Mombasa was nominally subject to the state of Oman, the Mazrui were practically undisputed masters of the town. This lasted till 1822, when Sayyid Said, the greatest of the long line of Imans, resolved to exercise the powers that he had nominally inherited from his ancestors. His designs were discovered, and the Mazrui appealed to England for protection. This was provisionally granted, but almost immediately withdrawn. A fierce struggle then ensued and lasted for five years, ending only with the capture of the fort by treachery. Mombasa then finally lost its independence. It was at first ruled by an Arab governor or "Lewali," in the name of the Iman of Oman. After the death of Sayyid Said at sea in 1856, his sons quarrelled. The question was referred to England for arbitration, and the award of Lord Canning made Zanzibar independent of Muscat. The African possessions of the Iman were assigned to Zanzibar, to which town Mombasa has ever since been subject. In 1875 the people struck another blow for freedom, and their revolt was only suppressed by British interference. It was probably the strength of the revolutionary party that led Sayyid Barghash, who was then Sultan of Zanzibar, to offer the late Sir William Mackinnon in 1877 a lease of all his possessions on the mainland. The offer was then refused owing to the timidity of the English Foreign Office, and it was not till German competition had roused English fears, that the British East African Association was able, in 1887, to accept a lease of the country. Next year this Association became the Imperial

British East Africa Company, with Mombasa as its head-quarters; and though the Arab Lewali is still nominally Governor, and the blood-red flag of the Sultan of Zanzibar still waves over the fort, the Company's Administrator became the real ruler of the town.[1]

The Administrator's residence is on the south side of the island, on a cliff commanding a fine view over the broad harbour of Kilindini and up the long estuary of Port Reitz to the distant summits of the mountains of Shimba. This fine sheet of water was named after J. J. Reitz, who surveyed it for the Admiralty in 1824, and died at the completion of his task. The house is connected with the town by a narrow gauge railway, along which garrys are pushed by native runners. Mr. Piggott kindly invited me to stay with him, and to his home I followed him after I had settled up affairs with the men.

The ride across the island was delightful; the narrow line "switchbacked" up and down hill through park-like scenery, of which the beauty was the more fascinating by contrast with that of the monotonous scrub-covered plains I had previously seen. We dashed through groves of lofty cocoa-nut palms, past rugged baobabs and massive shady mangoes, between which were glades radiant with meadow-flowers and orchids and festooned by creepers and vines of the india-rubber plant (*Landolphia*). Rushing down hill in a sharp curve, we startled a pack of monkeys who were playing on the line: they fled barking to the woods as the garry rattled past them. Mr. Piggott met me in his garden, and took me for a ramble along the shore to examine some beds of coral limestone, so crowded with garnets and fragments of emerald and beryl, as to resemble an ancient metamorphic rock or "calciphyr," rather than a recent coral rock.

It did not need any contrast with camp life on the Tana to make my stay in Mombasa enjoyable. The scenery of the island is most fair, the air from the ocean is fresh and bracing, and the members of the European colony were most

[1] The efforts of the Company were not fortunate in Africa, and therefore were not appreciated in England, and in 1895 the country became a British Protectorate. Had the Company cared less for the natives and more for itself, had it been more selfish and less patriotic, had it endeavoured rather to secure dividends for its shareholders than the freedom of the slaves in its dominions, its efforts would probably have been estimated with approximate justice, and the judgment passed upon it by public opinion might have been approximately fair.

entertaining and hospitable, and soon made me forget that the only clothes I had fit for Mombasa were, or rather were supposed to be, coming on a dhow.

I went into Mombasa on the day after my arrival to discuss my plans with the different members of the European colony. The result was not encouraging. I was too limited both in time and money. I had only six months' leave of absence from England still left, so I must be back at the coast in time to catch the mail at the end of August; and I felt that I could not afford the expense of a caravan of more than thirty men. My friends in Mombasa declared that both the time and force were insufficient for an expedition to Baringo and across the plateau of Laikipia to Kenya. The two most experienced men then in Mombasa were James Martin and C. W. Hobley. The former had accompanied Joseph Thomson in his expedition to the Masai country in 1883, and had been employed in caravan work in East Africa ever since that date. He had managed Sir John Willoughby's shooting party to Kilima Njaro and Sir Robert Hunter's up the Tana, and had been to Uganda seven times. His experience of the country is therefore unrivalled, and I was naturally much depressed when he told me that he thought seventy men the minimum with which I ought to try to reach Baringo, and that even that number would be insufficient on Laikipia. Both Martin and Hobley told me a good deal about the trade goods required on the journey, and I returned to Kilindini feeling more grateful for their information than for their advice.

Harris arrived next day, and Benett-Stanford the day after that. The former had marched overland from Melindi, and the latter had come down by dhow from Kipini, at the mouth of the Ozi. They left immediately for Zanzibar to take the Somali back to Aden. They both very kindly warned me not to forget my recent illness before deciding to go inland again. When I told them I had quite made up my mind to do so, unless recalled by the mail due in a few days' time, they most generously invited me to take whatever I wanted out of the stores of the expedition. I parted with Harris and Benett-Stanford with deep regret. The three months we had spent together had been a time of worry and disappointment and bitter vexation of spirit, but I had never had a single unpleasant word from either of them. They had treated me throughout with the greatest kindness and consideration. When I was ill they nursed me with womanly tenderness, and during my convalescence they insisted on my dropping my share of the camp work until my strength was quite restored. Whatever help I wanted in my natural history studies they were always most ready to give. It was no fault of theirs that the expedition was then being disbanded at Mombasa instead of in camp on the shores of Basso Narok.

Until the arrival of the English mail I could do nothing. My services had been lent to our former leader to help him in exploration. It seemed to me quite possible that the authorities might take the view that as soon as I found he did not want me, it was my duty to return home. I did not, therefore, feel justified in beginning to enlist my men till the mail arrived. I could, however, have done nothing in the interval, as I was prostrated by an attack of fever. As a result of this I went into Mombasa to get my letters, feeling more inclined to attach value to the warnings of my Mombasa friends. I looked again through a list showing the cost of different expeditions, ranging from Count Teleki's at £30,000, Mr. Jackson's at £25,000, down to more moderate ones such as Mr. Joseph Thomson's at £3000. Even over our expedition about £7000 had been expended. When I contrasted my limited resources with such figures as these, and thought of the history of previous efforts to explore Laikipia and reach Kenya, I began to feel that perhaps after all it would be as well if I were recalled home.

The letters came. They showed that the news of our leader's departure for Uganda was known in London before I knew it at Ngatana. There were, however, no instructions, so I felt at liberty to decide for myself. I rechecked my estimates of time and cost. I could find nothing wrong with them, so I sent off a message to Omari to say that if he were ready to go with me to Baringo and Kenya with a caravan of forty men, he was to meet me at the Transport Office at eight o'clock next morning.

He was there at the appointed time and was quite ready to go, so I engaged him at once as headman or "munipara." We talked things over, Hobley kindly acting as interpreter, for Omari could not speak a word of English.

The selection of my headman was a very important step. The success of the expedition and the harmony of its elements depended to a large extent on his nerve and trustworthiness, and also on his interest in the work. Several men were recommended to me on the ground of their knowledge of the country I wished to explore, but, for one reason or another, I did not take to any of them. One looked lazy, another sly, a third promptly destroyed his chances by expressing doubts as to the safety of the journey with so small a caravan. A famous old headman, named Wadi Bunduki, was the best of those recommended. His knowledge of the country around Baringo was unrivalled, and would have been most valuable. He seemed a most charming man, and thoroughly to deserve the estimate conveyed in the designation of "dear old fellow" which everybody gave him. I did not, however, regard the two features of dearness and age as the leading qualifications for the post. I resolved, therefore, to take Omari ben Hamadi. He had been the headman of the Zanzibari contingent in the previous expedition, and I had then some experience of his personal qualities. His conduct had won my respect. He seemed just the sort of man I wanted. He was young and ambitious, of immense physical strength and inexhaustible energy. He had a splendid record, having served under Stanley on the Emin Pasha Relief Expedition, and had been headman to the British East Africa Company's expeditions up the Tana and Juba rivers, besides having taken part in several less important journeys. The fact that the last expedition on which he had been engaged had failed was distinctly a point in his favour. The failure was due to no fault of Omari's. Nevertheless he had been so much chaffed about it by jealous rivals on the coast that he might be trusted to do his very best to prevent returning after a second breakdown. The chief objection to him was that he did not know the country: beyond Machakos, in Masailand, on Laikipia and at Njemps, he was as complete a stranger as I was. This, however, did not seem to me altogether a disadvantage. I thought it would make him all the more anxious to see the country, because the experience there would be of great value to him in the future. Moreover, his ignorance would give me some power over him; in districts that had been mapped I knew more of the country ahead than he did. This doctrine was regarded in Mombasa as altogether unsound. It was impressed on me that the first requisite in a headman is knowledge of the country through which the caravan is going to pass, and of the people dwelling there. I was very reluctant to act in contravention of local advice, but after pondering the matter carefully I resolved to take Omari, and well indeed did he justify the choice.

I went through the list of the men of our abortive expedition with Omari. I had watched them carefully, and so could tell him the names of those I wanted and of those whom I did not want. I said that I would not take a Somali at any price, though all of those of our former expedition then in Mombasa begged to be allowed to come. I was sick and tired of the very sight of the people. I gave Omari also a list of the languages for which I wanted interpreters, and told him that, if possible, he was to engage two or three of the men who had been with Count Teleki. If none could be found in Mombasa, they must be obtained from Bagamoyo or Zanzibar. As I wished to leave in a week, there was no time to lose. We set to work at once —he to hunt up recruits and I to collect stores. Mr. Piggott kindly lent me thirty-five Snider carbines, and I also took five of our old long Sniders for the headman and the four Askaris.

After a week spent in purchasing and packing, everything was ready. I was grieved to have to start with so incomplete a geographical equipment, but the necessary instruments could not be obtained. At first the porters came in very slowly. This was partly because they did not like Omari's manner —prompt and business-like, but unsympathetic—and partly because the great Mohammedan festival of Ramathan was at hand, and the men did not like to lose the fun. It was not till the 22nd of March that the last man was enlisted and the last load packed. I originally intended to start on that date, but as the mail from Zanzibar was a day later than was expected, I had to postpone our departure until the 23rd.

After a revision of my list of trade goods, my "ways and means," by Mr. Martin, who had returned with the mail, we read out the roll-call. Three men were missing, and we had to send out some of the Company's police to bring them in. Another porter, whom I knew and did not want, had given himself a new name, and signed under it. I refused to have him, but by the kindness of Hobley one of the Company's men took his place. These delays made it eleven o'clock before the last load was tied up and Omari was able to report that all were present. The men then formed line outside the Transport Office; each had a load assigned to him; arms and cooking pots were distributed. At last all was ready, and I was able to give the much-desired order to start. The "kiringozi," or guide, decked in a gorgeous Busoga goatskin, took his place at the head of the line and unfurled the flag of the expedition. A couple of drummers hired for the occasion began to beat their noisy tomtoms. Then the men struck up a marching song, and amid the cheers of a crowd of natives the caravan started on the journey. Shortly afterwards we despatched some men with ten loads of food and two of ammunition, which I preferred not to distribute until we were on the mainland. To have given them out before would have increased the temptation to desert, or at least to sell the cartridges. I still had to settle some accounts, and, in accordance with the Brussels Convention, to sign a guarantee not to part with arms or ammunition, and receive a license to carry them. Not until four o'clock could I get away, by which time the porters were already in camp on the mainland. My own pleasure in the start was lessened by an attack of fever, and I could not get up an adequate appreciation of the kind efforts of two of the Company's officials to cheer me by a parting song continued as long as I was in sight. The refrain of the song—and it seemed all refrain—was the question, "Will he nae come back again?" I walked across the island to the old fort of Makupe, which was built to guard the ford to the mainland in the days of the long struggle between the Suahili and the Portuguese. I crossed in a dug-out canoe, and at once began the ascent of the slope that leads to the great inland plateau.

On the way I picked up a few fragments of some ammonites, which occur in the clay that forms the lower part of the slope. These were the only fossils I expected to find throughout the expedition. So in spite of their fragmentary condition I stowed them away in one of my capacious pockets. I strolled on to camp feeling grateful to these, for they brought pleasantly to my mind, for a few minutes, memories of the charming valleys in the Cotteswolds where I had first collected specimens of them. But thoughts of the past soon gave way to the needs of the present—one of which was rest.

At daybreak next morning Omari served out nine days' rations of food and ten rounds of cartridges to each man. We read over the roll-call once again and I counted the men. Happily of the forty-one there was no absentee. There was the head-man, four Askaris, two of whom had previously been headmen and had had considerable experience, thirty-three porters, a cook, and a tent-boy. These were the forty, and to them I added myself. I had endeavoured to get a European to accompany me. Especially I wanted to find a sailor from a man-of-war, who could shoot game, skin animals, mend the guns, drill the men, take his turn as sentry on the sentries every other night. But no European could be obtained of any kind or on any terms, so I had to start alone.

The caravan was certainly small for the work before it ; but it looked as if it meant business. For its size it was very complete. We had interpreters for the languages spoken in the Taita Mountains, in Ukamba, on Kamasia and at Njemps, and several for the languages of the Masai and the Kikuyu. We had men who had served in most of the caravans whose routes I wanted to connect with mine ; we had one of the two men who had been with Count Teleki to his highest point on Kenya, and one who had been with the British East Africa Company's expedition as far as it went on the same mountain. Some of the men were almost new to caravan life, and two or three I would not have taken could I have made up the list without them. But most of the men were reliable, and we had a good proportion of veterans, including several of Stanley's Emin Pasha Relief Expedition.

Allowing myself but one last look back over the island of Mombasa, I followed the men out of camp on to the path, and commenced for the second time, but now with feelings of greater confidence, the march toward that remarkable feature of East African geography—the Rift Valley.

> "Whither, with what haste
> The weight we must convey with us will permit."
>
> *Ant. and Cl.* iii. 1.

CHAPTER V

ON THE UGANDA· ROAD

"At half-past five's Revelly, an' our tents they down must come,
Like a lot of button mushrooms when you pick 'em up at 'ome,
But it's over in a minute, an' at six the column starts,
With its best foot first
And the road a-slidin' past,
An' every bloomin' campin'-ground exactly like the last."

KIPLING.

THE east coast of Africa has a notorious reputation for unhealthiness. This it has acquired mainly from the reports of the earlier expeditions into the interior, which entered the country by routes to the south of the British possessions. There, an inland plateau is separated from the sea by a broad belt of marshy and malarious lowland, where fever and dysentery are rife. Explorers, therefore, had to encounter the most unhealthy part of their journey when they were least prepared for it, and at that date the exact nature of the diseases and the most efficacious remedies were unknown. West of Mombasa, however, the hills come so close to the sea that the worst of the fever belt is absent, and the first day's march ends on the edge of a high plateau. As a matter of fact, the difficulty of the Mombasa road is usually lack of water. In the dry season it is sometimes necessary to make marches of thirty-five, or even fifty-four miles, from one water-hole to the next. This difficulty, encountered at the beginning of the journey, when the loads are at their heaviest, and the men are fresh from the idle life and relaxing climate of the coast, is a severe trial to a caravan.

But I anticipated that we were more likely to be troubled with an excess than with a scarcity of water, for the great rains were due. That we should not have long to wait for them was indicated by the dense clouds that hung in the west, and the weather-wise on the coast predicted a "masika" or rainy season of exceptional severity. I had, therefore, made no special arrangements for carrying water, and felt a distinct satisfaction when, the day after we had left Mombasa, the rains began.

We struck camp very leisurely, as I had promised that we would that day go no farther than Maseras or, as it is otherwise known, Gonjeni. This place the men assured me was five hours' march inland; but I knew it ought to be reached in three. I consented to so short a march ostensibly as a favour to the men, though I felt it would be quite as much as I could manage, for my fever was worse and had entailed a breakfast-less start. The path almost immediately joined a narrow gauge railway, which was laid as the beginning of the projected Uganda railway. It now ends, however, at the foot of the hill of Mazeras, only seven miles from the coast. Even for this short length it is not intact; in places the line is overgrown by shrubs; here and there the white ants have removed the foundations of the sleepers, rain has breached the embankment, and the natives have at intervals stolen a few lengths of the rails. The track passed over undulating, richly-wooded country, with clearings in which were groves of cocoa-nut palms and papaws, orchards of mangoes and cashew apples, plantations of banana, maize, and millet, and fields of cassava (manioc or arrowroot), and of the plant the seeds of which yield simpsin oil. This district is the granary and market-garden of Mombasa, as it was in the days of old Ibn Batuta. The ground rises gradually to the foot of the steep hill of Mazeras, the summit of which, 600 feet in height, we reached just before a heavy storm of rain burst upon us. I at once took up my quarters in a comfortable iron hut, that had been erected for the manager of a mine once opened here. The workings show a thin vein of galena or sulphide of lead, which was prospected by the British East Africa Company; the ore is, however, in thin strings, and the specimens I saw appeared to be very poor in silver. An hour after my arrival, while I was taking my frugal dinner of a cup of arrowroot, I was startled by a visit from Mr. Edmonds, the missionary who was stationed at

Golbanti at the time of my first visit there. His residence on the Tana had shattered his health, and he had been dangerously ill since I met him; he had, therefore, been transferred to this station, as it was supposed to be healthy. Both he and his colleague, however, still suffered a good deal from fever. He took my temperature and said it was 106° F., and therefore kindly tried to persuade me to rest for a day or two. The fear of my men deserting if allowed to stop so near Mombasa weighed far more heavily on my mind than did my temperature; so I felt compelled to go on. I tried in vain to buy a donkey, but the missionaries kindly lent me one for a couple of days.

Early next morning, after a sleepless night, I was lifted on to the donkey, and followed the caravan, which had already started. The road ran for some distance along a turf-clad ridge which commanded splendid views to the south and west. The prospect to the south was especially fine, as it included the broad estuary of Port Reitz, with the rounded hills of Shimba at its head. Thanks to the rain, the grass was green and the whole country looked fresh and luxuriant. The rocks in this district are much older than those lower at the shore, in which occur the ammonites. They here consist of coarse-bedded sandstone and grits with layers of pebbles; the series appears to have been formed as a shore deposit on the slopes of the old gneiss plateau that extends inland. I did not see any fossils, but I could not hunt properly for them; there were some promising-looking shaley layers, which it was hard indeed to pass unsearched.

After leaving Mazeras we saw no more villages or signs of cultivation, and entered the great barren "Nyika" (see chap. xii. p. 223), inhabited hereabout only by some few families of Wa-duruma. After a march of six miles we reached a small stream bed, a series of holes filled with yellowish-green and filthy stagnant water. I was too ill to go farther, so we camped. Half an hour later Edmonds arrived to go on with us for a couple of days and bring back the donkey. It was a good thing he came, for I was delirious all the afternoon, and even in the lucid intervals could not read the labels on my medicine bottles. Next day we made a better march of ten miles through a drenching downpour of rain. The fever had subsided, but it had left me very weak, and for the last two miles I had to be held on the donkey. We camped on a hill-top amid some flat slabs of carboniferous sandstone, beside some pools of water. From these the place derives its name of "Maji Chumvi" or "Salt-water"; they are usually brackish, though after the heavy rain of the last few days they then tasted quite fresh. Next morning the weather was finer, and camp was struck by daybreak. Edmonds then had to return; he gave me on parting a good deal of somewhat startling medical advice, which I simply noted as not to be used, but which I fear he practised on himself. He then jumped on to the donkey and rode off, shouting back to me the promise of a hearty welcome on my return. The welcome was never given: the next news I had of my poor friend was of his death.

This day's march took us well out on to the Taro Plains; these are part of the barren plateau that, throughout Equatorial Africa, separates the edge of the coastal slope from the foot of the volcanic mountains of the interior. Before us lay a vast expanse of uninteresting-looking country, rising slowly to the west. It was covered by a dense scrub of flat-topped, umbrella-shaped acacia trees; the turf of the coast zone was replaced by tufts of dry grass; huge aristolochias, spiny sansevieras, and aloes formed the most striking features in the undergrowth. The only things that broke the monotony were the massive trunks and knobbed branches of huge baobabs that rose through the scrub, and rugged peaks of gneiss that towered above the plain on the western horizon. The country did not promise much of interest to the naturalist, and even if it had done so I was not well enough to collect. The fever had left me, and my temperature had even fallen two degrees below normal; but I was terribly weak. This was not surprising, as for three days I had taken only a little arrowroot. I stepped out as vigorously as possible for the first two hours, and then suddenly collapsed. The caravan had gone far ahead, and I had to follow it; I could, however, only drag on for a few hundred yards at a time, and then lie down to rest. It was the most dismal experience throughout the expedition. At last I could go no farther until Omari sent on to the camp, had a kettle of tea made and brought back to me. It was in this district that many of Thomson's men broke down for want of water; he, with some of the stronger porters, had to go on to get water and return with it to the rest. While lying on the ground waiting for the tea, I could not help contrasting my feeble condition with Thomson's rescue of his men, and thinking that African explorers should be made of tougher stuff.

In the evening a party of Wa-kamba passed with some donkeys, which they were driving to Mombasa to sell. I sent Omari to try to buy one, but the natives demanded the preposterous price of half a load of cloth, a load of iron wire, and twenty shaddas (or 2000 strings) of beads. We declined the offer, and charitably wished they might get their price in Mombasa.

Next morning I was sufficiently better to achieve the march of nine miles in a little over four hours. Camp was pitched before my arrival beside the famous "Ngurunga," or water-holes of Taro. These Ngurunga are a series of holes in sandstone, and they are often the only source of water in this part of the Duruma country. They resemble in form the "pot-holes" formed in boulders and in the rocky beds of mountain streams by the swirling round of a stone resting in a hollow. The Ngurunga have certainly not been formed in this way, and the exact method of their origin has often been discussed. They appear to be due simply to the gradual deepening of slight depressions in the surface of a rock; water collects in these, and its solvent power is increased by the organic acids derived from the decay of vegetable matter; the cement between the sand grains is dissolved, and the grains thus loosened are removed by the wind when next the water has evaporated from the hollow. A continual repetition of the process results in the formation of a deep hole. Mr. Joseph Thomson suggests that man aids in their formation by scraping away the sand from the bottom when drawing the water. This, no doubt, has often happened; but Ngurunga as large and deep as those of Taro occur in districts and in positions where human agency cannot have aided in their erosion.

At Taro begins the worst march in this stage of the journey. From this point there is often no more water to be had until a pool is reached on the summit of Maungu, thirty-five miles away. The water-hole there is sometimes dry, and then a caravan has to drag on for twelve more weary miles to

Ndara. Fortunately I was so much better that it did not seem necessary to delay the commencement of this march. I was cheered by the capture of a lizard about two and a half feet long (*Monitor albogularis*, Daud.), and began to take sufficient interest in things in general, to record the few meteorological observations which my imperfect outfit allowed. In order to lessen the fatigues of the journey, however, it was decided to start at midnight and make three short marches next day. It poured with rain all night, and the clouds made it so dark that we had to postpone our departure till the morning. As soon as the first streak of dawn appeared, we filed out of camp on to the muddy track. Anxious to avoid exposure to the mid-day sun, I went on ahead as quickly as I could. Just after daybreak, on turning a sharp corner, I surprised a herd of five sable antelope (*Hippotragus niger*, Harr.), which dashed off at once through the jungle. Had I known that this was the only time that I should have seen this, the noblest as well as one of the rarest of African antelopes, the temptation to follow them would have been irresistible. A few paces farther on was the fresh "spoor" or track of a lion, which appeared to have only just crossed the path. Probably it was stalking the herd when I came up, and had occupied the attention of the antelopes, so that I was allowed to approach close to them before they noticed me.

A sharp march of ten miles brought us to Butchuma, where we rested till two in the afternoon. Then we started off again, to try to reach a camping-place at the end of the road cut through the jungle. At six o'clock, however, we found a pool, and, as Omari was far behind with a sick Askari, I decided to camp there. We started again before dawn for a march, which the men said would take nine hours, to the camp at Maungu. Soon, to our great regret, we reached the end of the road; we had then to follow a native path, which twisted and twined in every direction through the scrub. This is here so dense that it is impossible to go straight ahead without cutting a way. When, therefore, the natives began to traverse the district, they had to follow the game tracks; and the path thus made has since been rigidly adhered to. We could often see our goal, the peak of Maungu, rising through the scrub, apparently near at hand. Sometimes it was on our right, and as often on our left; occasionally we were walking directly away from it. But we had to follow the trail in all its twining, for every attempt at a short cut ended by an entanglement in thorns and creepers.

Shortly after eight o'clock we reached three pools, known as "Mawiza Matattu." We halted there for a short rest, and then plunged again into the maze. This was unwise. The midday heat was too much for me, and brought on an attack of bilious fever. I had to rest till the afternoon under the shade of some trees. When I started again I could only go slowly, and the path through the scrub seemed interminable. At length, however, we began to ascend the hill. Great candelabra-shaped euphorbias and rock exposures became numerous. The path rose to a depression in the ridge, and then descended to the camping-ground. This is situated on a terrace at the height of 2200 feet, about 500 feet above the plain, across which it commands a glorious view. Forty miles to the north, on the other side of the river Sabaki, is the long low line of the lava-capped plateau of Yatta; sixty-five miles to the east-north-east the rounded dome of Mangea can be discerned upon the skyline. A few miles to the west the "Nyika" ends abruptly at the foot of the Taita Mountains, whose rugged crests and precipitous slopes form a pleasant contrast to the monotonous uniformity of the plain. But the sun was setting, and I could not afford to waste the few minutes of daylight even on such a view. I went on to camp as quickly as I could to read a round of angles with a roughly improvised plane table.

The view at sunrise next morning was weird and grand; the whole plain was covered in mist, through which rose the ridges of Taita, tinted for a few minutes with a ruddy glow. The wind whipped the upper surface of the mist into wave-like undulations, and tore the crests of the billows into shreds, or cut them off, and carried them away as streaks of cloud. We waited till the sun had dispersed the mist, and then, in a steamy atmosphere, crossed the last twelve miles of the plain, and pitched our camp beside a mountain stream at Ndara.

In the afternoon I scrambled up the hillside, and collected some specimens of liverworts from the rocks below a waterfall. Among other products of this excursion were some specimens of a wild banana. In the evening, on the plain, I got a puff-adder of the same species (*Clotho arietans*, Merr.) as that met with at the Cape, but which here attains a much larger size. Next day I added to my collection a Zanzibari, who had deserted from some previous expedition. I caught him quite by accident. He was watching the caravan from a hiding-place in some bushes about a quarter of a mile from our path. I came up behind him after a branch excursion to some rocks. I took him prisoner, and placed him in charge of my boy Philip and an Askari, to whom I promised the reward paid for a captured deserter when the man was handed over to the Company's officials at Tzavo. Later on, however, he succeeded in making his escape.

During this day's march we had to ford the Voi, which, as it was in full flood, was rather exhausting to the porters. The river is but a small one, but it flows through a swamp so choked by vegetation as to be impenetrable, except along the courses of the various branches of the stream. We worked our way along the channels, on both sides of which rushes, papyrus, and sedges rose as a wall ten feet in height. In places the plants met overhead, and then we had to creep along the tunnel thus formed, cutting away the lower stems so as to raise the roof and allow the porters to carry the loads through. In some places the floods had piled a barricade of vegetation across the channel, and this we had to clear away. As most of this time we were standing up to our shoulders in rapidly flowing water, we were very glad when we reached the northern margin of the swamp.

The men wanted to camp here, and as it was almost one o'clock I was quite prepared to do so. But Omari said it was unnecessary, and suggested that we should rest here for a mid-day meal and then go to a camping-place three hours farther on, at the foot of the mountain of Ndi. I thought this unwise; but, wishing to hasten, and not to check, the pace of the caravan, I did not like to discourage any excess of energy on the part of my headman. I was hesitating as to the decision when two of the porters insolently refused to go any farther, and began to "cheek" Omari. That settled it at once. Orders were given to the men to be ready to start for Ndi in an hour and a half's time. The decision proved to be a mistake. The passage of the Voi had been so fatiguing that an extra march ought not to have been attempted. The result was that some of the men did not get into camp till after nine o'clock, and they were all thoroughly tired out. It would not have done, however, to have confessed the mistake to the porters; so when some of the sulky ones came up to me to say that they were going back to Mombasa if they were to be marched like that, I merely remarked that they knew what they might expect when they arrived there. I advised them to go if they were very anxious to qualify for two years' hard labour in the chain-gang on the Uganda road. It was obvious, however, that some of the weaker men were too exhausted to proceed next day, and I decided to give the caravan a day's rest. This I did with the less reluctance, as I was very anxious to ascend the peak of Ndi, the highest mountain of this group, in order to compare the flora of its higher slopes with that of its base and of the adjacent plains.

At daybreak next morning I started from camp with a couple of the Askari; we followed the course of a small stream, which came plunging down the mountain side in a series of picturesque cascades. To force a way straight up the slope was impossible, owing to the denseness of the vegetation. We tried to do so, but were soon glad to take advantage of a path which wound up the face of the mountain to some Taita villages on the upper slopes. The path was well planned for security; it was arranged to lead occasionally across the face of an almost vertical cliff, where the foothold was reduced to a few knobs, or to a narrow ledge of rock. Many of the tracks in the Alps which are dignified with the name of "mauvais pas" are safe in comparison with these. That the natives can pass along them with heavy loads of food on their heads is a great testimony to their sureness of foot and steadiness of nerve. At first I thought that these rock-traverses were only short cuts, and that the main path ran elsewhere. But it was not so; the arrangement has been planned to enable the natives to keep their mountain fastnesses safe from the marauding Masai, who could not force them if defended from above by the natives with boulders, and with bows and poisoned arrows.

After a gentle ramble for two hours up the hillside, stopping here and there on the way to collect, we reached the meadows and valleys, in which are situated the villages and shambas of the Wa-taita. Only a few months previously the natives had quarrelled with a European caravan; they had dammed up the stream, and so cut off the water supply from the camp below. For this they had been so severely punished that I felt doubtful as to my reception. I found them, however, in a most friendly mood. They loaded my two men with presents of cobs of green maize, pumpkins, and sugar cane, and gave me a few eggs. The terrace beside the village commanded a good view of the upper part of the ridge, and I was thus able to decide on the best route to the summit. The headman of the village lent me two men as guides. A sharp walk soon brought us to the ridge, and we went southward along it, until at half-past eleven we reached the margin of the clump of trees upon the summit. The guides and my own men refused to go farther, as they said the wood was the abode of evil spirits and they dared not enter. I left them to light a fire, while I pushed on through the shrubs alone. This, however, was a waste of time, except in as far as it satisfied my climbing conscience. There was no view from the summit, and I had to make my sketches and observations from the edge of the wood. We boiled the water for the thermometers, and obtained data from which I subsequently calculated the altitude at 5640 feet.

I was interested to find a dense growth of the common English bracken on the higher part of the ridge. I did not like to accept it as the same species without a close examination, and tore up some feet of the long underground stem (or rhizome) to see if I could detect any difference in this part of the plant. While doing so I came upon a striking case of mimicry; an insect that lives among the dead leaves below the bracken so exactly resembled them in colour and in form that I should not have noticed it, had it not moved when I began to scrape away the soil.

During the descent we kept farther along the ridge to the north until we reached a col, which appeared white when seen from a distance. We found that both the depression and the colour were due to a reef of quartz which cropped out there. This was so brittle that it wore away more rapidly than the tougher rocks on either side. We were then compelled to beat a hasty retreat owing to the threatening aspect of the weather.

We reached our guides' village, and picked up our burden of provisions. I invited the natives to come with us for some return presents, and then rushed down the path to camp. But the storm broke upon us before we reached shelter; two minutes after the rain commenced we were wet to the skin.

Shortly after my return to camp I had the unexpected pleasure of a visit from Mr. George Wilson, a man well known to all readers of Captain Lugard's *Rise of our East African Empire.* He was then cutting a road to Kibwezi at the expense of the late Sir William Mackinnon, and had struck camp at Butchuma only the day before I arrived there. He had wisely travelled by a route to the west of the mountain group of Ndara, as he knew how troublesome the ford of the Voi would be after the recent rains. I had thus passed him on the road, and was afraid I should not meet him. I was delighted to see him, for I hoped to be greatly reassured by him as to the practicability of my plans. He was reported—and with truth—to be a man of such patience, tact, and good temper, that he is extremely successful in making friends with the natives. In this his facility for languages stands him in good stead, for it enables him to dispense with interpreters, and to communicate with the natives directly. He had had, moreover, great experience of the two tribes with which I was most likely to come into conflict, the Kikuyu and the Masai. He had lived for some time among the former at Dagoreti, and was a great friend of Tereri, the head of the Naivasha Masai. Unlike most men who have had much intercourse with the Masai, he has great faith in their intelligence and faithfulness, in addition to sharing the general admiration of their pluck and social organisation. So enthusiastic is he about the Masai that he is known in Mombasa as the "Masai faddist"—although a raiding party of this tribe had once attacked and routed a caravan of which Mr. Wilson was in command, and he only escaped after keeping the enemy at bay for a few minutes, by peppering them with buckshot from an eight-bore elephant rifle. I therefore hoped to receive from Mr. Wilson valuable advice from his knowledge of the people; and also encouragement as to the prospects of success, both on account of his friendship with the natives, and because, as a rule, the nearer danger is approached, the less it is reported to be. The advice I received, the encouragement he could not give. He told me, to my disappointment, that though he should not himself refuse to undertake such a journey with such a force as mine, he should regard it as a risky experiment, and that it was far more dangerous for a raw recruit, ignorant alike of the languages, the country, and the people. He said that in the present condition of the Kikuyu country, I should certainly have to fight my way through it. However, I consoled myself by the thought that Mr. Wilson's last recollections of the Kikuyu were most unpleasant. He had been in command of the station at Dagoreti in the Kikuyu country during a very stormy period. He had been besieged by the natives, and his communications cut off. He held the fort till his ammunition was exhausted, for the supplies sent him from the coast failed to reach him. His position was then absolutely untenable; so he sallied forth, and cut his way through his besiegers to the fort at Machakos. Having there obtained the ammunition that had been sent for him, he returned to Dagoreti, recaptured the position, and rebuilt the station, which had been burnt by the natives. Later on, however, he was persuaded to retire by the leaders of a caravan returning from Uganda.

But I obtained a great deal of most useful advice, and we sat up chatting in his cosy, artistically-upholstered tent till the early hours of the morning. Then I tore myself away from this mine of information, and went back to write out notes and press plants until, at dawn, we left our camp and marched again into the "barra."

Two further marches brought us to the station of Tzavo, a fort erected in 1890 by the British East Africa Company in order to stop Masai raids down the valley of the Sabaki. For this purpose its garrison of twelve Beluchi "Kiroboto" under a Goanese clerk was quite inadequate, even if the men had been trustworthy. These "Kiroboto" were part of a force of Asiatic mercenaries raised by the Sultan of Zanzibar. Their name was derived from that of the soldiers of the Sultan of Hyderabad, from whose army the nucleus of the force was obtained. Of the reliance that can be placed upon these men I had an illustration on the return journey, when the commander of the fort had gone away for a few days. Before his departure he had locked up all the ammunition, as the men were not to be trusted with such dangerous material. There was nothing, therefore, to prevent any Masai who might be raiding in the district from capturing the stores and spearing the defenceless garrison!

The fort consists of a group of huts, surrounded by a three-pile stockade and a thorn "boma" or hedge. It is prettily situated beside the ford across the Tzavo river, in a grove of Hyphæne palms (*Hyphæne thebaica*, Mart.), which were the first palms we had seen since leaving the coast.

In company with the Goanese commandant I spent a pleasant afternoon catching lizards and scorpions, and digging up the skulls of some Wa-kamba who had been killed by the Masai.

I had, however, the disappointment of losing my first man. I had been exceptionally fortunate in not having had a single deserter, and was now beginning to feel safe from this anxiety. The men had, however, suffered a good deal from illness, and a porter was found to be too ill to proceed: he was one of the two men in my force who had been with Count Teleki in his great expedition to Kenya and Basso Narok. I was therefore especially reluctant to spare him. He was, however, suffering from a disease that was incurable on the march, and it would have been sheer murder to have made him continue the journey in the rainy season. I had therefore to give him his discharge, and arrange for him to be kept at Tzavo till he was well enough to return to the coast with some passing caravan.

After leaving the station at Tzavo we had to proceed with greater care. Our road intersected those used by the Masai of the districts of Nyiri and Matumbato, to the north of Kilima Njaro, when marching to raid in the districts of Ukamba, on the Tana or near the coast; we had therefore to keep close together on the march, ready at any moment to repel attack. The Askari, Ramadhan Aperti, and I led the van, keeping a sharp lookout ahead for any sign of Masai. I had thus to keep to the path, and my opportunities for collecting were much restricted. More precious time was also occupied in teaching the men a simple drill. I had seen too much when with the former expedition, of the results of the attempt to train excitable natives as if they were British troops, to be inclined to repeat the process. It appeared probable that if we had to fight, that we should have to do so in forest or scrub, where formal operations would be impossible. So I made no attempt to teach the men how to wheel like a company of guards in a barrack square, or march in column like a girls' school; but simply trained them in skirmishing drill, outpost duty, forming square, and volley firing. As there was no ammunition to spare in practice, we had to check the men's smartness in the last exercise by the click of the hammers of the rifles. The men did not seem very courageous, and had an ingrained terror of Masai; but I was consoled by seeing that they had a good idea of bush fighting, and chose their positions with judgment.

In the evening Ramathan mounted the pile of goods and harangued the camp. He told the men that they must keep their guns ready beside them and their fires burning all night; and that any one who moved out of camp after dark, without a burning brand from the fires, would be shot. A party of Wa-kamba traders, returning to their homes, joined us here, and begged permission to accompany us as a protection from the Masai; to them we gave even stricter orders as to their behaviour during the night. Here I had to begin the practice of going round camp several times during the night to see that the sentries were awake. As breakfast was at five, I generally felt sleepy when I went to bed, usually about eleven. I therefore always drank a cup of strong cold tea just before turning in. This rendered my sleep so light that any movement in camp disturbed me. I soon acquired the habit of waking at regular intervals through the night.

That our precautions were not needless we saw next day, when we found some fresh footprints crossing our path in a north-easterly direction. Some deep square punctures made by spear points, and occasional marks where a heavy shield had rubbed along the ground, showed that the track was that of a party of Masai. But for the day's delay at Ndi we should have spent the previous night at the pools of Kinani, and should probably have seen something of this party. Double guard was kept next night. At three in the morning we were roused by a shot. We all sprang at once to our places, ready to repel attack. The sentry had seen a dark object sneaking up to camp, and had fired at it; it fled at once, and from its howl I concluded that it was only a hyena. Some of the porters, however, maintained that it probably was a Masai scout, and that he had imitated a hyena's cry to throw us off our guard. This was just possible, so we kept a sharp lookout till morning.

Our next camp was beside a dry river-bed, at a place of which the name has two different forms according to the two different hypotheses as to its meaning. One etymology makes it Mto wa Undei, meaning the country of hawks, while the other derives the name from that of the son of a great chief of the Wa-kamba, who was killed here by the Masai; according to this, the more probable theory, the name should be spelt Mtoto wa Ande.

We were here near the borders of the district of Kikumbuliu, the first inhabited by the Wa-kamba. Early next day we came to their plantations or shambas. We passed through these, and camped beside some wells dug by the Suahili traders, in a valley known as Masongaleni. As our food supply was again nearly exhausted, Omari proposed that we should rest here for a day to buy a fresh stock, instead of at Kibwezi, where it was said to be very expensive. We therefore fired a couple of shots as a signal that food was wanted, and as an invitation to the natives to come in and open a market or "soko." Women soon appeared with calabashes of flour and plaited baskets with grain. After a long discussion the price was fixed at fifteen strings of small pink beads for six kibaba, or about nine pounds, of beans. As each woman brought only two or three kibaba, the purchase of as much food as we wanted proceeded very slowly. We could not get enough that day, so I left Omari to continue the market, and went on with a few men to Kibwezi. I was anxious to get a day's rest there in order to see the experimental plantation of the East African Scottish Mission, and have a chat with the missionaries. The station is situated in some woods on an old lava sheet around the sources of the river Kibwezi.

The rain fell in torrents as we crossed a swamp and entered the Mission grounds. But this had become too familiar to diminish the pleasure which the sight of the fine timber trees of this oasis gave me. Any one who has experienced the delight of suddenly entering one of the chestnut groves of the Italian slopes of the Alps, after weeks among the monotonous fir-woods of the Swiss highlands, will understand the joy with which, after weeks in miserable scrub, I entered the forest of Kibwezi.

The Mission was founded in 1891 by a party of Scotch philanthropists, who were dissatisfied with the results of the work of existing missions. They therefore resolved to start a fresh one on a purely industrial and non-sectarian basis. Dr. Stewart, the head of the well-known Lovedale Mission in Bechuanaland, was sent into the country to find a suitable site. He selected Kibwezi, from its apparent healthiness, the excellence of its water, and its convenience as a place of call, it being about half-way from Mombasa to Machakos. No doubt the beauty of the situation was partly responsible for the choice.

The two principal objections to the site are that, owing to the luxuriance of the surrounding vegetation, it appears to be very unhealthy, and that an Industrial Mission ought to have been settled in a well-populated district. At Kibwezi the natives only number some two or three hundred, and they are not a promising set of pupils. For years they have been demoralised by Masai oppression and by intercourse with Suahili caravans. They appear to do nothing except for payment, and the children require higher wages for attendance at school than their parents do for work in the plantations.

At the time of my visit the staff consisted of three Europeans, of whom Mr. Watson was the only member of the original party of missionaries. The superintendent of the station, Dr. Charters, had arrived a few weeks previously. He will be known to readers of *In Darkest Africa*, for he was in charge of the mission-steamer, the *Peace*, which carried Mr. Stanley up the Congo from the Pool to the mouth of the Aruwhimi.[1]

The British East Africa Company had given the Mission a grant of 100 square miles of land around the station. This is being cleared, and an experimental farm has been established upon it. Mr. Pattison, a gardener, who came out with Dr. Charters, was then trying to rear the vine and various fruit trees, as well as testing the suitability of many different kinds of grain.

As Mr. Watson had been rather unwell, at Dr. Charters' suggestion he took a holiday next day, and we went off together for a day's fishing and collecting along the Kibwezi river. This

[1] Dr. Charters disappeared when out shooting near Kibwezi in the autumn of 1894; there is now practically no doubt that he was murdered by Masai.

first rises on the Mission grounds, and then plunges into the lava and flows through a subterranean channel for a couple of miles. It rises again in some pools. In these we caught some fish of two new species: one, a Barbel (*Barbus tanensis*, Gthr.), occurred also in the Tana and in the Guaso el Narua, south of Lake Baringo; the other, *Oreochromis niger* (Gthr.), I did not find elsewhere. In the woods we found some fungi, the remarkable case of insect mimicry, illustrated by the frontispiece and described on p. 71, and a black variety of a small mammal (*Macroscelides rufescens*, Pet.).

It was by far the most successful day's collecting I had had. More can be obtained here in a week than in a year in the open barra. In the forest clearings butterflies of most gorgeous colours occurred in great profusion. But as I had hurt my leg a few days before, I could not run after them. The porters sent out to collect them brought back a good number, but they were all terribly damaged owing to the men's clumsy methods of capture. As Mr. F. J. Jackson had made a large collection during his residence at Kibwezi, I threw all mine away in disgust.

Next morning the men started ahead of me for the march to the Mkindu river. I followed later with Mr. Watson. Just as we were starting I caught my only glimpse of the snowy dome of Kibo, the higher of the two summits of Kilima Njaro. I was within range of view of this mountain for several weeks, during the journey up and on the return march. This, however, was the only occasion on which I saw it. All the rest of the time the mists and clouds obscured it. It was then visible only for a few moments; the clouds closed round it again, before I could draw my prismatic compass from its case, though it was slung ready by my side.

The gravel path across the Mission grounds ended at the edge of a rough tract of lava, which showed the irregular ropy structure that forms only on the actual surface of a flow. It was, therefore, clearly more recent in date than any lava we had previously seen. Sir Gerald Portal[1] has graphically described the uncomfortable nature of the track across it. The rock was so rough and the edges so sharp that his boots were cut to ribbons, and it is therefore easy to imagine how severe a trial it was to the

[1] *Mission to Uganda* (1894), p. 64.

porters with their bare feet and heavy loads. The men went so slowly across this that I soon came up with them, and we marched on together over the gap at the southern end of the ridge of Bwinzau. The porters wanted to camp here beside some pools, though it was only half-way to the Mkindu river. Mr. Watson assured me that the latter was the regulation camping-place, so I hardened my heart against the men's entreaties, and told them they must follow me to the Mkindu. Mr. Watson returned, and I went on with an Askari and a boy, in the hope of getting some shooting on the way. I reached the camping-ground shortly before dark, and Omari came in soon after with the "fly," or outer sheet, of the tent. The river-bed was lined by a species of date-palm (*Phœnix*), after the native name of which (Mkindu) the river is called. The bed was quite dry, but this did not cause us anxiety, as a heavy storm was threatening, and began very soon after our arrival. The locality has a bad reputation, as a Masai warpath follows the course of the river. We examined this, and found that a party of Masai had passed along it that day. We therefore did not dare to light a fire. Omari and I had to take up stations at the opposite ends of the path that ran through the belt of thorn scrub in which the camp was situated. The Askari and the boy Philip, who were the two prize cowards of the caravan, were in such a fright that they were useless. It poured with rain, and the hours during which I stood in the narrow path were the reverse of pleasant. At eight o'clock the cook came up with the news that the porters, at the instigation of the Kiringozi, Wadi Hamis, had resolved to throw down their loads and desert. As I knew, however, that Omari would never have come on had there been any chance of this, I dismissed it as "bluff," and continued my sentry duty. At nine o'clock, however, I became seriously alarmed, for I thought the porters might have met Masai, and prepared to start back with Omari. We were just drinking some cocoa, stirred up in rain water, to fortify ourselves for the march, when we heard a rustle in the palms on the opposite bank of the stream. We stood to arms. To our intense relief we found it was a porter, Fundi Mabruk, who was hastening to tell us that the rest of the men were close at hand.

They soon arrived, cold, wet, miserable, and most of them ashamed of their fit of sulks. It was too dark to arrange camp properly, and we could not light a fire, so we put up the fly of the tent and huddled together beneath it. To cheer the men I told them stories, my boy helping out my limited stock of the native language; the ever merry Mwini Mharo sang us hunting songs, while Omari described the miseries of the march through the forests of "Darkest Africa."

At daybreak I paraded the men, and scolded them for their misconduct. Then through a raw cold mist we left our dismal resting-place, and struck out again into the barra. When near the next camping-ground, beside the Mto Kiboko or Hippopotamus river, I let the men go on and stopped to collect. I found some chameleons, and was bottling them, when my attendant pointed out a snake who was eyeing us viciously. Its head was just thrust out of a hole in the side of one of the mounds made by white ants. I fired at it with my shot-gun, and the snake glided swiftly on to the ground, and I thought it was coming for me.[1] But it stopped, writhed its body into knots in agony, and then darted again for its hole. Anxious to secure it, I sprang forward and seized its tail. I held on till I could place the butt end of my gun on it. I got my hand free none too soon, for it came up again as soon as it found that it could not escape. I broke its back by a blow with the cleaning-rod of the gun, and its powers of mischief were destroyed. The two men with me refused to touch the snake. They said it was "Uchungu, *uchungu* sana" (Poisonous, very, very poisonous). As it was still wriggling, I had to empty the collecting gear out of my satchel, put the snake into it, and carry it on to camp myself. Dr. Günther tells me that the snake (*Dendraspis polylepis*, Gthr.) is one of the most poisonous of known reptiles, and is worse than the Indian cobra. It usually lives on trees, and has not been met with before in this region of Africa, or known to live in holes. It has probably acquired this habit, owing to the absence of trees from the plains in which it lives in this district.

The camp that night was an especially dangerous one. Readers of Sir Gerald Portal's *Mission to Uganda* may remember that he describes the massacre of a Suahili caravan at this place

[1] Mr. C. W. Hobley tells me that a snake, which, from his description, is probably the same species, charged him in the Mumoni district some 100 miles to the north: he only stopped it by blowing off its head with his rifle.

only six months previously. We kept an especially vigilant watch; but as it rained continuously from four in the afternoon till eleven o'clock the next morning, I felt that we were safe, for Masai are reported to lose most of their energy when cold and wet.

After the "Kiboko" we had to cross the "Salt River," and then make the last two days' journey over the "Nyika." The country began to rise more rapidly. We had fine views to the west across a lava plain to the craters of the Kyulu chain, and before us rose the sharp ridges of the Iveti Mountains. The first of these was a conical peak named Ngu. Owing to its position far out on the plains it forms a splendid watch-tower, and on it Wa-kamba scouts are always stationed, so as to give timely warning of the approach of any raiding parties of Masai. We soon found further proof of the activity of these freebooters. Scattered over the path were broken boxes and bottles, which were being sent to Sir Gerald Portal's party in Uganda. We heard at the next station that a caravan, laden with reserve stores for the Commissioner, had been attacked here a few weeks before and all the loads destroyed. Thrice during the past three months the Masai had routed caravans along this section of the road.

Two days later we reached the ridge of Nzaoi, and camped on a platform at its base. Nzaoi is a narrow ridge of gneiss, sloping gradually to the north, but facing the south with a precipitous cliff nearly 2000 feet in height.[1] Lugard well describes it as "the massive sentinel that guards the gate to the heart of Africa." We were all delighted to pass through the gate from the barren wastes of the Nyika into the fertile valleys of Ukamba. Magile, the "Mzee" or chief of the district, called to see me and ask for a present. This I promised him if his people would bring us food for sale. He said they had none, as Sir Gerald Portal's caravan and one that had recently returned from Uganda had eaten the district bare. He told us, however, that neither of these caravans had purchased any food at Kilungu, a long day's march farther to the north, and there I could buy as much as I wanted. He said he would do his best to get us a little, but his efforts were not

[1] The camping-ground is at the height of 3700 feet; the summit of the peak at about 6000 feet.

sufficiently successful to gain the present he coveted. Some eggs only were brought. I was away at the time, vainly trying to boil the thermometers in a hurricane of wind and rain on the summit of the peak. When I returned, wet and weary, I was cheered by the promise of omelette. With the promise I had to be content. The natives seem to regard it a sinful waste to eat eggs, as there is so much more meat on a chicken. So they try to hatch all that are laid, and no egg's chance is spoiled by lack of patience. If nothing happens to it in three weeks, it is allowed three months. Eggs which remain obdurate are at length removed and stored up to await the arrival of a caravan. My cook had accepted the guarantee that the "nest" eggs offered had been laid yesterday, and bought the lot. Their contents, however, were more solid than savoury, and no addition to our food supply.

Unable to refill our empty sacks at Nzaoi, we rushed on next day to Kilungu. After our arrival in the land of plenty the chief came to see me, and showed me a copy of a treaty he had made with the British East Africa Company, by which he was pledged to supply food to passing caravans. Though it was then but little after noon, for we had made a very early start, the native potentate said it was impossible for his people to bring us anything that day. He left us, promising that we should have as much as we wanted on the morrow. After this not a single native came near us. Early the next afternoon, as the people still held aloof, I resolved to go and interview them. It was now obvious that the wily chief had beguiled us into stopping there, so as to prevent our going to some more friendly villages farther on. My men were very angry, and clamoured to be allowed to go and seize food. I should, under the circumstances, have been quite justified in doing this; but I was anxious to avoid fighting anywhere, and it seemed to me that the people at Mombasa would not think much of my leadership, if I could not carry a caravan up to Machakos, without quarrelling with the natives on the road. But I considered I ought to go and protest. So I took twenty men, crossed the river, and marched to the nearest village. On the way we captured an old man, who was too infirm to run away. He told us that the natives had plenty of food, but would not sell it to white men; he informed us that they had used the last twenty-four hours' delay to remove all supplies from the villages into hiding-places in the hills. We let him go, and then resumed our march to the largest village. The natives beat their war-drums ; the women and children fled to hide in the dense scrub ; the men, armed with bows and poisoned arrows, took their places beside the loop-holes through the walls of the village. As we passed beside it we could see the warriors with their arrows ready strung, and it was only by threatening to shoot the first of my men who fired, that I managed to prevent a fight. The natives made no effort to stop us as we entered their shambas ; these we crossed to the foot of the highest peak in this group of mountains, which was occupied by a crowd of natives. I was very anxious to get into communication with the people, and as a sign of friendliness, left my men behind and climbed up to a shoulder, accompanied only by an interpreter and my boy. Two elders came to meet us ; we told them that we wanted to be friends, and scolded them for the behaviour of the chief. We were soon surrounded by a crowd of natives, armed with knives, clubs, and the primitive hoes, which are their only agricultural implements. The three of us kept back to back, and thus we held our "shauri." As my gun and the two men's rifles were loaded with buckshot we were perfectly safe ; a single discharge would have knocked over a sufficient number of the natives to have frightened off the remainder. So I jeered at them, told them we were going to the top of their biggest hill, to show that we did not care a jot for the lot of them, and assured them that if we wished, we could help ourselves to all the food in the district. I asked two of the elders to come up with us in order to tell me the names of the neighbouring hills. We then continued the ascent, accompanied by a crowd of about a hundred men, who were not allowed, however, to approach within twenty yards. We rested for half an hour on the summit, while I boiled the thermometers and made a sketch map of the surrounding country. The former observations gave results which fix the height of this peak of Etwa at 6120 feet.

The elders offered me a goat as a present, and begged me to wait till one could be brought. For this generous offer I was not as grateful as they expected. I explained I had no intention of being caught on their hills in the dark, and was therefore going back at once. I promised, however, that if they brought their present down to camp, they should be suitably rewarded. Of course it never came. The idea was simply to delay our return till dark, when we should have been at their mercy.

As we had ascended by the south-east ridge, the return was made down the south face. We rejoined the porters and marched back to camp. On the way I let the porters help themselves to some sugar-cane, as a protest against the behaviour of the natives. The men grumbled at not being allowed to loot a village, for they said sugar-cane was "water and not food." Had I then known the whole story of the past treatment of European caravans by these Wa-kilungu, I might not have been contented with the mere seizure of a few sugar-canes. I only knew that the people here had quarrelled with a caravan, and been severely chastised in consequence by Mr. Ainsworth and Captain Nelson. I therefore thought that the natives were indulging in a not unnatural fit of the sulks, and hoped, by leaving them alone, to help them to a happier frame of mind. So we fasted once again, and next day, with belts drawn one hole tighter, marched on to a more friendly clan.

We obtained abundance of food at Zuni, where the natives told us the people of Kilungu were not Wa-kamba. Mr. Ainsworth subsequently informed me that they were reported to be Kikuyu who had been driven from their own country and had settled at Kilungu. They now speak a Kikamba patois, but their features are far more of the Kikuyu than of the Kikamba type.

At this camp an incident occurred which very literally clouded my prestige among the men. It was their Christmas Day, and they were all eager to get the first glimpse of the new moon. An eclipse of the sun was due the same day, as predicted by the Greenwich Almanac. On the west coast it was a total eclipse, but in this region it was only partial, as the sun set half an hour after the moon's shadow began to creep across the solar disc. I had said nothing about it, as the weather was so bad that I expected we should not be able to see it. The afternoon of the eclipse, however, was so bright and cloudless that I was tempted to prophesy. So when the men told me about their new moon, I said a piece of the sun would be bitten out of its edge a little before sunset. Just before the time of the eclipse, a great bank of cloud rose from the western plains and completely hid the sun from view. I did my best to explain that the piece had been bitten out all the same, but that the cloud had prevented our seeing it. But the men thought this a very lame excuse. I overheard one man say to another: "If Mpokwa[1] could bite a piece out of the sun, he could have kept those clouds out of the way." For some days after this, whenever the sun went behind a cloud, the porters would ask if I were not going to bite a piece out of it.

Our next camp was on the site of one occupied by Count Teleki on his return march to the coast in 1889, on the summit of the pass of Kwazome (4750 feet), which is named after Kwathome, the chief of the district. We had had a long march, and I therefore made my usual afternoon excursion alone. I climbed to the summit of Tututha (6050 feet), the ridge that forms the eastern boundary of the pass. From this I had a glorious view to the north-east, over the plain which here runs up to the foot of the Iveti Mountains.

Next day I explored the ridge of Givoni, which forms the western side of the pass, and had a pleasant climb to the summit (6780 feet) over some rather difficult rocks. I joined my Askari again at a col upon the ridge, which he reached by a path, as he did not care to follow me up the rocks. We then struck straight across country toward the hill of Machakos, swimming two flooded rivers on the way. At one of these, a party of Wa-kamba women was waiting beside the ford for the flood to subside. I modestly went up stream to find a more secluded point at which to cross; but as they did not every day have the chance of seeing a white man swim a river, they resolved not to lose their opportunity, and so followed up the opposite bank. So there was nothing for it but to waive prejudice and cross in front of them. It was lucky that they came, for one of my boots fell into the river as I was throwing it across, and had not one of the shameless damsels rescued it, the priceless treasure would have been lost. I almost lost my best interpreter, for my attendant at first refused to cross, as he said the river was full of crocodiles, and that the current was too swift. As he could swim quite as well as I could, I

[1] *i.e.* "loaded pockets," my native name.

advised him to stick to one excuse in the future, and continued my way to Machakos. Afraid of being left alone, the Askari at length plunged in; but he swallowed some water, lost his head, and was swept down stream, until he was rescued by a young Mkamba warrior.

Late the same evening, in a tremendous downpour of rain, we reached the British East Africa Company's station at Machakos. The porters had arrived there several hours previously, and had given Mr. Ainsworth, the officer in charge, such a remarkable account of my habits that he was puzzled as to who was coming. He invited me to stay with him, and at once gave me some hot tea, which was very welcome. I was wet to the skin, and as I had forgotten to take any beads with me in the morning, all the food I had been able to get during the day was some arrowroot ("mhogo") cooked in "ghee" (a kind of native butter), which a Mkamba had given to me in exchange for my pocket-handkerchief.

The fort of Machakos was originally founded in 1889. In the following year Captain Lugard recognised the value of the position, and greatly enlarged and strengthened the station. It is situated in a rich food country, and the meteorological returns show that it is better adapted for European colonisation than any other part of British East Africa. On the neighbouring hills there is a dense population, and the people are enterprising and industrious. In the early days of the fort they were hostile, and twice attempted to storm it. At the present time, however, their attitude is most friendly, thanks to the tact and skill Mr. Ainsworth displays in his administration of the district. Most of the work of the station is now done by Wa-kamba, who, under Ainsworth's influence, show that they are willing to work and able to learn. The mail is carried to Tzavo more expeditiously and economically than when Zanzibari runners were employed. The fort is garrisoned by a company of Wa-kamba, whose fidelity and fighting capacity have recently been tested. In 1894 the Masai made a sudden attack upon the fort, but were defeated with heavy loss. Mr. Ainsworth is enthusiastic about the Wa-kamba and the possibilities of their country; and this is not due to an indiscriminate love of aborigines, for he dislikes the Somali more than I do, which is saying a good deal.

My host at Machakos is a keen topographer and a good draughtsman. He has prepared an admirable map of western Ukamba, a reduced copy of part of which he kindly gave me on my return. I was glad to find that the names on his map agreed with those which I had obtained. African place-names are very confusing, though they probably do not vary as much as one would expect. It is true that the people on the opposite sides of a range of mountains often have different names for the peaks, as is also the case in the Alps. This has frequently led to confusion, for a traveller hearing two different names for a mountain or a river has concluded that there are two mountains or rivers, and marked them accordingly on his map. In many cases, therefore, the native names have been ignored, and the places called after eminent people at home. As these names are unknown in the district they are useless, and merely burden geography with a set of synonyms. In uninhabited regions, where there are no names, it is necessary to introduce them, and it is open to the first traveller who visits a country to propose what terms he likes. But where there are native names, the advantages of using them are so great that geographers now condemn the old practice. The European names are falling into disuse and disappearing from maps, where those of the natives are being substituted in their stead.

CHAPTER VI

ACROSS THE LANDS OF THE KIKUYU AND MASAI

"Maneno madzo gausa ndzovu m'ndani"
(Fair speeches turn elephants out of the garden patch).
Giriama Proverb (W. E. TAYLOR).

I SHOULD have been glad to accept Mr. Ainsworth's proffered hospitality for a few days' rest; but I was now at the entrance to the country I had especially come to explore, and nothing could tempt me to linger on the threshold. Up to this point the work had not been of special interest. The geology was monotonous and dull, and the geography was fairly well known. But a change in the geological structure of the country was at hand. In the best existing sketch map of the geology of East Africa, a large tract of country to the west of Machakos is coloured as recent alluvium. I expected this would prove to be an old lake basin or a desert of wind-borne drift. From the hills crossed on the way to Machakos, I had caught through the clouds occasional glimpses of a vast level plain, and these had strengthened this expectation. Great therefore was my surprise, on reaching the summit of the last ridge of the Iveti Mountains, to see to the west an undulating prairie instead of a level plain, and that this was composed not of alluvium or sand, but of a hard, dark-coloured rock. Its extent also was greater than I had expected. Here and there in the foreground, bosses and ridges of gneiss, such as Lokenya and Koma, rose above the surface; a few dark lines of trees marked the courses of the rivers. Except for these, we could see only a vast expanse of rolling grass-land, extending westward and southward as far as the eye could follow it. The rock of this prairie ended abruptly at the foot of the old gneiss ridge on which I stood, but it followed its outline, running up the valleys, round the spurs, and into the hollows of the mountains, just as the water of a lake adapts itself to the irregularities of its shore. In this, as well as in other ways, the view reminded me so much of that of the Snake River lava plains of Idaho, as seen from the western flanks of the Rocky Mountains, that I felt sure that this was a plain of lava and not of alluvium. I walked quickly down the slope to the nearest point where the rock could be seen, and found, as I expected, a lava, a coarse trachyte with very large porphyritic crystals of sanidine.

The impression that the whole plain was of lava was confirmed during our march across it, and I was mentally braced up by a crowd of new and perplexing problems. But the work of the caravan was a terrible tax upon my time. The day after leaving Machakos the box containing most of my sporting cartridges, and all the plants collected on the way from the coast, was dropped into a stream. I spent four hours in the effort to dry them; but as they had been soaking in water for several hours before I reached camp and the accident was reported to me, nearly all the plants were ruined. I was consoled for the loss at the time by the vain hope that I could collect another set on the return march.

Our camp that night was a rather risky one. Major Eric Smith of the 1st Life Guards had had an encounter with the Masai here, and its name was marked with a double asterisk on the list of dangerous camps that had been given me. Moreover, it was here that Mr. Jackson and Dr. Mackinnon had seen a "herd" of twenty-three lions, and the place was reported to swarm with them. Hence the sentries had been ordered to keep a specially careful watch. Nevertheless, when I went round camp at a quarter past three in the morning, the sentry was sound asleep in his tent. I called out "Askari" twice without rousing him, so I woke Omari and showed him the sleeping sentry; we knocked over the man's tent, and I kicked the culprit round the camp. The noise roused the porters. When they heard the cause of the disturbance, they created a tremendous uproar. The poor sentry had a very bad time of it. One porter declared he saw some lions just sneaking away from camp, and some others that they had seen some Masai. So I knew camp was safe for the rest of that night, and leaving the porters to hold an indignation meeting and abuse the culprit, I turned in and slept soundly until daybreak.

I had hoped to get some antelope shooting in this district, for it is one of the richest game fields in Africa; but the Athi river was in flood, and it was obvious at once that all my energies must be devoted to getting the caravan across it. Mr. Ainsworth had warned me that we should probably be delayed here, as a small caravan that had just gone on to Fort Smith had been obliged to stop beside the river for a week. The ordinary ford was quite impassable. Omari and I managed all but the last channel, which was only fifteen yards wide; but through this the stream was rushing with such force that we were obliged to give it up. Omari went off in one direction, and I in the other, to try to find a practicable ford. My search was unsuccessful, but Omari found a place a couple of miles down stream, where the river widened out into a shallow. Here we crossed next day. The passage, however, was troublesome, especially over the flooded land on either bank; for amid the sharp angular rocks were shrubs, whose still sharper thorns broke off in our bare feet.

Later on the same day we had to cross the second branch of the Athi, which was not so broad as the first, but deeper and swifter. The men said we could not possibly cross, as this branch of the river was always worse than the other. When I joined them on the bank, after an ineffectual attempt to stalk some giraffe, the porters seemed determined not to try the ford that day. Omari, however, said it must be possible to cross, so we plunged in and raced to the other side. The current was more powerful than it looked, and twice knocked me off my feet. Omari went back for the rope while I sat on a rock and exhorted the porters. Fundi Mabruk and the Askari, Ramathan Jumma, next tried the passage, but as the latter was carried off his feet and dragged ashore half choked with water, the others preferred to wait till a rope could be fitted up across the river. A length (80 ft.) of Buckingham's Alpine rope went about half-way across; we fastened other ropes to this and tied them round trees on the banks. Ramathan Jumma and I then took up stations in shallows, equi-distant from the bank, and held on to the rope to relieve the strain. Omari and two or three of the strongest porters then carried across all the loads, and the rest scrambled over along the rope. By great good fortune we crossed without accident, but the four of us on whom the brunt of the struggle fell were too tired to go farther that day. I had been in the water for an hour and a half, during which my feet and legs were numbed with cold, while my shoulders were being sun-blistered.

Next morning we continued our march across the lava plains. During the day we were joined by a caravan of ninety men, under a famous old headman, Wadi Bunduki ("The Son of the Gun"). They were going to reinforce and carry mails and stores to the garrison at Fort Smith, and had been hastened forward so that we might traverse together the country inhabited by the Kikuyu. We camped a little distance from the forests, keeping double guard all night. Next morning we met twenty of the Company's soldiers, who had been sent by a night march to guide us. Our united forces therefore amounted to 150 men. The forest paths were flooded, so while wading along these we were safe from attack; but when we reached the plantations there was more danger, and crowds of natives appeared and watched us. We marched in a compact body, with skirmishers thrown out ahead to examine the clumps of bush before we approached them; and we kept ready at any moment to repel attack from the natives who hung upon our flanks. But they made no attempt to interfere with us, not even at midday when we rested for a few minutes in a yam-field and helped ourselves to the potatoes.

In the afternoon we reached Fort Smith, the British East Africa Company's station in the Kikuyu country. I found that Mr. Purkiss (who has since died) was then in command, for Captain Nelson, one of the officers of the Emin Pasha Relief Expedition, had died two months previously. The history of this station is short but eventful. The Company's first settlement was about four miles away at Dagoreti, which was built by Captain Lugard and Mr. George Wilson in 1890. That station, however, was twice destroyed by the natives, and the existing fort was built by Major Smith. The defences are so strong that there is no fear of its capture except by surprise. Of this, however, there is some danger, as the Zanzibari are such unreliable sentries that Mr. Purkiss told me that his last thought every night when going to bed was that if a well-planned attack were made, no one would know of it till the enemy was in the fort. The whole country was in an extremely unsettled condition, and no one could go half a mile from the fort except under escort. One of the first things Mr. Purkiss told me was that, according to regulations, no one was allowed to go anywhere in this district with less than twenty armed men; and he requested me while his guest to obey that regulation. In the November of the previous year there had been a big fight with the Kikuyu; but as at that time the garrison was reinforced by the caravan of the Railway Survey, by one under Major Smith, then on its way to Uganda, and by another under Mr. Martin, then returning to the coast, the natives were easily defeated and severely punished. This had not, however, prevented them from attacking the fort early in 1893, when they had besieged the station for a week and killed ten of the garrison.

The farm attached to the station has a splendid kitchen garden, and most of the best English vegetables, including potatoes, tomatoes, turnips, carrots, radishes, cauliflower, lettuce, water-cress, and spinach, and all of excellent quality, were growing there. The altitude of the station is estimated by the Railway Survey at 6350 feet; but as Mr. Purkiss told me that they had a mercurial barometer in the fort, I hoped to get a more exact determination. The instrument, however, was an ordinary ship's barometer, of use only at sea-level. The mercury was several inches out of sight down the iron tube.

Mr. Purkiss invited me to stay with him for a couple of nights, and I did so, to rest the men and make my final preparations. For, after leaving Fort Smith, I did not expect to see another European until my return to Machakos or the coast. I had also to buy enough food to last as far as Baringo, and for this needed nearly a ton. To help to carry the additional load I bought three donkeys. The food supply was further increased by the purchase of a small flock of sheep, and by a sackful of vegetables, a welcome present from Mr. Purkiss.

The day after my arrival one of the new recruits for the garrison "ran amuck" through the fort, and tried to brain

Purkiss with a beam of timber. He was soon disarmed, tied to the flag-staff and flogged. He was one of the Somali who had given me a good deal of trouble at Ngatana.

After leaving this hospitable fort we struck away to the north-west over forest land, intersected by numerous deep valleys. We were watched by a crowd of Kikuyu, who howled and jeered at us, but did nothing more alarming. On arrival at our camping-place I found that two boxes had been left behind by mistake, and so had to return next day with thirteen men, while Omari held the camp with the rest. Our reappearance created temporary consternation, for the rumour had come in the night before that a powerful force of Masai was marching northward, in order to join the Kikuyu in an attack upon the fort. Some friendly Kikuyu had had a fight with the Masai outposts and killed five of them on the previous day. Purkiss therefore thought that I had come across the Masai and been driven back by them. I was glad of the warning, and so, having got the boxes, started back at once for camp. We moved this on a little farther, and then built our thorn boma in a steady downpour of rain. As some Kikuyu were sneaking about watching us, we prepared to resist an attack. In addition to the ordinary sentries, Omari and I both kept guard all night.

We were off betimes next morning, anxious to leave the Kikuyu forests and to reach the Rift Valley. We crossed a swamp, which must be one of the head springs of the Tana, and then entered a valley, the character of which is different, both in structure and scenery, from any that we had previously seen. Its eastern wall was a straight precipitous cliff, due to a dislocation of the type known to geologists as "faults." At the northern end of the valley we should be on the edge of the Rift Valley, so we hastened along it. For five weeks I had been looking forward to the view that I expected to get from this point. My disappointment may therefore be imagined when, just before we reached the summit of the pass, a dense cloud settled down upon us and completely blotted out the view. We descended a few hundred feet, and then a wonderful prospect burst upon us. We were on the face of a cliff 1400 feet in height, broken only by a platform 500 feet above the floor of the valley. From the foot of the cliff a level plain extends thirty miles to the west, to the foot of the scarp of "Mau." Most Zanzibari have an eye for beauty in landscape, and to many of the men, including Omari, this view was new. So we sat along the path and enjoyed it. Now and again the cloud-banks that floated up the valley settled round us and blotted out the prospect; but a friendly gust of wind would cleave a passage through them, and give us a glimpse to the north of the great cone of the volcano Longonot, or to the south of the breached crater Doenyo Suswa, and the newer cone rising within it. Sometimes the clouds would lift for a few minutes and reveal the plain, with its patches of green swamp and glittering sand, and the dark sinuous line of flat-topped acacias that mark the course of the Guaso Kedong; while far to the west we could see the long, dull gray scarp of the plateau, which forms the western boundary of the valley. We stopped there, lost in admiration of the beauty and in wonder at the character of this valley, until the donkeys threw their loads and bolted down the path. We soon caught them and started again down the steep slope, intending to go on to a place known as "Martin's Camp." But I noticed on the cliff face, a little farther to the north, what appeared to be some old shore lines. This was a point that had to be settled; so we camped on the platform already mentioned, 500 feet above the floor of the valley, and I devoted the rest of that day, and most of the next, to an examination of these terraces. The dense vegetation, sodden with moisture, was a great hindrance, but sufficient evidence was obtained to show that these strange straight lines along the hill-sides were old lake terraces. I at once named the lake that had made them after Prof. Ed. Suess, whose work has thrown so much light on the geological structure of this region.

Late on the following day we moved down to the floor of the Rift Valley, and in the dark, during a deluge of rain, pitched camp in a deserted Masai kraal.

We had now left the Kikuyu behind us, and as there seemed no natives in the immediate neighbourhood, I thought it was no longer necessary for me to spend the whole day with the caravan. The men were so overladen with food that no one could be spared to go with me, except my boy, who was so slow a walker as to be an encumbrance. But the

No. VIII. THE EASTERN WALL OF THE RIFT VALLEY, WITH THE TERRACES OF LAKE SUESS. *Page 94.*

geology was so tempting that I went off alone. By this time the men were accustomed to my going by myself, for I did so whenever the country was safe and the next camping-place easy to find. These solitary rambles were to me the most delightful incidents in the expedition. Free from the bother of the caravan, I could climb a mountain, track a river, visit a neighbouring lake, chase butterflies, and collect plants as careless as a schoolboy. But that day I nearly had to pay a high price for my pleasure.

The Kiringozi, who alone knew the way, told me that the next camp was at the foot of "yonder hill." This seemed very simple, so I said *Kwaheri* (Good-bye), and struck westward through the scrub to examine a mountain—Doenyo Nyuki—which appeared very different from the others in the Rift Valley. I found that it was the denuded remnant of an old volcano, traced the bands of ash and tuff that encircle it, examined the sections cut by ravines through the parasitic cones on its lower slopes, and climbed its four highest pinnacles. From the summit of one of these I thought I could see the camp in the place where I expected it to be; I therefore finished my study of the mountain at leisure. Afterwards I struck off to the supposed camping-ground, but could find no one. I shouted, but only an echo from the cliff answered. I climbed a tree, but could see no smoke. I made a circle round the point at a distance of about half a mile, but could find no track that answered to the men's. So I went back to Doenyo Nyuki, and after some trouble picked up the trail of the caravan. I ran along it for three miles as quickly as I could go, but then the sun set and I had to slacken my pace. It was soon dark and I lost the trail; it began to rain, and before long I was wet through. As I floundered on, hopelessly lost in the thorn scrub, I remembered that it was here that one of Thomson's men, who had straggled behind, was killed by a lion. I thought I could see a lion under every bush, and regretted that I had not stopped at the point where it first became so dark that I could not recognise the trail. I ought to have done this, made a "lean-to bivouac," and waited till the men came to look for me. But I was too proud then to admit that I was lost, and it was now impossible to find my way back. I thought, however, that the camp must be near the river, so I struck off due east until I reached it and then followed up its course. The river forked, and fortunately I took the wrong branch. Before I had discovered that this was only an insignificant tributary, I came across a trail in the mud. I went on my knees to examine it, and found the footprints of men, sheep, and donkeys; so it was probably that of our caravan. By going very slowly, often on hands and knees, I managed to track it for about half a mile when I saw the flicker of a fire. I crept cautiously towards it, fearing it might be that of a Masai kraal; but as I approached I could see a tent, and I knew I was "home." I called out to inquire, as I approached, "Mpokwa rudisha?" (Has "Loaded pockets" come back?). But in spite of my effort to disguise my voice it was recognised, and the men came rushing out to greet me, shake my hands, and lead me into camp. I found that Omari and some of the men were out on the search, and that since four o'clock there had been great anxiety owing to my non-appearance. I ridiculed the fear of my losing the way, and told them what a charming day I had had; but I did not think it necessary to add that the feelings of self-reproach and fright that had possessed my soul, from the time when I realised that I was thoroughly lost, until I picked up the trail half a mile from camp, had more than outbalanced the charm of the scramble over the peaks of Doenyo Nyuki.

The next day I climbed the cliff above the camp to conclude the examination of the lake terraces, but it rained all day, and the jungle was so sodden with moisture that I could not do as much as I wished. At noon I found a deserted Masai kraal, and crept for shelter into the only hut that still retained a roof; but it was already occupied—by the bodies of two Masai in an advanced stage of decomposition; so I ate my frugal meal in the rain. This was my first experience of the Masai, and it was not a pleasant introduction. A short time before reaching camp I saw some more members of the tribe, and this second meeting at one time promised to be livelier than the first. Unexpectedly I came across a number of them tending a herd of cattle. Some of them drove the animals away, and the rest walked towards the path as if to intercept me from camp. They waited about fifty yards from the path, and we tried to talk; but as neither side knew a word of the other's language, the conversation was not in-

No. IX LONGONOT FROM THE SUMMIT OF DOENYO NYUKI. *Page 97.*

structive. The Masai knew, however, that holding up a tuft of grass was the sign of friendship; they appeared also to understand the powers of a revolver, for they watched mine nervously. They followed at a distance of about a hundred yards for over a mile, and then as I approached camp they withdrew to their cattle.

The camp was situated on some open moorland, on the summit of the "Longonot Pass," on the ridge that separates the basins of the Kedong and of Naivasha. The pass is at the height of 7200 feet. It is bounded on the east by a rock mass named Kajabe, while to the west rises the great crater of Longonot. This was ascended by Mr. Joseph Thomson in 1883 as far as the crater rim, but the actual summit of the mountain on the western wall of the crater had not been climbed; this I resolved to attempt. It was very cold in the morning, and so wet and cloudy that it was eight o'clock before we were able to start. The lower part of the mountain consists of a series of platforms or terraces of lava. The rock is a black trachytic pumice, containing a good deal of obsidian. The mountain is very uniform in character, and lava forms a large proportion of its mass. The cone is in the main composed of lava, instead of fragmentary materials, such as ash and tuff, as is the case with most volcanoes. A sharp scramble of an hour and a half brought us to the rim of the crater, which at the point where we reached it has been worn by zebras into a broad cinder track. The floor of the crater is a large and fairly level plain, covered with acacia scrub; the walls are usually precipitous, but a descent could easily be made on the southern side. The great surprise was the discovery of a large steam vent on the inner face of the north wall of the crater.

The actual summit of the mountain is on the western side, and is 1800 feet higher than the rim of the crater at the point where we reached it. We started for it along the northern wall. At first the way was easy, but, as the edge of the crater rose, it became jagged and densely covered with scrub. To avoid some of the teeth, we worked across the inner face of the crater. After a weary scramble of an hour and three-quarters we reached the foot of the final pinnacle. A narrow sharp ridge of volcanic ash led up to it. This was very slippery, and it sloped rapidly down on either side to cliffs of such height, that the two Zanzibaris refused to traverse it. I cut steps across it and reached the dense bush that covered the summit. For a moment I doubted whether after all I should be able to force a way through the bush, for the foothold was precarious, and a slip would have been easy and disastrous. To return, however, when so near the summit was not to be thought of, so I hewed a way through the scrub with a sword bayonet. After sketching the surrounding country from the summit I returned to the men, and we boiled the thermometers, thus obtaining data which indicate for the highest pinnacle a height of 9350 feet. We then raced back to camp. Civilisation proved its superiority to nature, for thanks to boots and an alpenstock I arrived there at 2.15, an hour before the men.

Camp was struck at once, so that we might reach the shore of Naivasha that evening. We started at three o'clock, as soon as the two men who had been with me on Longonot came in sight. Our route lay over rich turf, on which roamed many antelope and enormous herds of cattle guarded by Masai. I had been told that the terrible havoc wrought by cattle disease had annihilated the vast herds once possessed by this tribe; I was therefore surprised to see so many. I was also rather sorry, for the abundance of cattle showed that there would be a large force of Masai near the lake. But my men assured me that they would be quite friendly, and not interfere with my plans. I was very anxious to examine the structure of the Mau scarp to the west of Naivasha, in order to discover if its structure corresponded with that of the eastern wall of the Rift Valley; and also to make a collection to illustrate the flora and fauna of the lake. I had brought with me presents for the Masai chiefs; for I intended to make friends with them, to leave most of the porters on the lake shore, while with fifteen others I made a branch excursion to the west.

Great, therefore, was my disappointment when, on our arrival at the camping-ground, a band of insolent young warriors came crowding round us and forbade my men erecting the tents. We had unwisely divided into two parties; Omari, the Askari, and eight porters were with the donkeys, while I had hastened on with the other men to get the camp into order before dark; so for a while I was bound to temporise. The attitude of the El-Moran was insolent in the extreme; they

No. X. THE CRATER AND STEAM VENT OF LONGONOT. *Page* 98.

bullied the porters, who were in a cringing state of terror, but fortunately they were rather in awe of me. One of them stepped up to the door of my tent and wanted to enter it. I asked in a polite tone "El Moru?" (Are you an elder?), and when he cheekily replied "El Moran" (Warrior), I kicked him away from the tent and told him at once to leave the camp, an order which the other Masai made him obey. At last, to our intense relief, Omari and the men with the donkeys came in, and a temporising policy was no longer necessary. I ordered two El-Moran to go and fetch some of the elders, as I wanted a "shauri." Two elders came at the head of a powerful body of El-Moran, and the shauri began. I could see that the idea of stopping at Naivasha for a week was absolutely impracticable; I therefore simply stated that I wanted to go through the country on my way to Baringo. I asked for permission to rest there for that night, to buy some firewood and water from the women, and next morning to continue my march to the north. The reply was short and emphatic. I was not to go through their country; I had no right to pitch my camp in it without permission; and I was to pack up at once and return by the way I had come. Omari and the interpreter Ramathan both looked as distressed as I was disappointed at this reply. They begged for a peaceful answer, but I had been advised on the coast that the right way to manage Masai was by "bluff"; so I replied that we were not going back, that we intended to stay where we were for that night, and next morning march on to Baringo; and that we should leave it to them to decide whether we were to be friends or foes. They then altered their tone, and said I might go on if I paid them "hongo"—a kind of toll for passage through their country. Hongo seems to me a very fair tax, at least when a caravan uses the paths and wells made by the natives; I had therefore no *a priori* objection to paying it. But the amount demanded exceeded my whole stock of goods; I therefore pointed to the stack of loads and then to the men, who with rifles ready were standing round the camp, and told the Masai that whatever hongo they wanted, they had better come and take. They did not seem to like the tone of the invitation, and left with threats as to what would happen next day.

As soon as the Masai had withdrawn we devoted all our

energies to strengthening the camp, cutting down any thorn bushes that would have afforded shelter to the enemy, and adding them to our thorn zeriba. Omari then came into my tent to discuss matters. He said the Masai were certainly present in great force, and for some reason or other were determined not to allow us to proceed. The interpreter Ramathan said he had passed Naivasha several times before, and had never seen so many kraals, such great herds of cattle, nor experienced such a hostile reception. With his usual cowardice he begged me to return. Omari was also very ill at ease, but with his usual pluck said that whatever I decided on should be done. Everybody on the coast had told me that if we met Masai parties on the war-path, we should have to fight them ; but that, except on Laikipia, the people of the villages would be friendly. The porters only the day before had said that we should find the Masai here most amiable ; they would come up to camp, we should shake hands and spit on one another ; they would sell us food and trinkets, and we should go in and out of their kraals like brothers. We were therefore all of us much disconcerted by our reception. The Masai never offered to shake hands, much less did they spit on us ; and until these two expressions of peace and goodwill had been made, we knew we could not trust the people. We could not quite understand it, so we made every preparation to repel attack, and stood to arms all night. It poured with rain, and this added to the discomfort of our situation. We kept the fires burning, and piles of wet grass beside them, so as to smother them in case of an attack.

Rain was still falling heavily at dawn, and it was useless to think of continuing our march until it left off. To encourage the men, we served out an extra ration of food and let them prepare breakfast. Some Masai children were sent out to watch us, but we saw no one else till about eight o'clock. The elders and the El-Moran then visited us, and once again the game of brag and bluster was resumed. They repeated their taunts as to the weakness of our numbers, and boasted that they had massacred caravans of twenty times our strength. They said they had done this with much smaller forces than they could now bring into the field, for they claimed to have 9000 warriors on the shores of Naivasha. Any hesitation or offer of hongo would have been mistaken for weakness, and probably have completely ruined the whole expedition. There was nothing for it but to put on a bold front and answer defiance with defiance. So, as soon as they repeated their refusal of permission to proceed, and again ordered us to return to the coast, I told them that we were going on, even though we had to fight our way through the lot of them. I warned them that if any harm befell us it would be the worse for them, for a great caravan of brave soldiers was even then approaching their country from Uganda. I said that the men in this caravan were more in number than there were papyrus stems around Lake Naivasha, that they would sweep upon their country, kill all the El-Moran, eat up all the cattle, and drive the elders and women and children out into the deserts, where not even their slaves, the Wanderobbo, could manage to live. This little speech had been translated by Ramathan during the night, and I had learnt it by heart. The Masai replied by driving their women and cattle back to the kraals, and it looked most uncomfortably like a fight. At about a quarter to ten the rain stopped, and a few minutes later we started north across the plain. When the elders saw that we were determined to go, they became more friendly. As the first man moved out of camp the chief came up and for the first time held out a "knobkerry," cut from a rhinoceros horn, for me to shake. I of course shook it, glad of this expression of friendliness, though quite aware that if it stopped at this it meant nothing. But as I walked down the slope from the camp after the men, he came up again, and this time not only held out his knobkerry, but as I shook one end, he shook the other. After walking a few yards he repeated this more vigorously. Then at last he held out his hand; we shook hands, at first coldly, and then more cordially. Finally, after we had walked a couple of hundred yards, we repeated the process and the chief spat upon me, a salutation which I returned with perhaps unnecessary vigour. I had been warned that whenever Masai retire from a conference without spitting the spit of peace, squalls may be expected. I was therefore much relieved when this friendly rite had been performed.

We marched northwards over the plain beside the lake. We were followed by a crowd of El-Moran, who seemed at first disposed to be fairly friendly. But as soon as they had passed out of sight of the first kraal, their love of mischief became too much for them. They drove up some donkeys towards us; these brayed to ours, which tried to dash off to join the others. One broke through our line, threw its load, and succeeded in reaching the herd. The El-Moran tried to pillage the load, but Omari and I reached it as soon as they did, and by striking the butt end of our rifles on their naked feet, sent a few of them limping away. As it was obvious that there was nothing like bluff for impressing the Masai, I ordered the El-Moran to go and catch the donkey. I said they were trying to steal it, and that if they did not bring it back I would seize three other donkeys and burn the kraal. They soon brought back our pack animal, and after this behaved much better. They made, however, another attempt to seize a load while we were crossing a stream. As we had to march in square, and were much annoyed by the Masai, we only made four miles that day. We had to pitch our tents amid some rocks about a mile from the north-eastern corner of the lake. There was no thorn scrub with which to make a zeriba, so we arranged the tents in a small circle and filled up the interspaces with boulders and baggage.

Late in the afternoon a powerful force of Masai in full war costume marched up to camp and said they had been sent to dance to us. I put some marks on the ground at a little distance from the camp, and said they were not to come nearer to us than these. I insisted on a Masai elder taking a seat on a box before me, and saw that all the men were at their posts with their rifles ready and the reserve ammunition boxes in a convenient place. Then I gave permission for the dance to begin. It was certainly extremely picturesque; the men were all armed with great shovel-headed spears, with heads varying from 30 inches to 3 feet in length, and spikes a little longer; they carried thick oval shields 4 feet in height, which were coloured in various patterns. They had rattles on their ankles and their arms; their fingers were protected by bright iron knuckle-dusters; their heads were adorned with aureoles of feathers. At first they simply leapt into the air, throwing their heads violently backwards and forwards, and making a series of guttural grunts. Then they marked time with their feet while they hurled their spears upward, giving them a twist which made them flash in the sunlight. Meanwhile they shouted a kind of song without words, to the accompaniment of the music of their rattles. Next they marched and ran in Indian file, twining in and out in a series of complex evolutions, and finally arranging themselves so that the designs on their leather shields were symmetrically disposed. Then they executed the last dance performed before starting on a war raid; they followed this by a series of sham rushes at the camp. As they came on roaring and shouting, their bodies covered by their shields, and their long spears raised over their heads, they presented a weird spectacle. We watched them with breathless anxiety, expecting every moment that a sham rush would become a real one. My revolver covered the head of the Masai elder, while the men prepared to close the breeches of their rifles, which were kept not quite closed, so that if a trigger were pulled accidentally no harm would be done. After this, the Masai gave us the dance with which they celebrate victory; I told them I wanted to see that, but that they need not trouble to perform their dance after a defeat, as, if they did not behave better than they had done the day before, they would soon dance it in earnest. At the close of this "Wild East Show" I distributed among the dancers a couple of shillings' worth of beads, with which they went away apparently contented.

The Masai elder offered to stay with us in camp, as a sign of friendship and to protect us from interference by any of the El-Moran. I said I should be most happy to entertain him for a night, but that our guns were our protectors, and to them alone did we trust. I had a long chat with our guest in the evening, and found out from him what I imagined to be the explanation of the hostile attitude of the usually friendly Masai of Naivasha. He said they were preparing for a great war raid, but would not tell us against whom. They appeared to think that the garrison of Fort Smith had heard of this, and that I was being sent on to hasten the caravan returning from Uganda, in order to stop the raid. Major Smith, however, who knows the Naivasha Masai very well, afterwards told me that they are sometimes very bumptious and troublesome; that they planned an attack on the Railway Survey, and tried at night to surprise the camp of Mr. Newman, the famous rifle shot, when on his way back from Uganda. It is therefore probable that it was the small size of my caravan that tempted them to display their innate love of bullying and thieving.

Although after my conversation with the elder I felt more at ease, I still did not deem it safe to go to sleep. I sat wrapped up in a blanket, with my revolver, shot-gun, and rifle ready loaded on a table beside me. Every half-hour I went round the camp to see that the sentries were awake, and that the fires were burning. We had fortunately lighted fires at a little distance from the camp, and these illuminated a considerable space around us. Shortly after midnight, just after returning to my seat beside the fire, I saw the cook spring to his feet in such haste that I guessed there was something wrong; I threw off my rug, seized my rifle from the table, and looked round. By the light of the outer circle of fire, we could see a party of about twenty Masai creeping up to camp. They were carrying their spears, but not their shields, apparently lest these should make noise that might betray them. The cook and I shouted "Masai," and this, with my order of "Bunduki tiari" (Guns ready), roused the camp.

The men took their places quietly and quickly, and looked as though they meant fight. Even my tent-boy Philip, whose cowardice was the joke of the camp, went to his station inside my tent and stood ready to serve out ammunition. Several of the men under the Kiringozi, Wadi Hamis, had been told off to climb the small cliff above the camp, to guard that approach, and to be able to fire down at a foe upon the plain. I forgave that surly old grumbler many of his sins, for the pluck with which he led his men to that important post. He found another party of Masai approaching from that side, and at once opened fire, while I discharged four shots over the heads of the men on the plain. As the Masai now saw that there was no chance of a surprise, they turned and fled. I sprang at once to the shelter tent where the Masai elder was supposed to be. But the savage had escaped unnoticed, leaving his goods behind him. He certainly left before the attempted attack, as I had set two men to guard him at the first alarm. His behaviour thus lends some support to the idea that the attack had been planned, and that it was not a mere attempt by a batch of thieves to rush through the camp and seize any objects that they could lay their hands on in the confusion.

We stood on guard, expecting at any moment an attack in force by the main body. But as this was not delivered, and I knew there would be no lack of sentries for the rest of that night, I turned into the tent and enjoyed my first three hours of continuous sleep since I had left the protection of Fort Smith.

At four in the morning I had breakfast, and the men packed up the loads. We waited anxiously for dawn, for we dared not start in the dark, lest we should walk into an ambush. The moment the light was strong enough to enable us to guard against surprise, we marched northward across the plain. Unnoticed by the Masai we waded the swamps beside the streams that enter the north-eastern corner of the lake; but as soon as we lost the shelter of the papyrus in the swamps and the scrub on its borders, we were discovered by some shepherds. They took the news to the nearest kraal, and a party of El-Moran came out to watch us. On the open grass-land they dared not attack, so they followed at a respectful distance. A few miles farther on we were stopped by the river Malewa or Murendat. To avoid passing near some kraals we had bent our course far to the west, and thus, instead of striking the river at the ford, we reached it where it flowed through a deep sinuous cañon. We marched along this to find a place where we could descend to the river, so as to be able to get water even if we could not cross. We found a track which led down to a ford, but the river was impassable. The flood, however, was subsiding rapidly, for the bank was still wet for more than a foot above the level of the water. We placed some notched sticks in the river to mark its rate of decline, and camped. For twenty long hours we sat beside the ford, watching successively the river, the Masai, and the clouds on the hills to the east, and feeling probably much like the Israelites, when they had the Red Sea in front of them and Pharaoh's hosts behind.

In the evening, as the river had fallen several inches, I tried again to cross; I reached a shoal in the middle, but the last channel was too much for me. A school of hippopotami was playing in the pool below the ford; as I had been carried down into it in the morning, I did not think it worth the risk of adding to their sport, and swam back to the southern shore.

Twice during the night we had false alarms of the Masai; our rest was also disturbed by the attempt of a pair of hyenas to get at our donkeys. Fortunately the night was rainless, and by the morning the river had fallen sufficiently to permit us to cross. We did so, though with difficulty. Omari and two or three of the stronger porters carried over the loads, while the feebler members of the party pulled themselves across, hand over hand, along a rope. Some Masai watched us, but made no attempt to interfere. A short march farther brought us to the Gilgil river, the waters of which were unusually low. There were no Masai in the district, and nothing to lessen the luxury of a feeling of safety and peace. The porters were delighted to get into an uninhabited country, and I was glad to turn again to the subjects of geology and botany, after my experience of the over-exciting occupation often presented by African anthropology.

CHAPTER VII

ALONG THE RIFT VALLEY TO BARINGO

" Upaci, upaci,
Hatta Baringo.
Mbali kidogo
Tukafika Uganda."

(Hasten, hasten,
As far as Baringo.
But a little farther
Then we're at Uganda.)

Zanzibari Marching Song.

DURING the march from Naivasha to the Gilgil we had risen slightly in elevation. The lake stands at the level of 6200 feet, and in the two marches we had only ascended 200 feet. After crossing the river the ascent became more marked; and, at the height of 6710 feet, we reached the summit of the ridge that separates the basin of Naivasha from that of Lakes Elmetaita and Nakuro. To the south was a long slope, covered with turf and loose " lelesha " scrub (*Tarchonanthes camphoratus*), and in the far distance the crater of Longonot. To the north, a cliff descended abruptly to a plain on which lay Lakes Nakuro and Elmetaita, leaden-gray in contrast to the intervening tracts of glistening sand and salty desert. We scrambled down the cliff, leaving to our left a group of extinct volcanic cones, and made our way towards the southern end of Elmetaita, where we camped. There, on the banks of the Kariandusi river, we found a powerful caravan under Major Eric Smith and Captain (now Major) Williams, R.A., who were returning to the coast after the evacuation of Uganda by the British East Africa Company. They gave me a great deal of most valuable advice and information. This included, as a matter of course, the usual warnings against the attempt to cross Laikipia with so small a force. Captain Williams advised me to strike westward from Njemps to the Victoria Nyanza, and along its northern shore to Kampala; there I could get whatever trade goods were necessary, and thence proceed to Ruwenzori. Food was abundant, and the natives friendly all along the line. The scheme looked attractive when compared with the risks of the march across Laikipia, and for a while my resolution of "Kenya at any price" was shaken.

Next morning the Company's caravan proceeded on its homeward journey. I spent the day in an excursion eastward to climb one of the Dondole Mountains, and in walking westward across the plains to a group of broken-down volcanic craters. The morning climb was rewarded by a view of exceptional beauty from a peak of 7650 feet in height, which, owing to the form of its flat, lava-capped summit, I named Kilima Meza (*i.e.* Table Mountain).[1] Of more practical value than the view was the collection of some interesting plants, including a handsome *Gladiolus* and a new species of *Lagarosiphon* (*L. hydrilloides*, Rdle.)

The lake itself I visited next day. It receives two rivers, the Kariandusi and the Guaso Nagut, but has no outlet. Its level is being lowered by evaporation, and it is now much smaller than it once was. The water is bitter, and the only signs of animal life in it were some insect larvæ and small crustaceans (amphipods), though the empty shells of a new water-snail (*L. elmetaitensis*, E. A. Sm.) were abundant on the shore. Huge flocks of pink flamingoes (*Phœnicopterus roseus*, Pall.) waded in the shallows or swam on the surface, browsing on masses of alga, which in places imparted a deep green colour to the water. The simultaneous discharge of both barrels of my shot-gun secured six flamingoes, and yielded for some days afterwards a pleasant addition to our bill of fare.

After leaving Elmetaita the journey, for a few days, was as easy and restful as a picnic. The route lay over smooth turf steppes; water and firewood were abundant; there were no natives to worry us, and the rain did not begin till late in the afternoon. But this did not last long. After crossing the

[1] See *Alpine Journal*, vol. xvii. (1894), p. 91.

ridge (6730 feet) which separates the Elmetaita and Baringo basins, the valley descended rapidly to the north. We entered some old lake basins, in which the grass was so rank that marching became slow and wearisome. We left the ordinary track, which was well known, in order to try and find a new route to Njemps along the eastern side of the Rift Valley.

The next week was a succession of misadventures. In order to reconnoitre the country ahead I climbed a mountain, which, as it lies exactly on the equator, I called " Equator Peak." From the summit (6150 feet) I could see a lake to the north-east which I resolved to visit. I sent the Askari in one direction while I went in another to find the best track by which to lead the caravan next day. We were caught some miles from camp in a storm that was simply terrific. We were used to heavy storms, but this was a hurricane. Even before sunset it became so dark that it was impossible to see the way or even to read a compass ; only the wind gave any clue to direction. We both spent some hours stumbling about in the darkness in the thorn scrub, and it was not till late at night that we regained the welcome shelter of the camp.

This was only the first of a trying series of accidents. The men at first objected to going by the route I proposed, as they said they had no Kiringozi who knew it. I replied that I would be Kiringozi, and they would have to follow me. We marched north across the plain to the foot of Equator Peak, and then along the eastern base of the ridge, of which it is the highest summit. After a fatiguing march through dense jungle, we reached the shore of the lake. One of my men said he had previously visited it with a Suahili trader, and that it was known as Zewi (or Lake) Kibibi. This is apparently only a Kisuahili name and means " Little Ladies," which may be a playful allusion to the swarms of mosquitoes that are said to infest it. As it poured with rain throughout the whole of the fourteen hours during which we camped beside the lake, we were not troubled with these pests. We fished in the lake, but without success. Next day we continued our way northward across a country so intersected by ravines, and choked with scrub, that progress was very slow. We had to march in single file, chopping a way through the jungle and trampling the undergrowth into a path. Several times during the day we were charged by rhinoceros. They lay asleep until awakened by the noise we made, and then, frightened and muddled, they charged wildly in all directions through the scrub. Twice they broke through the line of porters, but the men managed to dodge them, and the only damage done was to the loads.

In the afternoon we emerged from the valleys on to a plain where the scrub was thinner, and I went on ahead to try to find water and a camping-place. Suddenly, without the slightest warning, I found myself on the edge of a precipice 1900 feet in height. For some hundreds of feet the cliff was absolutely vertical. A few yards away from where I stood it actually overhung, as the wind had cut away the soft beds of ash below the lava that formed the summit. The change was so startling that for a moment it made me feel giddy. When the porters came up they were as surprised as I had been, and we all stood along the edge of the cliff and admired the extraordinary view before us. At our feet, at the base of the precipice, lay a long narrow lake, in shape something like Windermere. The opposite shore is formed of a series of steps and terraces which rise one above another to the summit of Doenyo Lugurumut. Beyond this ridge is the valley along which runs the trade route to Njemps, and beyond this again are the undulating foot-hills and the dark gray scarp of the plateau of Kamasia. The view was certainly the most beautiful I had seen in Africa. As a rule, except at sunrise and sunset, the colours were disappointing; but here the lake itself was of an exquisite blue, broken by the green of dense growths of alga, or by pink where vast flocks of flamingoes floated on the surface. The colour effects on the shore and on the islets were as striking by the contrasts as by the brilliancy of the tints: here and there a glittering tract of sand, or a dazzlingly white deposit of sinter around a hot spring, interrupted the sombre brown hue of the acacia scrub.

We were all so fascinated by the landscape, with its wide expanse of view, its striking contrasts of colour and form, that it was some time before practical considerations forced themselves upon our minds. Then we realised that the view was not without its drawbacks. The precipice that gave it to us barred our progress and drained the plateau dry. Water was absolutely necessary, so we started off in different directions to search for it. Of all hunting there is none so exciting as that for water, none which exacts sounder judgment and greater patience; and on the edge of a cliff 1900 feet high no quarry is more elusive and difficult to stalk. Once or twice my hopes were roused by finding some damp soil on the bed of the brook I was tracking, but I returned to camp unsuccessful. A porter had, however, found water in some rhinoceros foot-holes, and these yielded us a limited supply. We hoped the usual evening's rain would enable us to fill up our bottles; but though a terrific storm of wind nearly blew the tents over the cliff into the lake, not a drop of rain fell. I had in consequence to make a breakfastless start. We continued northward along the edge of the plateau, in the hope of finding some place at which we could descend to the lake. At half-past ten we reached a gully which had cut back the edge of the plateau and offered us a chance of descent. I started down it to prospect. I found some interesting plants and a most instructive geological section, but after a climb of some 800 feet, a vertical cliff of lava formed an absolute barrier to further progress. I tried again in two or three other places, but in vain, and I had to return to the plateau foiled and disheartened. An enterprising porter, Fundi Mabruk, had, however, gone off on his own account, and towards evening succeeded in finding an old game track that led down to the shore. It was then too late to descend that night, so we camped in a hollow, and served out the last ration of food, which, however, could not be cooked owing to lack of water. We started at dawn next morning, hoping to reach the lake early, breakfast there, and then hasten on to Njemps.

But the descent was a longer and more difficult task than we had expected. The porters and sheep scrambled down easily enough, but a good deal of ledge-digging and bush-cutting had to be done to render it practicable for the donkeys. It was therefore nearly eleven o'clock before we reached the shore. Then, to our horror, we found that the water was salt and sulphurous. The first man who reached the lake returned to us making hideous grimaces and groaning, "Dowa, dowa, hapana maji" (It's medicine, medicine, not water). And very effective medicine it proved to be; its emetic properties acting on the man's empty stomach brought on such a violent attack of retching that it was necessary to give him some cocaine. The sight of the water drove the donkeys almost mad, and they made most desperate efforts to reach it; we dared not let them drink, for it would probably have killed them. Bitterly disappointed we resumed our march along the eastern shore of the lake. In our thirsty condition we soon began to feel the heat. The whole place seemed to have been planned as a sun-trap. The black precipice and the bare lava at its foot became hotter and hotter as the day drew on, till the glow from them became almost intolerable. The cliff, moreover, screened us from the refreshing breeze that generally softened the midday heat. Hence the air was stagnant, and there was nothing to carry away the moisture that rose from the surface of the lake, or the putrid odour of the vegetation that lay rotting in the submerged meadows along its margin. Here and there we had to cross stretches of sand, raising a cloud of salt-covered dust which added to the pangs of our burning thirst. Occasionally we had to wade for short distances through the lake, to avoid dense thickets of bush or rocky headlands, or to cross meadows now submerged owing to the high level of the water. During these we had to tie sacks over the donkeys' heads to prevent them from drinking; as we could not take the same trouble with the sheep, they drank what they wanted, and two out of the three died.

I dragged in the lake for shells, but could get none, nor any trace of aquatic animal life. It was more barren than Elmetaita, for the water of that lake, though bitter and salt, was clear and pure, and yielded a few insect larvæ and amphipods; but the putrid sulphurous water of Lake Losuguta seems fatal to life. Some trees that stood a few yards from the shore were dead, though, as their leaves were still attached to them, their submergence must have been recent. The grass was yellow, and whatever the water touched it seemed to kill. The only exceptions were a green alga and some pink flamingoes, and in the absence of competition these throve exceedingly. The alga grew in such dense masses that it often coloured the water green; while the number of the flamingoes was such that when, towards sunset, they rose from the lake and flew northward, one of the kite-shaped flocks must have measured 400 yards in breadth and a mile in length. But these alone seemed able to touch the waters of the lake and live, and we saw neither birds nor insects on the shore. The day's zoological collection was represented only by a water-scorpion (a *Nepa*) which we found in some half-dried mud, and a small snake (*Rhagerrhis tritæniata*, Gth.) An effort to add a rhinoceros nearly resulted in a disaster. We saw a pair, and as we were desperately hard-up for food, I told Fundi to follow me, and started off to stalk them. They bolted into the bush at the foot of the cliff, and it seemed impossible for them to escape. I soon found one of them, but could not get a safe shot at it, as it stood facing us, and its head was over its chest. A small stream bed, 4 feet deep and about 6 feet wide, ran towards the animal. We crawled along this till I was level with the rhinoceros. I was taking a steady aim at its brain when I heard a shriek from Fundi of "Kwea, bwana" (Climb, master). The other rhinoceros was also in the bed of the stream, and, having scented us, was charging down upon us. We both sprang up into the thorn scrub, while the rhinoceros passed beneath me and stopped, as if it intended to turn. I jerked my rifle to my shoulder and fired at the animal's neck. The effect was startling to both of us. The mouth of the barrel had scraped against the side of the gully, and become half choked with sand; the recoil was so heavy that it threw me back into the thornbush, while the bullet, instead of cutting through the backbone, tore its way through the muscles of the neck. With a grunt of pain the rhinoceros rushed on along the gully, while I rolled into it. Fundi picked up his rifle and ran to help me, shouting to the others that I was hurt. He thought the rhinoceros had caught me, and was much relieved to find that it was only the gun that had knocked me over. My right shoulder, however, was so bruised and battered that for some weeks afterwards I had to shoot from the left. Later on the same day a rhinoceros threatened to charge the caravan, but after a minute's reflection it walked slowly southward. I suggested to Fundi that we should go and stalk it. "Bus faro leo, bwana" (Enough rhinoceros for to-day, master), he replied. I agreed with him, and so we made no attempt to intercept its retreat.

About five in the afternoon we passed round a headland to a wooded gully cut into the face of the cliff, in which we hoped to find water. It was our last chance that day. But, to our intense disappointment, it contained nothing but sand. Never before did I realise so fully the sad truth of the line in the missionary hymn—

"Where Afric's sunny fountains roll down their golden sand."

All the fountains we passed that day rolled down nothing but sand. To add to our annoyance, we could see rain pouring down on the other side of the lake. It had rained every day but two since we had left Mombasa, and we therefore hoped that it would yet do so; but, as I feared, the hot cliff beside us kept the rain-cloud away. By this time the porters were too exhausted to go farther. Several of them had already fallen far behind, and some of these had to be almost carried on. We were too tired to pitch the tents, so our camp that night was a very dismal one. After dark it became cooler, and the pangs of thirst became more tolerable; but later on it became intensely cold, and several of us were seized by an attack of fever. The men clamoured for medicine, but as there was no water they could not have any, and we spent a miserably feverish, restless night.

Dawn came suddenly; a dim tinge of grayish pink tinted the western sky, and almost before I was certain what it was, the sunlight caught the summits of the Kamasian hills. We prepared to march at once; every man who had the strength to do so picked up his load; I nodded the order to start, and without a word being spoken we resumed our march. The scrub here came down to the water's edge, and we had to chop a way through it. Fundi and I, with an Askari, led the van to cut the path; but every jerk sent such a pang through my shoulder, that I had to let the others do the work. Progress was very slow; the slightest extra exertion produced an awful feeling of sickness and weakness. We could stagger along on the level fairly well, but if we had to climb a few feet to cross a ridge, the men would fall exhausted upon it. Rests became longer and more numerous; at ten o'clock we found the heat of the sun unbearable, and it seemed as if the whole caravan were on the point of complete collapse. A little way ahead there was a gap in the jungle, so I let the men drop their loads and hasten on to this. I intended to put up the fly of the tent and let the men rest under it during the heat of the day, while I went on with a few others to search for water. Just as we were preparing a slight shower of rain fell; we caught as much as we could and shared it. Omari refused to take his share, and gave it to a porter. The drink, small though it was, refreshed us somewhat, and we went on again. I started ahead with Fundi and an Askari, and at length we found some water in a swamp at the north end of the lake. We fired a shot to announce the good news, and the two men went back with water for the others. The porters soon arrived, many of them without their loads, and we were soon revelling in the luxury of abundance of water, after forty-one hours' total abstinence.

We had found water, but we had found more than we wanted. Through the middle of the swamp there flowed a river, which was too deep to wade, and too swift to swim. It ought not to have been there, for we were now close to a well-known road, and neither of the latest maps showed any sign of a river at the place. But we did not trouble about this: we drank what we wanted, and let the rest run by. We could not understand whence the river came, and at first we did not care; but when, a few hours later, we proposed to resume our march, we *had* to care. We found that the river flowed from a gorge in the mountains and entered the northern end of the lake; we were therefore compelled to cross it. Omari and I spent a couple of hours wading along the hippopotamus paths, that ran like tunnels through the reeds, in search of a place where we could cross. The only possible ford was at the edge of a waterfall; but the rocks were covered with slime and algæ, and a slip would have plunged us into a pool and into the close company of some crocodiles. It is not as a rule advisable to allow the fear of crocodiles to interfere with plans, but in deep muddy water they are dangerous, and neither of us chose to risk a fall into the pool below. We had at last to recognise that it would be necessary to bridge the river. We came to this decision very reluctantly, for there was no food left, and the delay was serious. A clump of acacias stood beside the camp, and men were set to fell timber and hew it into a convenient shape. The men worked in relays by firelight through the night, and in the morning the beams were ready. We dragged them through the swamp to a point below the waterfall, where the river ran between steep banks of clay, which afforded a good foundation. We threw a light trestle bridge across the river, and crossed by it in safety. An hour later we passed a huge sycamore, close by one of Teleki's old camps; beyond this we forded a stream (the Mudoleto), flowing northward into the swamps to the south of Baringo.

Soon afterwards, on crossing a ridge, we saw steam rising from the ground, and knew that we had reached the hot springs of "Maji Moto." The men ran forward singing, while I stopped behind to collect and make some observations. When I came up to them, I found them bathing in the springs and enjoying themselves thoroughly—larking, indeed, like a lot of schoolboys. The ordinary caravan route passes beside the springs, so we knew that there were no further obstacles between us and Njemps. We felt assured that there a supply of welcome food awaited us, and some days of equally welcome rest. I could not resist the temptation to ask my hungry men the precise delicacy which each hoped was in store for him. I told them they could follow on at leisure, and in the hope of having food ready for them on their arrival, I started with Fundi, the interpreter Ramathan, and my boy Philip, at the best pace we could achieve, in hot haste for Njemps.

We took at first a north-westerly course round the lower flanks of three "fault ridges" of lava; we kept upon these as long as possible, for the lowland was one vast swamp, but when they trended too far to the west, we struck off across the plain for the woods around Njemps. We soon entered the shambas, which are of enormous extent and watered by hundreds of irrigation channels and ditches, which we had to wade or jump. The crops consisted in the main of dhurra or millet, but only the stubble of the previous season's crop remained upon the ground. While collecting some plants I startled, and was startled by, a family of ostriches, but Philip had my rifle, so I lost that chance of food; and later on I wasted a cartridge in a flying shot at a small antelope belonging to the genus *Madoqua*. At length we left the fields and entered a forest of acacias, with an undergrowth of mimosa scrub; the path had been worn by the feet of many generations into a hollow, that served as a channel for a sluggish stream often 3 feet in depth. But having to wade along this did not lessen our sense of the welcome coolness of the shade. The paths became more numerous, and, after a long wade—which was rendered even less pleasant than usual by the odour that arose as our feet churned up the slimy ooze at the bottom—the ground began to rise, and I guessed that we were approaching the river. A stream 3 feet wide was jumped by all except the fat boy, who preferred to wade, and was in consequence knocked over by the current, which was rushing down like a mill-race; but I was too excited to stop and help him, and shouting to the others to take care of the rifle, I walked quickly up the path through the wood. A family of monkeys fled screaming through the trees, a green parrot shrieked its shrillest in rage at being disturbed from its midday nap, and flew across the river. A few steps more and a sharp turn in the path brought me to the steep bank of the Nyuki. On the opposite side was a cluster of beehive huts, densely grouped in the shade of some lofty acacias, and protected by a strong double thorn stockade. It was the village of Njemps Mdogo. The river ran swiftly in a deep channel that it had cut through the red sandy alluvium. A fallen tree formed a rude bridge a little distance up stream, and I preferred to scramble across its slippery trunk and through its irregular branches, rather than to cross the ford opposite the village. I paused for a minute on the bridge; for a water-snake was gracefully wriggling up the river, and the temptation was irresistible to flatten a revolver bullet on the head of a crocodile that was basking on the bank. As it was two o'clock in the afternoon, no one was about; an air of drowsy peace and security seemed to rest upon the village, as about an English hamlet on a summer Sunday afternoon. So with visions of a week of rest, with a rich harvest of precious specimens to be gathered in the daytime, and unbroken sleep at night, I walked gaily up the path to the low narrow gateway, and fired off a couple of shots to announce my arrival. A native crawled along the passage cut through the otherwise impenetrable hedge of thorns to see who I was. He was a tall Njempsian, with a breadth of shoulder that told of former fine physique; but he was terribly emaciated, and stooped from sheer weakness. He looked at me for a moment with his dreamy, hollow, shrunken eyes; then with a sad smile held out his hand and said, "Yambo, yambo." His aspect was that of a starving man, and filled me with dismay. I knew from his salute that he must know a little Kisuahili, so I eagerly asked him if there were much food in Njemps. "Jocula, tele jocula?" he repeated, and his sickly smile faded into a look of bitter despair. "Jocula *hapana*" (There is *no* food), he said with an emphasis that his appearance made superfluous. Yet I assumed, or at least tried to assume, a sceptical aspect, and pointed to the village, and asked him what the people all lived on if they had no food. In reply he only glanced up at the trees, tore off a few leaves from a branch he carried in his hand, and voraciously munched them. I pointed again to the highlands of Kamasia, rising like a wall on the western horizon, and asked if there were food there. "Jocula hapana," he repeated; but after a pause he turned to the north-west, and waving his hand several times in that direction said, "Mbali, *mbali*, jocula *kidogo*" (Very far, very far away, there is a *little* food).

Fundi and Ramathan then came up, having been delayed in pulling Philip out of the stream: I at once set the latter to cross-question the native in his own language. But he could only tell us that last year they had had no rain, and so the crops had failed, and this year the rains had been so heavy that the crops had been washed away with the soil on which they grew. I staggered across the path and sat down on a fallen tree-trunk, while visions rushed through my mind of the disappointment of the caravan, how Wadi Hamis would grumble, how each of the Wakame would look the image of despair, and Omari would become even sadder and more thoughtful than he had lately been. I rummaged for a whetstone in one of my capacious pockets, and began to sharpen my already supersharpened hunting-knife. When this mechanical motion had dulled the sickening sense of disappointment, I could contemplate more at ease all that was meant by the simple statement: "There is no food in Njemps."

CHAPTER VIII

THE STAY AT NJEMPS AND EXCURSIONS AROUND BARINGO

"Ay, now am I in Arden; the more fool I; when I was at home, I was in a better place: but travellers must be content."—*As You Like It*, ii. 4.

NJEMPS is the district on the floor of the great Rift Valley, at the southern end of Lake Baringo. It is inhabited by a section of the tribe of the Wakauvi, who are related to the Masai; they are, however, agriculturalists, having abandoned the pastoral nomadic habits of their ancestors. The first European who succeeded in reaching this country was Joseph Thomson in 1883. He gave the natives a character for trustful friendliness and simple honesty, which has been confirmed by every subsequent traveller. So peaceful are the Njempsians that one can walk about their country unarmed and unattended, or chase butterflies or stalk zebras with as little fear as if one were rambling through English lanes. In contrast to the caution necessary in the land of the Masai, this feature makes Njemps a welcome haven of rest for the weary. The country, moreover, is usually rich in food, and caravans refill their empty sacks for the journey northward over the foodless wastes that must be traversed in order to reach the ivory-yielding districts of Karamoyo and Samburu.

I had been warned both by Captain Williams and Mr. Martin that it would not be safe to rely on getting much food at Njemps, but that I could doubtless procure enough to take my small caravan on to Kamasia, where it could be obtained in any quantity at a small price. Captain Williams advised me to make assurance doubly sure by having at least one day's food in hand when I reached Njemps; and, but for the disturbance of my calculations on Lake Losuguta, I should have had this. I had, however, distributed the last load with the less hesitation, as I felt persuaded that in a district where there was sufficient food to support two such large villages as Njemps Mdogo and Njemps Mkubwa, enough could surely be obtained somehow to ration forty men for a couple of days. If this could not be done, I resolved to leave the loads and sluggards at Njemps, and push on to Kamasia by forced marches. I had never doubted that I could get as much as I wanted there, and the news that the famine extended not only over Njemps, but also over the rich plateaux to the west, was a blow as staggering as it was unexpected. I had always regarded the failure of supplies at Njemps as a possibility, but as one that could only be a nuisance. Now, however, it was obvious that the extension of the famine to Kamasia might easily mean a disaster to my expedition.

My plan of action was soon settled, but I was still sitting on the log, wondering how I could most easily appease the wrath of the caravan, when I was interrupted by a soft voice giving me the Arab greeting "Subulkairi." It came from an elderly Arab dressed in a spotless white robe or "kanzu," which would have looked eminently respectable in any bazaar in Zanzibar. The usual dialogue of salutations followed, and he made me a long speech of welcome in an easy flow of Kisuahili. I was too charmed by its musical rhythm to bother about the meaning, and told Ramathan to make an appropriate reply. The Arab led me by the hand into the village and introduced me to his friends, a company of Mombasa merchants; they were waiting at Njemps for the return of the parties they had sent off in different directions to purchase ivory. They asked me if I were hungry, and I confessed I was; I believe I looked it. They gave me a bowl of delicious bean-flour "potiss," or gruel, and a cupful of sour cream; a plague of flies lessened the comfort of the meal, for in spite of the efforts of two of the Arab's slaves to keep them off by flourishing over me zebra and eland tails, a crust of drowned and drowning flies made the food look like a black-cap pudding. At first I dived for flyless handfuls—my spoons had not yet arrived—in the deeper strata of the mass; but as a few of my tormentors were always picked up between the basin and my mouth, I ceased to struggle against the inevitable, and completed that meal on the principle of "Open your mouth and shut your eyes, and see what fate will send you." My new friends had not heard from the coast for fifteen months, so I told them all the news that I thought would interest them; such as that Mr. Piggott was now "Bwana Mkubwa" at Mombasa; that the Sultan of Zanzibar was dead and another was reigning in his stead; that the British East Africa Company had withdrawn from Uganda, and that Sir Gerald Portal had gone there to take it over in the name of the British Government. They were delighted to receive news, and when I asked about food for my men, a load was at once given me; I was told, moreover, that if I had cowries and "kiketi," I could buy as much as I wanted in Kamasia. So, fully recovered from my fright, I strolled down to the ford to meet the porters. I saw at once that the men had heard the news of the famine; they had been told of it by some women whom they had met on the road. Wadi Hamis marched first with a look of despairing triumph on his surly, pock-marked face. "Jocula hapana, bwana" (No food, master), he growled with an I-told-you-so sort of air. The attitude of the three Wanyamwezi, who followed him, was very different; their sympathetic looks showed that they felt rather sorrow for me in my perplexities, than regret at the prospect of suffering for themselves. So I hastened to relieve their anxieties by telling them in as off-hand a manner as I could, "Upaci fannya motu—Jocula tiari" (Make fires quickly—the food is ready). The news ran back along the line of men across the ford; the shouts of joy cheered on the stragglers, and camp was soon pitched under the shade of some acacias on a sandy plain near the western gate of the village. We bought firewood from some women, rations were served out, and the men were soon merrily cooking their food and fraternising with the porters of the trading caravan.

Kizizi, the chief of Njemps, was now in Kamasia, but was due back next day, and it was advisable to make friends with him. The Arabs offered me the loan of 500 lbs. of flour, and said if necessary I could pay for it in Mombasa; so there was no immediate necessity to go on, and I decided to rest next day and see Kizizi.

The traders visited my camp in the morning, and Philip

acted as interpreter to a shauri that began most amiably. At last Timami, who was the chief spokesman and president of the caravan, asked me to sell them some cartridges, as they had used all theirs shooting food. This I did not believe, for I knew they were not cannibals, and one of the party had previously confessed they had used up all their ammunition in a fight with a tribe six weeks' journey to the north-west. I had given a solemn pledge, in accordance with the terms of the Brussels Convention, not to "give, assign, or sell to any person or persons" either arms or ammunition. As natives are not allowed to have breech-loading rifles on any pretence whatever, I had been surprised to find that the caravan had both Sniders and Remingtons, and had therefore taken precautions to prevent the porters selling either cartridges or the powder out of them.

From the moment of my final refusal to sell cartridges, the attitude of the traders changed; and, with the exception of the two Beluchi, they did what they could to hinder instead of help me. They suddenly discovered that the Njemps' statements were true, that there was no food in Kamasia, and it was useless for me to go there. I replied that if there was none there, I would go on to Elgeyo or Kavirondo. They then tried the effects of intimidation, and told me the story of the massacre of Bishop Hannington and his men, embellishing their account with as many unnecessary horrors as they could remember or invent; they assured me that his caravan was a much larger one than mine, and solemnly warned me against going into such a district with so small a force. I laughed at their fears, and explained that I regarded Bishop Hannington's fate as due to negligence, as his men were apparently undrilled, and the natives had been able to take him prisoner without a shot being fired. I pointed out that, like the Bishop, I was a man of peace, but if necessary was quite ready to fight anybody and everybody on the road, and that my camp would never be taken by surprise. As they saw such intimidation as they could try had no effect, they fell back on entreaties; they said there was a little food in northern Elgeyo, but since the time when Dachi-tumbo (Count Teleki) had raided there and massacred the natives, they would not sell to Europeans. I was assured that if I persisted in going, the natives would at once burn their villages and fly to the hills. The Arabs suggested that I should stay behind in

Njemps, and send Omari into Elgeyo to purchase food, while one of them would act as guide to him; I asked which of them would go, and they pointed out the very Arab whom they had only just told me had that day come back from this district at the request of his porters and slaves, as the natives mistook him for a Mzungu (European) and would not sell him food. I called their attention to this little inconsistency, and was then shamelessly told that this incident had happened twelve months before, and that the man was now blood-brother with all the chiefs. I thought at first that they really only wanted to make me buy my food from them at exorbitant rates, and so determined to go on; but, with an unfortunate excess of caution, before breaking up the conference, I obtained from Timami as exact a description as he could give of the place where Teleki was said to have raided. I was quite certain that Teleki had never even been there, and that his great fight for food took place much farther to the north; but the description tallied precisely with the locality where Dr. Peters had performed one of his marauding exploits. I attached no importance whatever to the string of lies and contradictions I had detected in the traders' statements, for they were but Orientals, and it would have been absurd to expect in them a Teutonic sense of truth. I was dealing with people among whom, as Mrs. Humphrey Ward would put it, "the capacity for testimony has not developed." I had to remember that their lies were possibly not lying lies, uttered with an intention to deceive, but mere dots and dashes honestly inserted to carry conviction of an argument, otherwise truthful. So, though I had demonstrated the falseness of their two premises, by showing that Teleki had never been in the district in question, and that a man who had just failed on his own account would not be likely to succeed on mine, this did not blind me to the fact that in modern Arabian logic the truth of the conclusion is in no way connected with that of the premises. So, accepting with Mrs. Ward that "the witness of the time is not true, nor, in the strict sense, false; it is merely incompetent, half-trained, pre-scientific," I broke up the conference in order to talk things over with Omari.

I found that he also disbelieved the traders' statements of facts, but feared there was some truth in their main argument, and that Teleki was bearing the blame, while I had to face the punishment, of Karl Peters' misdeeds. Later on, one of the traders to whom I had rather taken a fancy—a Beluch, named Jumma ben Abdullah—came back privately to beg me not to go; there was a look of sincerity in the man's face, and I resolved to be frank. So without any effort to soften my words, I told him that he knew they had been telling me lies, and asked him why they did not want me to go. For a moment he hesitated and then said: "We don't want you to go, because we are dependent for our food supplies on that district, and we fear you will spoil our market; you have *kiketi* (large blue beads much prized there) and cowries, and we have not. If you go, you will either spoil our market, or the natives will fear you and run away, and we shall get no food for our return journey to the coast."

This seemed to me reasonable; and though it was equally possible that they had been raiding up there, and did not want me to follow on their trail, it was hopeless for me to attempt to decide as to which was the true motive. So after talking the matter over with Omari, I determined to send him to Elgeyo with twenty-two men to buy as much food as they could carry back; I could meanwhile explore Baringo and try to find a route northward to Basso Narok (Lake Rudolf) intermediate between the two used by Teleki. After we had got the food, I thought I could go on to Kamasia to examine the gneiss outcrop reported there by Thomson, and draw an east and west section across the Rift Valley. The latter was really the main object of the expedition, and I resolved to carry it out at any cost or hazard.

The Arabs seemed much relieved when I announced my decision, and that I had arranged to leave two men with the donkeys, sheep, and most of the loads in the village. I told Omari that I must have Fundi Mabruk, as he had been with Teleki, and a porter named Alli, as he could speak a little of the Masai language, but I would leave the selection of the other men to him. He asked me to take Wadi Hamis, Mwini Amiri, Abdullah, and Stahabu, as they were the four worst men in the caravan, and he feared that if they went with him, they might get into trouble with the natives and ruin the chance of buying food. Ramathan Jumma was chosen as my Askari, as he was less to be trusted with an independent command, and it might be necessary for Omari to send back a party to Njemps with messages and food.

Loads were packed, the number of cartridges in each man's possession recounted, and early next morning Omari started for Elgeyo. The reserve goods were stored in a hut which Kizizi lent me. I had to wait for a short time until the boys had collected some spiders, and these had spun lines of web across the cracks around the door. I was then asked to take particular notice of the arrangement of these, so that on my return I could be sure that no one had been into the hut to steal my goods. Then after giving last instructions to the men left to look after the sheep and donkeys, we started northward across the red sand plain on the way to Baringo.

I had asked Kizizi for a guide who could be trusted, who knew the native place-names, and was a good walker. He picked out for me a primitive-looking savage named Lomweri, who satisfied these three requirements so well that he proved a perfect treasure. I had often heard of the "simple Ethiopian," but I doubt if any of the natives whom I had previously met with could be truly described as men of no guile. But Lomweri was as simple an innocent as one could hope to meet. His character, as well as his costume, reminded me greatly of Rudyard Kipling's hero, Gunga Din—

"The uniform 'e wore
Was nothing much before,
And a little less than 'alf o' that be'ind,
For a piece o' twisty rag
And a goatskin water bag
Was all the field equipment 'e could find."

The main difference in the costume of Gunga Din and my guide Lomweri was the absence of any piece of rag, twisty or otherwise. When we started I gave him a few "hands" of cotton cloth by way of prepayment, and also in the hope that he would have considered our feelings sufficiently to have worn at least part of it. But Lomweri was far too prudent. As he afterwards pointed out, if worn it would get dirty and crumpled; moreover, the thorns would tear it, while neither dirt nor scratches would matter on his skin. Nor did he need clothes for the sake of pockets, for he carried his impedimenta, a toothpick and a quid of tobacco, in the expanded lobe of his left ear. So after the cloth had been duly stroked, and patted, and licked, it was tied up into a bundle and left safely at home in Njemps. His only weapon was a bow and arrows, but he was such an appallingly bad shot that he could not have carried these except to charm his lady friends and frighten his foes. His intellectual attainments were as simple as his personal adornments. His vocabulary was most limited; he could only count up to five, and had no idea of distance; everything that was not in sight was "loqua" (far off), a word he used so much that we gave it him as a nickname. His appetite was insatiable, and nothing to him was unclean, at least when away from the eyes of the Mrs. Grundy of Njemps; and when properly fed, he was a splendid walker and simply did not know the meaning of fatigue. Within a narrow range, his knowledge of the country was surprisingly exact, though he knew nothing of the Masai-haunted countries beyond one day's march to the south and east. He had the morals of a child of five; he begged for everything he saw, and was never abashed by refusal. He would eat three roast duck straight off before my eyes, and then declare with tears and lamentations that he was dying of starvation. He could be terrorised by the men into saying anything, and while they were listening would lie by the yard. But for all this, he had a simple instinct of duty which nothing could shake; as far as it went, he followed it with a blind faithfulness and carelessness of consequence that won my admiration and regard.

Under the guidance of this psychological curiosity we crossed the open plain, until the scrub thickened and passed into the forests beside the Tigirish. We made very little progress that day. The men, except Fundi, were all as angry as they could be; they seized every possible excuse for delay, and it was past noon before we reached Njemps Mkubwa, or "Njemps the Larger," and crossed the Tigirish to the north of the town. The ford was deep, the current swift, and the banks were steep and formed of slippery greasy clay, while the abundance of crocodiles did not add to the pleasure of the passage. I shot a couple of the brutes, to the intense delight of the natives.

The men objected to going any farther that day, as they

declared that the guide said there was no water or camping-ground ahead for at least eight hours. This was manifestly untrue; so with the few words of Kikauvi that I had picked up, I questioned Lomweri, but could only make out that we must stop there for the night, as Kizizi was coming to see us in the evening; and I could see that the guide had been so bullied by the porters that he would say anything. I felt a certain amount of sympathy for the men, as they had had a very rough time of it lately, and had doubtless been looking forward to a long rest at Njemps. There was therefore some excuse for their bad temper, and I did not like to be hard on them, so I consented to pitch camp at once. Kizizi came in the evening, and I also had a visit from a native of the town, who had been with Teleki to Reschiat and back to Mombasa. He told me that the name of the great salt lake to the south of Njemps is Pirias, but I could get no recognition of this name from any other native of the district and so distrust it. He was an intelligent man, and spoke Kisuahili very fairly, but with a curious, and even comic, guttural accent. He possessed also three other accomplishments as a result of his contact with civilisation—the ability to put on clothes, to smoke, and to swear. The last he had acquired very thoroughly.

We had been annoyed all day by swarms of flies. They blackened the roof of my tent, and turned my basin of pea-soup into a dipterous decoction. The flies, however, were harmless in comparison with the mosquitoes, which rose like a mist from the marshes immediately after sunset. Mosquito curtains were far away with civilisation, and so I could only wrap myself in my blanket-bag. Its texture was sufficiently impenetrable to allow me complacently to pity the porters, whose thin cotton sheeting was easily pierced. But mosquitoes are bed-fellows that facilitate early rising; the men were eager to be off in the morning, and asserted that they had never experienced such a night before. We reached the south shore of Lake Baringo fairly early, and as I was anxious to communicate with the natives, we camped opposite the island of Lukrum and fired a couple of signal-shots. Meanwhile I started off to explore the course of a river called the Ndow, which enters the swamps at the mouth of the Tigirish. I was at first much puzzled in trying to reconcile the maps of von Höhnel and Thomson of this part of the lake, either with one another or with the facts. Von Höhnel omits the Ndow, but I found out later that he must have passed a little distance to the west, on the platform that forms the lowest of the foot-hills of Kamasia, under which the Ndow flows by a subterranean course. Where we forded it, the river was 30 feet broad and 3 feet deep, and flowed with a powerful current between banks raised above the level of the surrounding plain.

Next day I intended to camp at the north-west corner of the lake, while the men seemed determined not to march at all, and it was only after some delay that a start was made. I pointed out the position where I wanted camp pitched, and then struck off westward to examine some old lake beaches, formed when the level of Baringo was much higher than at present, and to look for fossil shells and living antelope. Farther to the north I returned to the shore to pick up the trail of the men, but could not find it, so I sent Lomweri back with a note to Philip ordering them to hurry on, while I climbed a lava crag to sketch, and take a round of angles for the map. An hour later the guide returned with a somewhat impudently-worded letter from the boy, telling me the men refused to come on, and asking me to return and hear what they had to say. I rushed back at once and found that the loads had been dropped in disorder, the microscope smashed, a lot of preserving spirit thrown away, and that the men had bolted, leaving the two boys to guard the goods. Later on the Askari came sneaking up to us, but he would say nothing except that the men were hiding in the swamp, and were going back to Njemps as soon as it was dark. A little before sunset the others came near and sent in a message to ask if I would hold a shauri, which I was only too glad to do. Wadi Hamis acted as spokesman, and a very insolent spokesman he proved. He said they had seen Wasuk, that it was not safe for so few men to travel alone in that country, and that they would not go on or obey orders any more. I had no trouble in pulverising these arguments, for the guide confessed that Sukuta—the country of the Wasuk—was some days' journey to the north, and that he had seen no signs of any strangers in the district; and, as I reminded the men, both Kizizi and the Arabs had said in their presence that we

Loel Divis. Langari. Doenyo Lersubugo. Lukrum. Doenyo Nelesha.

No. XI. THE SOUTHERN END OF LAKE BARINGO (LOOKING EASTWARD).

Page 128.

should be perfectly safe. I taunted them with their cowardice in leaving me for some time alone and unwarned, on the very hill where they said the Wasuk were in ambush. The men were as insolent as they could be, except Fundi and the Askari, who were a picture of despair and grief. The shauri was throughout a stormy one; Wadi Hamis at last pushed a cartridge into his Snider, seized his sleeping mat, kicked over his load, and said that I might go where I liked, but that he was going back to Mombasa to appeal to Judge Jenner. "Then as you have appealed to Judge Jenner, to Judge Jenner you shall go," I replied, as I angrily swung round on my heel; "but to-night you shall all stop here, and to-morrow I will go on alone."

The poor guide had been sitting through the shauri open-eyed and open-mouthed, intensely puzzled by the whole performance. By the aid of signs and gestures I explained that I was going on alone, but he could come with me or go back with the men, whichever he pleased. He replied at once that he must keep with me. After our return to Njemps, when I had the assistance of an interpreter, I asked him why he did not go back with the others, as they had tried to make him do. "I hoped you would go back," he replied, "for if the Wasuk had caught us, they would have killed us; but if you went on, I had to go on too, for what should I have said when Kizizi asked me, 'Where is the White Man I sent you with, to show him the way with the fewest thorns, to wherever he wanted to go?' If I had told him, 'The White Man has gone on and I have come back,' then Kizizi would have killed not only me, but my wives and children, and wiped my family out of the tribe."

My own course of action was perfectly clear; to have returned to Njemps would have been a sheer waste of time, as I could do nothing until Omari had come back from Elgeyo. I was, moreover, extremely anxious to examine the passes leading north from Baringo, especially since I had discovered how much higher the level of the lake had formerly been; for I now wished to determine whether the lake had ever had an outlet to the north. It had been believed that Baringo was one of the sources of the Nile, and though Thomson had shown this is not so at present, it seemed quite possible that such may once have been the case. I calculated that I could easily complete my survey of the lake in four or five days, during which time the shot-gun would supply ample food; I had slept out in the open far too often to be in any way concerned about the absence of a tent. Storms of rain were certain to occur, but a bivouac that could be constructed in half an hour would afford fair protection against these; my only serious regret was at having to go on without collecting apparatus.

I packed up a few medicines and filled a pocket with cartridges, which was all the luggage I intended to burden myself with, and having made these preparations, sat up writing, as I feared if I went to sleep the men would bolt during the night. Shortly after midnight I heard some one walk quietly up to the tent. I coughed to show that I was awake and indicate my exact position, which, it is barely necessary to add, I immediately changed. I was relieved to hear Fundi say, "Bwana, may I come in?" He asked if I were really going on, and hearing that I was, told me how twice before, when with enormous caravans, he had been through terrible fights with the fierce Wasuk, who had killed more of his comrades than the whole number of men in my caravan; "But if you go, I must go on too," he concluded.

I simply thanked him, and assured him that if I did not fear to go on, he ought not to do so either. But I was very glad to get him, for the fact that he had joined the mutiny was the bitterest constituent in this very unpleasant pill. I neither liked nor trusted any of the other men, but Fundi was a favourite.

The youngest of the porters, a mere boy named Baron Abbas, then came in and reminded me how I had nursed him when he had been ill on the Tana, and said he must go where I went. A third, a three-quarter-witted porter named Alli, and the two boys then said they would come on too; so I could now look forward to the rest of the excursion without the feeling that it was going to be a case of work all day and watch all night.

Next day the mutineers returned to Njemps, where they were stopped by the absence of food and presence of the Masai farther to the south; the rest of us continued our way northward along the shore. We crossed a number of lava sheets, broken by parallel faults into a series of cliff-faced terraces; one of these, rising directly out of the lake, compelled us to go a little distance inland. Elephant tracks were abundant, but they were at least two months old, and we saw no game of any kind. I pointed out a suitable place for the camp on the northern shore, and told the men I would join them there at sunset; I followed the raised beaches of the lake across the northern watershed, at a point where Lake Baringo must once have had an outlet to the north. The pass is dominated by a massive, straight-faced, flat-topped lava hill named Lobat; I intended to climb it, until I saw that time could be more profitably spent on the pass than on the peak. The view from the former, however, was very fine; it embraced a long stretch of desert that sloped northward between the peaks of Mesuri and Chibchangnani, that terminated the escarpments of Kamasia and Elgeyo on the west, and the ridges of Weweini and Subugu Loluko, that formed outliers of the Laikipia plateau on the east.

At the other end of this basin, less than ninety miles away, was the southern shore of the great Basso Narok (Lake Rudolf). It was mournful to have to turn back when so near it, but it was useless to go unless I could get some time there for scientific collecting. To have dashed across the desert simply for the fun of dashing back again and saying that I had been there, would have been an unjustifiable waste of time and energy, and a needless risk. More than two months of my five had already gone, and there was nothing for it but to lament the waste of time on the Tana, and go back to camp. This it was now high time to do, for the sun's lower rim was already resting on the western plateau, and Lomweri was going through a pantomimic show to remind me that when the sun had set, the lions would rise. "Kampi lokwa, Pokwa" (The camp is far off, O Bulging Pockets), he had been repeating for some time past, making as good an effort as he could to pronounce my nickname, which he had learnt from the men. After a last look over the northern desert, at the rose-tinted slopes of the eastern hills, and the dark shadow that was creeping from the frowning cliffs of Kamasia across the sparkling salt steppes, we turned to go. We recrossed the Lobat

Pass in the twilight, and in the darkness scrambled over ridges, and forced our way through the thorn scrub down to the northern shore of the lake. I fired a signal-shot, and the reply guided us into camp.

The remainder of this excursion was most delightful and restful, at least mentally, for Lomweri and I did our twelve hours' walking every day. There were no natives to worry us, and the men were so ashamed of their part in the row that they did their work without a grumble. Wild duck were abundant; I had therefore no trouble about the commissariat, as I often brought down ten at once, with a right and left. The absence of vegetable food was unwholesome, but the ducks roasted on spits over the fire were simply delicious.

The scenery, too, was most charming. A view of Elgeyo would occasionally suggest doubts as to how things were going up there, but thinking about this could do no good until our return to Njemps; so the subject was resolutely kept in the background, and not allowed to disturb my enjoyment of the ever-varying panorama of quiet bays and rocky headlands. The northern shore of the lake is especially picturesque; for the lava sheets that form it have been broken by faults into a series of troughs with vertical walls, arranged as regularly as if they had been drawn with a parallel ruler. The summits of the ridges and the floors of the valley are flat, and slope gently southward toward the lake; in this, the former end either abruptly as cliff-bounded promontories, or as jagged ribs of rock continued by lines of islets. The ground is covered with rich green turf, studded with gorgeous meadow flowers, while here and there are patches of scrub. The shrubs of which these consist are not dreary, spine-leaved acacias, but are clumps of *Dombeya*, with masses of dense green foliage glowing with an outer crust of brilliant yellow bloom; there is the willow-like Lelesha (*Tarchonanthes camphoratus*), which, like the olive, when the silver gray of the under sides of the leaves is seen, as the branches sway on the breeze, "turn all hoary to the wind." Where the valleys reach the shore-line they form bays fringed by dense growths of reeds, rushes, and papyrus, and containing acres of the light green cabbage-like rosettes of *Pistia stratiotes*, amid which rise the bright blue flowers of the Lotus. In the open waters beyond sport schools of hippopotami—a small, dark-coloured variety—which grunted at us as we passed. Flocks of wild fowl browsed on the weeds and algæ, fish-eating birds dived after their prey, and every now and then the sharp crack of the jaws of a crocodile would tell us that the tables were turned, and that the birds were food as well as feeders. The crocodiles occur in enormous numbers, especially among the bays that are almost choked with vegetation; there they lay like logs amid the rushes, preying on the ducks that swarm on the pools. Bates [1] tells us that in some of the upper waters of the Amazon the alligators occur in huddled, jostling crowds, and are as thick as tadpoles in an English ditch. Until I saw Baringo I thought this wild exaggeration. In such places I never got all the birds I shot, and to secure anything like a fair share of the bag I had to plunge in at once. Fortunately the crocodiles are as cowardly as voracious, and shouts and splashes readily drove them away. One evening in the dusk I almost tumbled over one that I had not noticed, and though it fled at once, if possible more frightened than I was, I never waded for duck again. After that I always sent in the men on the excuse that the wet spoiled my boots and clothes; if none of the porters were with me I left the duck severely alone. In this work Lomweri was absolutely useless; his ingrained terror of crocodiles was so abject that it would have been pitiable had it not been so comic. He would stand on the margin of the swamp wringing his hands in agony, and imploring us to come back; but when we brought in the birds he would dance with glee, pat his stomach, and smack his lips, and when the ducks were roasted, he would beat any of us in the pace at which he ate them.

The one disappointment of the return journey was the failure to reach a small archipelago of islands in the southern half of the lake; I wanted at first to visit them to see the Wakwafi who inhabit them, but the islands themselves acquired an interest, when I learnt to recognise in them the broken remnants of a volcanic cone. The natives, however, held aloof, and I postponed building a raft till I could return to the lake later on. Lomweri and I once surprised a couple of the natives on the shore, but neither the guide's assurances

[1] H. W. Bates, *A Naturalist on the Amazons*, 5th ed. p. 299.

of friendship nor my offers of strings of beads would tempt them near us. We could easily have cut off their retreat to their skin coracle, but forced friendship would have been useless, and we let them get into their tub and paddle off.

Our march back along the east shore was rendered unpleasant by the vast extent of swamp through which we had to wade ; the route to Njemps twice followed by Teleki was now so flooded that it was impassable. The deviation to the south that this rendered necessary, and a delay caused by a hunt for Philip, who had lost himself, to my great annoyance lost us a day. We had to keep along an old lake terrace on the eastern side of the Rift Valley, and wade a swamp that ran up "Summuran Bay" into a fault valley into the hills ; we crossed the lower slopes of the terraced ridges of Lolbogo, forded the swift and muddy Mudoleto, and finally struck the path from Maji Moto, by which we had originally reached Njemps.

I was welcomed back most cordially by Kizizi, who said he had been seriously alarmed as to my safety ; he had arrested the porters, kept them inside the village, and seized their loads. As they had not been able to get any food, they were now starved into tameness and submission. Though Omari had promised to send back a native, there was no news from him. Sokoni, the chief of the guides at Njemps, had returned from Kamasia the day before and came in to see me. I had messages to him from Mr. Martin, and we had an interesting chat. He remembered Mr. Thomson's visit very well, and had acted as Teleki's guide in the march to Basso Narok ; he told me several stories of the hardships of that expedition, during which he had several times abandoned all hope of return. The information he gave me about Elgeyo and Kamasia was most disquieting, and some of his remarks increased my distrust of the coast-traders, and regret that I had allowed Omari to go off alone. I therefore resolved to follow him, and told my "stalwarts" they could rest in Njemps, and ordered the rebels to be ready to start next day for the hills. Wadi Hamis had the impudence to protest against my going with only five men against "such Waschenzi" as the people of Kamasia. As he was speaking, there came back to me the memory of a scene in a geyser basin in the Rockies, at the end of some hours of waiting for an eruption which did not come off ; my guide had guaranteed a performance, and was very angry at the failure of his pet geyser. As we turned to go, he made one last effort, apostrophising the geyser, and in the name of all the gods of a Yankee stable-boy adjured it to come forth. I felt at the time that his language was a thing that could never be forgotten and ought never to be quoted. When, however, Wadi Hamis finished his protest I did quote a little of it. I was ill with fever at the time, and that is my only excuse for the outburst and the confusion of the names of some Eastern deities with those of the gods of that western cowboy. But when at dawn next morning, in answer to my inquiry if the porters were ready, Ramathan's soft voice replied, "Dio, Bwani: yote tiari" (Yes, Master, we are all ready), I felt thankful that that language had not been as wasted upon my porters, as it had been upon the demon who disturbs that boiling Wyoming spring.

It is unnecessary to follow in detail the excursion to the west, and I am glad to hasten through with the rest of the dreary, anxious period of my stay at Njemps. As I was foiled in my endeavour to return to the lake, that I might collect representatives of its fish-fauna, and to visit its islands, and as I only again saw it in the distance, a few notes on the geography of Baringo may be here inserted.

Lake Baringo[1] is the best known of the members of the lake-chain which I was able to visit. Its existence was first made known to Europeans by the reports of native traders. The old estimates of its size were greatly exaggerated, probably as it was confused with Basso Narok. It became at once a bone of contention among geographers. Thus Livingstone, in accordance with his axiom that all fresh-water lakes must have an outlet, which led him into his most serious geographical blunders, connected it with the Nyanza. Burton welcomed Baringo as another argument in favour of the dismemberment of Speke's Nyanza into a lacustrine heptarchy. Others, of course, claimed it as the source of the Nile. The first European to reach the lake was Joseph Thomson in 1883, and he showed that its size had been greatly exaggerated, and that it had no connection with either the Nyanza or the Nile. The maps of the lake were, however, erroneous in many respects, and the circumambulation of it enabled me to correct these,

[1] See inset map, on Map II.

and to determine the character of the basin. (The topography of the lake is described in the *Geographical Journal*, vol. iv. 1894, pp. 311-313.)

Baringo and Njemps both lie on the floor of the Rift Valley, the walls of which closely resemble in their structure those of the region around Naivasha. Here, however, there were no Masai to hinder exploration, so I could hope to work out the geological structure of the country more conveniently and thoroughly. The study of the eastern wall I intended to postpone until the return march, and I was determined to make the most of the present opportunity of studying the western plateau, in order to obtain materials for a geological section right across the valley. The eastern wall is in places a single face of rock, over 2000 feet in height, the foot of which is 7 miles from Njemps. The western wall, formed here by the great fault-scarp of Kamasia, is 14 miles from the village, and a series of foot-hills intervenes between the two. From the descriptions of Thomson, I expected to find that while these foot-hills were composed of volcanic rocks, the plateau was a great island of the old rocks (gneisses) rising up through the lavas and forming the backbone of the Kamasian ridge.

We left our camp at Njemps on the 28th May and crossed the plain towards the base of the foot-hills. Our progress, however, was soon stopped by a messenger who recalled me to the village; there I was detained till it was too late to do more than reach and cross the river Tigirish, before approaching nightfall arrested us. We had no guide, as, probably owing to the instigation of the Arabs, none of the natives would go with us. One of my men said he knew the way by which Omari had gone, and I trusted him to guide us; he misled us, no doubt wilfully. We got into very bad country, a district intersected by deep ravines. We had to cut our way through dense acacia scrub from one ridge to another. We trudged for miles over sandy river-beds, through gorges so sinuous that no breeze could work its way along them to refresh us, and with walls as bare and black as if designed expressly for a sun-trap. The men were frightened; alarm begot anger, and they dawdled as much as they dared. Illness made me irritable, and irritability made me energetic; so at last I threatened to flog them, one and all, for I was determined not to have a repetition of their behaviour on Baringo. The geology, moreover, was very tantalising. It was only too interesting, but it was complex, and the conditions of the country were not favourable to a hasty unravelling of its story. There were clearly two series of lavas of very different ages, but they were so much displaced and mixed up by faults that I could not make out their relations one to another. Occasionally a gully would promise an instructive section, but when a way had been laboriously cut through to it, the sides were found to be hidden under a wash of the "talus" of disintegrated rock, or were obscured by a dense growth of scrub, or else a vertical face of rock barred approach to the desired spot. At last I gave up the attempt to settle the relations of the two sets of lavas by mapping them, and so plodded through sandy gorges, or forced a way through ravines choked with a dense growth of jungle, until I found a series of old lake deposits, buried beneath the later set of lavas. The upper part of the lake deposits had been baked by the lavas into an intensely hard and exquisitely white porcelain; but below they contained some beds of gravels, all the pebbles of which must have been derived from the older set of lavas. This gave me the clue I wanted, so I weighted the men's loads with a fair assortment of the pebbles, and then went merrily on. We marched for a couple of days westward over the rest of the foot-hills, until we entered the inhabited region at the foot of the main plateau of Kamasia. The natives were in great force, and very suspicious of our objects. I did not, therefore, think it safe to take the men farther, though I was resolved to go on myself. So I posted the men and baggage among some rocks, in a position which five men with guns could hold against an army of timid spearsmen, and then started, with the Askari and one porter, to attempt the ascent of Doenyo Lubikwe. This mountain rises above the plateau of Kamasia as a fine conical peak on the end of a well-rounded ridge; its bold outline towered above the long flat edge of the plateau, and lured me upward with a fascination that I could not resist. We nearly reached the top, but alas! the start had been made too late in the day, and we had the annoyance of having to return, when the summit was well within our reach. We dared not delay, for the red-painted natives, who were watching us in crowds at a respectful distance, were timid and suspicious. Had we been caught away from camp after nightfall we should probably have seen more of them than we cared about. The sound custom of not walking abroad at night and my hope of a second attempt next day had to yield to more pressing needs, for bad news awaited me on my return to camp. There I learnt that the quarrel regarding which Sokoni had warned me, had led to an outbreak of hostilities between the Wa-njemps and the Wa-kamasia. A party of the latter was said to be already on the march to Njemps, and to have cut off communications with Omari. My reserve supply of goods and ammunition were in the threatened village, and an immediate return was necessary to save them. It had taken us three days to reach our present position, but I resolved, in spite of the men's protests, to do the return in one. I had my dinner and we took down the tent; we stacked up the fires in such a way that they would keep alight all night, and thus prevent the natives having any suspicion of our flight; then, an hour before the moon rose, I crept stealthily out of camp to reconnoitre. I could not detect any natives, so I went back for the others and led them out along the ridge at some distance from the path, and then we stole quietly through some fields out on to the uninhabited region of the foot-hills. At first we all kept together, for there were only seven of us, and we had to be very careful; but when we had left the danger from the villages far behind us, and had only to fear that ahead at Njemps, I took my rifle from the Askari, wished the men "Kwaheri" (Good-bye), and hurried on. Dawn found me alone on a basalt platform near the eastern margin of the foot-hills; I rested on the edge till it was light enough to take my bearings, and to decide on the easiest way down to the floor of the Rift Valley. The sky was absolutely cloudless, and in that dry, cold air the sun rose over Laikipia with a sudden blaze of light, that flashed upon the cliffs of Kamasia with the glare of a search-light. In two or three minutes its rays had reached the foot-hills, and only the Rift Valley lay black and dark before me. Then ridge after ridge within it suddenly sprang into sight, till the sunlight caught the tracts of sand upon the floor of the valley, and "thrilled her black length burnt to gold."

A few hours later I was in Njemps, listening to the head Askari's tale of troubles in Elgeyo, and trying to calm the complaints of the ten porters who had come back with him, and who for two days had not had a morsel of food. There was nothing for it but to make an effort to appease their hunger. I gave one of them my rifle and gun to carry, crossed the Nyuki, waded for nearly an hour through the swamps, and scoured the plain beyond in search of game. The bones of buffaloes lay rotting on the ground, but the work of the plague [1] had been only too complete; there was not a beast to be seen, and I returned to the camp with only a few brace of guinea-fowl and some wild duck. This was better than nothing, but it soon disappeared before twenty-five hungry men. So I turned into bed, after fixing up a stick and telling the men to wake me when the Southern Cross came in the line of sight with it. Alas! when the sentry called me I was too weak to stand. For the flesh diet and the malaria of the swamps, possibly aided by overworry and overwork, had brought back my old foe—malarial dysentery.

On the following day I was better. I went off to shoot, but soon broke down and had to be carried back to camp. Fortunately one of the traders, the Beluchi named Jumma, was a good shot and a keen sportsman, and he offered to make good use of half a dozen cartridges. He kept his word, and soon sent back for men to carry in a "Swara mdogo," as the Suahili call the bush-buck (*Tragelaphus* sp.) Later on he got a zebra, but this was shot so far from camp that the porters sent for the meat could not reach the place until dark. This involved the loss of the food, for the porter who had been left on guard would not stay out alone after sunset; the men could not find the place, but the lions did, and during the night they devoured the carcase. Next day I was better, and went off in one direction, while the Beluchi started in another. We bagged a couple of zebra, and on the way back, by a most fluky shot, I bowled over a fine female water-buck (*Kobus ellipsyprymnus*). Next day we were still more fortunate, for though game was really scarce we killed two zebra and three mpalla (*Æpyceros melampus*). In the afternoon we tracked a lion to its lair; but after an interesting stalk through dense scrub he bolted, just as

[1] See p. 266.

we thought he was going to charge, and we lost him. In spite of its having been unsuccessful, this lion hunt was the most pleasant incident in the dreary stay at Njemps. I am not an enthusiastic sportsman under the most comfortable of circumstances. As this shooting involved starting at three in the morning, stumbling for nearly two hours in a swamp, and then waiting wet and shivering till, with the first streak of daylight, the zebras came down to drink, I thought it very poor fun. Moreover, when the first bullet was not fatal, as was generally the case, a furious obstacle race was necessary, across rough country and through thorn scrub, till a second could be placed with better effect. As most of this work came at a time when I ought to have been snugly tucked up in bed, and fed on hourly doses of arrowroot or rice gruel, such violent exercise was especially objectionable. A friend once asked if I were not sorry for the zebra, but on these occasions I was far too sorry for myself to have any sympathy to spare for anything else. Once or twice I sent out some of the porters, who could be trusted not to sell their cartridges to the traders. But the Zanzibari had even less of the spirit of Nimrod in them than I had; they wasted some precious ammunition, and probably spent the day asleep under a bush. They thought that as I had led them out into the desert, away from the fleshpots of Mombasa, it was my duty to feed them. There was so much reason in this, that, in spite of the discomfort and loss of time, I went on with the hunting.

The days thus dragged wearily on, and there was still no news of Omari. The threatened war was confined to a few attempts at sheep-raiding in Kamasia, by small parties of the Njempsians. The active hill-tribes, however, were too much for the starved natives of the valley, and succeeded in massacring every party sent against them. My own efforts at zoological collecting were almost as unsuccessful as those of my friends; for after my return to camp from shooting, I was always too tired to attempt any work. I had a shed built beside the river, and used to lie in it, watching for a crocodile to show its head upon the surface, when I at once planted a Martini bullet in it. After a few days, however, this amusement was played out; but we could cross the ford in the morning, feeling more at our ease.

The one stroke of good luck that befell us at Njemps was enlisting a man who had just crossed Laikipia, and consequently knew the position of the Masai kraals. I had announced the offer of a reward to any one who would bring me a guide. For so long there was no answer to the advertisement, that I began to fear I should have to be my own guide. This was not a very cheerful prospect. One evening, however, a native of Njemps Mkubwa came in to say that an Mkwafi of Laikipia had just arrived in that town, and that he should be brought to me on the morrow. The Njempsian told me that the man was the sole survivor from a kraal of Wakwafi, the members of which had been massacred by the Masai.

Shortly after my return from shooting next day, the proposed guide was brought into camp. My first feeling when I saw him was one of bitter disappointment. I had been told he knew the district extremely well, and that it was only owing to his familiarity with the game tracks that he had succeeded in finding his way to Njemps. But he looked absolutely useless; I thought he was dying. He was comparatively a young man, certainly not past middle age, but his hair was gray. He was terribly emaciated and seemed to consist of nothing but a mere bag of skin hung over a framework of rather massive bones. When erect he stood six feet four inches high, but he was then bent forward, leaning on a staff. His mouth was open, and his huge bloodshot eyes were fixed on the body of an mpalla that had just been carried into camp. The man who brought him and I both spoke to him, but he did not heed us. I took hold of his wrist, apparently without his being conscious of it. The feeble spasmodic pulse confirmed the suspicion that he was in the last stage of starvation. He told me afterwards that since his escape from Ndoro nine days before, he had eaten nothing but leaves and a few berries. We gave him a seat upon a box, and as solid meat would have killed him, we fed him at short intervals with small doses of beef-tea. In the afternoon he looked better and fell asleep. Next day we told him that we wanted him to guide us over Settima to Ndoro. I offered to feed him on the way, give him as much food as he could carry for the return journey, and pay him well. He said that he *could* not go with us, that the Masai would kill him, and that he would rather die than go on to Laikipia again.

I promised him full protection, and to let him return as soon as we saw Doenyo Egeri (Mt. Kenya), and he could point out to us the position of Ndoro. He refused to go, so I had sternly to tell him that he had better return to Njemps Mkubwa. He begged for food; but I could only say that I had not enough for my own men and had none to spare for him. With that the shauri ended, and I went back to my tent knowing that the pangs of hunger would be a more powerful argument than anything I could say. Every now and then during the day the man's piteous cry of "Nyama, nyama" (Meat, meat) rang through the camp. He persisted in his refusal, however, till the evening. Then, when the porters cooked their food, the savoury odour of the roasted antelope meat was too much for him; he decided that the risk of death by a Masai spear was less awful than the certainty of slow starvation. Terms were easily settled. I was to feed him, give him a supply of food for the return journey, and leave ten rings of iron and brass rod with the chief of Njemps as his payment. I offered to take him back to Mombasa and pension him, but this he declined. He signed articles by surrendering his bow and arrows, and then we gave the poor fellow his food. He seized it like a ravenous hyena, and nearly choked himself with it. We had to tear it from him by main force, cut it up, and give it him in pieces. For days afterwards the taste of food always drove him mad; at his meals one man had to hold him down, while another cut up his meat and fed him like a child. Mwini Mharo, the head Askari, was the most attentive to him. We were a stony-hearted crew, but there was not one of us who was not touched by the sight of the merry Mwini patiently soothing that wild, gaunt idiot as he doled him out his food. In a few days he grew stronger and became calmer; then he used to sleep all day and stand all night in the smoke of a fire beside my tent singing a doleful dirge, and waving about a flap of leather to drive away the mosquitoes. If he thus disturbed the mosquitoes as much as he did me, from his point of view this performance was a success.

Our relations with the coast-traders were of a very different character from those with the guide. Omari at length returned, but only brought 500 lbs. of food, which, with my store of zebra meat, would last us only ten days. The Arab guide had played us false. He had gone up under our protection, bought for himself as much food as he could carry, and spread abroad the news that Omari was buying for a European. Only once before had a "Mzungu" or "white man" visited that district, and the report that it was about to be honoured by the visit of a second sent the natives into hiding in the hills. The Arab guide well knew what a legacy of Mzunguphobia Dr. Karl Peters had left behind him for his successors. In consequence Omari could buy very little food, and we had to pay by days on half rations for Dr. Peters' misdeeds.

My position was now rather an awkward one. Owing to the unhealthiness of Njemps my time there was being worse than wasted, but I had not food enough to plunge with safety on to the great foodless plateau of Laikipia. I was very anxious, however, to get away from Njemps, for local politics there were getting mixed. The coast-traders, the starving Wa-njemps, and the hill-tribes of Kamasia were all at loggerheads, and agreed only in their suspicions of me. I did not want to get involved in their squabbles, and so I resolved to move elsewhere.

Four courses were possible. The easiest was to return along the Rift Valley back to Fort Smith; but a known road had no attractions for me. A more attractive scheme was that which had been urged on me by Major Smith and Captain Williams, viz. to strike west to Uganda, and, replenishing my stock of trade goods there, to march on to Ruwenzori. This mountain range had, however, recently been visited by Emin and Stuhlmann, and it did not seem to me worth going there without knowing what they had done. A third possible course was to go north to Basso Narok. Sokoni, the chief guide at Njemps, described to me the great fish which occur in that lake, and this tempted me to try to visit it. I had been within ninety miles of its southern end, and I had thought of trying to make a dash across the deserts to it, with a party of my six best walkers. But with such a scarcity of food, and such an uncertain basis of operations as Njemps, the risks were too great. The last plan was to cross Laikipia to Kenya. The objections to this were twofold: there was the possibility that Mr. Astor Chanler and Lieut. von Höhnel had already worked out the structure of that mountain; there was the certainty of trouble with the Masai if I met them. But against these there was the argument that this was my original plan. I had started intending to go to Kenya, and so to Kenya I resolved to go.

I told the Arabs that I was going, and explained to Azizi that, as I could not return the food, I must pay him for it by "chit" (or cheque) in Mombasa. He then refused to accept that method of payment, though when he had lent me the food he had himself suggested it. He now demanded all my "kiketi," a kind of large blue bead, and cowries, and a cheque for 300 rupees to be paid in Mombasa. The ordinary price for food at Njemps is one pice per kibaba; he wanted more than ninety pice for each measure, besides sufficient trade goods to buy the food five times over. After a stormy conference that lasted for four hours I consented to give him all the cowries and half the kiketi, and to pay his agent in Mombasa 160 rupees. Even this was a frightful swindle, but I was too fagged to fight the matter further.

At daybreak next morning (5th June) the camp was roused by the shouts of "Safari! Safari leo! Funga! Fannya tiari!" (A journey! A journey to-day! Tie up the loads! Make ready!) We were delayed for some time by the difficulty of catching the donkeys, which had run wild, and had to be broken in again. The natives would not help us, for they were holding a great war shauri, to decide whether they should make an attack in force upon the people of Kamasia.

We marched round the north wall of the town, and then eastward across the plain. We were soon stopped by the swamp of the Nyuki. When the guide saw it he said we could not cross it, and sat on the ground, folded his arms, shut his eyes, and refused to move. Fundi and I tried to get through but failed, and I feared the guide was right. When Omari came up he simply set his teeth and said, "Bwana, we *must* go through." Thanks entirely to his pluck and resolution, we did get through.

Beyond this swamp there was no water for a long distance, so we had to camp beside it. I spent the afternoon drying a load of plants that had been dipped into the swamp during the passage. This and two other incidents that happened at this camp considerably lessened the pleasure of the departure from Njemps.

One of the flour sacks was cut open during the night, and

some of the contents stolen. Our supplies were so limited that this was a serious offence against the whole caravan. We could not, however, detect the thief, though the sentry who was on guard at the time was accused by the men. He was of course responsible; but I hoped that he had committed the minor offence of being asleep on guard, and that the only reason why he was suspected was that he had served for several years in the Zanzibar police. Later on, however, he was degraded to the ranks for another act of theft, so probably the men's suspicions were not unjust.

Anxiety about the success of our march across Laikipia was increased by the behaviour of the guide. That I could find my way across without a guide, provided I had plenty of time, I never doubted; but with our limited supply of food delays might be fatal, and the usual rules of mountaineering were no help in finding one's way up such cliffs as those before us. Just before sunset I walked up to a terrace beside the camp, in order to sketch the scarp face of Laikipia. The guide came up. I pointed out to him a gap in the skyline, and asked by a sign whether that was the pass by which we were to ascend on to the plateau. He looked steadily at it for a few minutes; a cold shudder shook his whole frame, an expression of anguish came over his face, and without answering my inquiry he strolled slowly down the slope to camp. His look was one never to be forgotten. It told of a heart-broken submission to a cruel fate, like that of a dying antelope when it has realised the hopelessness of escape and the uselessness of further struggle. Whether his agony was due to some recollection of the past, or to a dread of the cruel fate which did befall him, I could not tell. But it indicated such disinclination to face again the perils of Laikipia that, in spite of our sympathy with the guide, he spent that night tied to my tent pole.

CHAPTER IX

ACROSS LAIKIPIA

"O'er the wide sierras and the high plateaux."

WALT WHITMAN.

LAIKIPIA is a plateau formed of volcanic rocks, situated to the east of the part of the Rift Valley between Lake Naivasha and the steppes of Sukut. Most of the plateau lies at an elevation of between 6000 and 7000 feet, and consists of rolling prairie. Its area is almost that of Wales. Bounded to the south by the forests of the Kikuyu country, it extends northward between the volcanic piles of Kenya and Settima, and the ridges of Doenyo lol Daika and Subugu, until it is cut off by the westward trend of the Loroghi Mountains and the eastward bend of the Rift Valley.

Considering the proximity of the country to the road to Uganda, it is surprising how little Laikipia has been explored. The enterprising Arab traders long ago found their way across it, along two routes. One of these started at the north-eastern end of the country of the Kikuyu, skirted the western foot of Kenya, traversed Ndoro and the open steppes to the edge of the Rift Valley above Baringo. The second route left the Rift Valley to the north of Naivasha, and thence continued along the eastern flanks of the wooded range of Subugu, enabling the traders to reach Basso Narok (Lake Rudolf) without crossing the country of the hostile Wasuk. These two routes were recorded in 1874 in the valuable paper by Denhardt, referred to in the first chapter. No further information about Laikipia was obtained until Joseph Thomson made his courageous and determined effort to cross it, to reach Kenya in 1883. His expedition had been fitted out by the Royal Geographical Society, and one of its main purposes was the exploration of Mount Kenya. Thomson entered Laikipia by the second of the above routes, and, working his way round the northern flanks of the mountain mass known as Settima, struck eastward toward Kenya. He soon, however, got into trouble with the Masai, who were present in force. Their attitude was so threatening that Thomson could not cross the Nyiro. He had to abandon his camp under cover of night and escape north toward Njemps, which he reached after a series of adventures, which are well described in his fascinating work *Through Masai Land*. Four years later Count Teleki, with his accomplished assistant Lieut. Ludwig von Höhnel, marched across Laikipia along the main Suahili trade route, along which they were guided by the great Suahili trader, Jumbo Kinameta. The Masai were then not very numerous, and Teleki's force was so powerful that it suffered no interference from the natives. Von Höhnel mapped the line of march with his usual skill, and made two branch excursions to the east, to track the course of the Guaso Narok and Guaso Nyiro, the two main rivers that drain the plateau.

Laikipia was crossed a third time in the winter of 1889-90 by the German Emin Pasha Relief Expedition under command of Dr. Karl Peters. The Masai, ignorant of the power of field-guns and repeating rifles in the hands of trained Somali soldiers, were aggressive. Dr. Peters' attitude was not conciliatory, and his caravan was accordingly attacked by hordes of infuriated El-Moran. Dr. Peters' victory was complete; he burnt most of the leading kraals, and captured great herds of cattle, which had no doubt been originally pillaged from other tribes. Some other attempts have been made to cross Laikipia, but without success.

With von Höhnel's map to guide us, it would have been comparatively easy to follow the Suahili trade route; but I resolved to leave it, and cross the plateau along a new line, partly for the sake of the exploration, and partly in the hope of avoiding the Masai. After the treatment they had received from Peters, it was certain that they would be thirsting for revenge, and that if we met them we should have to fight. The route I proposed to attempt was not only new, thus offering more chance of interesting geographical work, but it avoided most of the open steppes, on which concealment was impossible and defence difficult. The scarcity of food greatly increased the risks of the march, for nothing rouses the suspicions of Africans more than insistence on an immediate reply to a request for permission to enter their country. And we had no time to waste on the road in "shauri" with the natives, for our supply of food could not last until we reached the Kikuyu plantations.

The men understood the difficulties as well as I did, probably much better. But by this time they had also learnt that they had to do as they were told. Most of the older men, moreover, were pleased at the prospect of seeing new country: I had chatted with them round their camp fires in the evening about the great mountain, whose face was speckled with white; about the rivers and supposed lakes that formed the sources of the Tana. Fundi, moreover, had related exciting stories about the wonders of Kenya and the glories of the country at its foot, with its abundant firewood, pure mountain streams, vast herds of game, and freedom from mosquitoes. Omari had told the men it was their duty to go wherever I wanted, and so long as I did not fear to lead, they must not fear to follow. So between us we had roused general interest in the unknown land into which we were about to venture. As soon, therefore, on the morning of the 6th of June, as I gave the order to "Taenda," adding "Harako leo! maji mbali!" (We must hasten to-day! water is far off!), the men took up their loads and marched briskly across the plain to the foot of the forbidding cliffs of Laikipia.

We soon rose above the valley of the "Paragara" (as our guide called the tributary of the Nyuki by which we had camped), and then crossed the opening of a bay of alluvium which ran from the plain into the hills. We forded the Ngusagari, a small stream that drained this bay, and reached the foot of the cliffs, at the mouth of a narrow gorge half choked with vegetation. We pushed through this, walked along the bed of a dry stream over dark rounded surfaces of "andesite," and climbed over fallen boulders of black basalt. We had to scale a steep cliff on the north side of the river bed, hauling up the donkeys with ropes. We followed one valley to the north-north-east, and crossed a gap to another, and went along it to the south-south-west, until we came opposite to the "domo," or pass, which we had seen from the plains. A steep zigzag game track led towards it, and during the ascent of this we enjoyed a series of magnificent panoramic views of the Rift Valley and Lake Baringo.

We crossed the summit of the pass and descended to the springs of "Njoro Larabwal." Here we intended to rest for a couple of hours and then resume our march to the southern end of the basin. But a message came from Omari that the porter Jumbe had fallen ill and could not walk. I returned with a hammock, some medicine, and eight porters to fetch him. Carrying the invalid up the steep slope proved a difficult task, and when we reached camp it was too late to go farther.

Next day we crossed the basin of Larabwal, and had again to make a shorter march than we had intended. The guide was seized with one of his crazy fits, and nothing we could do would allay his frenzy, or rouse him from his subsequent sullen stupor. The great cliff of Doenyo lol Mwaru rose high above us, and to be caught at night upon its face involved dangers too serious to be lightly risked. We therefore camped beside the river, the Guaso el Narua, and nursed the poor Mkwafi till the morning. Some fish we caught in the stream consoled me for the delay.

We struck camp at daybreak, crossed a tributary to the river, and reached the foot of a wall of rock that seemed to bar further progress. The guide led us round the base of a spur, to a game track, which mounted by a very sinuous course up the face of the cliff. Numerous halts were necessary, but they were doubly useful; the ascent was very toilsome, and rests were welcome. The whole region of the foot-hills, the basins of Larabwal and Lake Baringo, lay spread out before us with the diagrammatic clearness of a map. After an ascent of 1400 feet the track became horizontal, and ran south along the face of the cliff to a gap at its southern end. We expected that this cliff was the face of the main Laikipia plateau, and that after we had gained its summit our course would be level. We found, however, that a deep ravine lay between us and the steppes beyond. At first it looked as if we should have to descend into this gorge, but the guide led us to a narrow ridge at its head, by which we easily crossed to the plateau.

The day's march had been very fatiguing, though it had only taken us about four miles onward. We hoped on the morrow to make up for this slow progress, for we were apparently on the western edge of a level plain, on which were occasional lines and patches of forest. We were again doomed to disappointment. A series of hidden ravines cut across the plain, and each of these had to be crossed. The largest, Longeyu lol Mwaru, was 350 feet deep, and its steep banks were so densely clad with vegetation that some path-cutting was required, and finally we had to camp on some open grass lands only five miles from our previous resting-place. There was nothing with which to make a "boma," but as there were no signs of inhabitants we did not trouble about this, and simply picketed the donkeys to stakes beside my tent door. The night was cloudless and moonless, and soon became intensely cold. The highest shade temperature observed during the day was 82°, but by eight o'clock the thermometer had fallen to 39° F., and the men crept into their tents or crouched beside the fires. When I went round camp at midnight, I noticed that the donkeys seemed restive and on the alert. A few minutes later one of them kicked against the tent ropes, so I went out to try to calm them. I stroked them and chatted to them for a minute or two, and as they seemed quieter, I told the sentry to look after them, and resumed my seat in the tent. Before I had finished arranging my blanket, the donkeys, with a wild snort, burst their bonds and fled, while a cry of "Mashimba" (Lions) rang through the camp. In spite of our circle of fires, two lions had rushed the donkeys. The camp was in an uproar in an instant. "If there are many, they will charge us," shouted Omari, and the men took up their stations, with guns ready, as if we had expected Masai. The lions, however, followed the donkeys; we heard a struggle a few yards away, and so, seizing brands from the fires, went to the rescue. We were just in time to see a lion sneak off from the body of a donkey it had killed, and send a shot after it. At the same moment we heard renewed excitement from the camp, with Omari's voice shouting orders to the men. We ran back, and found that one of the donkeys had escaped and returned to camp. It was trembling in every limb; we hobbled it securely and placed it in my tent for protection, and then waited for the lions to return to the body of the donkey they had killed. They followed the third, however, and as they did not catch it till it had run some distance from camp, they devoured it at leisure.

As a result of this disaster, we had to rearrange loads. It was not fair to add to the men's burdens, for we had many double marches ahead of us; so the skulls and many of the skins of the zebra and antelope shot at Njemps had to be left behind.

We marched to the southern end of the plain of Alng'aria, and then turned eastward to traverse the parallel ridges of Subugu (the Marmanett Mountains of von Höhnel). The highest pass proved to be at the level of 6950 feet, and from this we began the descent towards the undulating steppes of the typical portion of Laikipia.

Shortly afterwards the guide suddenly stopped, and pointing ahead said, "People!" He could not say whether they were Masai or Wanderobbo, so I went forward with him to scout, leaving the caravan to follow slowly. An hour later we came to the edge of the woods, and I could see in the far distance several columns of smoke. These, the guide said, were at Lari lol Morjo. I could see no other signs of natives, so we marched on to the plains and camped, waiting, however, till after dark before we lighted fires.

Next morning we enjoyed our first view of Kenya, which the porters called Meru. We saw it a few minutes before sunrise, just as we were preparing to start. It was eighty miles away from us, but it stood out sharp and clear on the eastern skyline. There was a look of fascinating mystery about it, as its dark jagged outline suddenly appeared above the dull gray mist. There was no time, however, for sentimental reflections or æsthetic rapture. I hastily traced its outline, and read the bearing of the central peak through my prismatic compass. Before I could repeat the observations, the mountain had disappeared in a gray expanse of mist.

The view cheered us all immensely. Some hills we had seen the day before had shown the men that we were now only level with Lake Kibibi, and had thus made very slow progress. Our food supply was running low, and we feared that we could place but little reliance on our crazy guide.

Now, however, we knew the position of Kenya, and if the guide failed us, we could brave the Masai, and march straight across the plains to the Kikuyu plantations on the southern slopes of the mountain.

During the day we passed close by some extensive prairie fires, which must have been lighted by natives, though we saw none. The guide led us between two of the largest of the fires, and we were all half choked by smoke. We reached the Guaso Narok fairly early, and forded it at a place where it was 20 yards wide and 2 feet deep. As our guide said we should come to some Masai kraals next day, he and I spent the rest of the day in finding a route that would avoid them. On returning to camp in the evening, I heard that the porter Jumbe was missing. We searched for him as thoroughly as possible, but without success. We could not delay, to continue the hunt in the morning, and had to abandon him to his fate. It was the man's own fault and due to disobedience, though he could not be blamed for this, as he had been in a condition of imbecility ever since a sunstroke on the shores of Lake Losuguta. The night had been intensely cold, and the exposure probably killed him. If not, it is to be hoped that the man soon met with either lions or Wanderobbo, for the physical pain of death from either would be merciful, in comparison with the mental agony of wandering lost in the wilderness, and dying slowly of starvation.

At this point I became much troubled about our route. According to the two maps of this district, if we followed the course of the Guaso Narok we should enter a broad valley between a range of mountains to our left, and two great peaks, Goyito and Kinangop, each 14,000 feet high, to our right. The topography, as represented by the maps, had led me to expect that a nucleus of old Archean rocks would occur in this mountain group; and I hoped to find on the flanks of these rocks some, intermediate in age, between them and the volcanic series. I therefore resolved to continue along the valley of the "Ururi" (the Upper Guaso Narok) until opposite the pass between the two western peaks, and then turn eastward across the Aberdare Mountains of Thomson, or Settima Kette (or Chain) of von Höhnel. This route seemed an easy one to find, and the only part of it that might have been expected to offer difficulties was the traverse of the "Settima Kette." But a track across this was marked by Mr. Ravenstein on the evidence of Denhardt's report of a journey by Ferhaji of Pangani. If a Suahili trader could cross, so could I.

As soon, however, as I tried to determine my position and the course of the next march, I was bewildered by being unable to find any one of my three guiding points. The Guaso Narok appeared to end off abruptly, and there was no trace of a double range of mountains to the south; all I could see there was one great and extremely eroded volcanic dome. Of the valley, five miles broad, which I proposed to enter, I could see no sign, and no help could be got from the guide. I asked him the name of the mountains to the south, and he said there was only one, Settima, the position of which he correctly indicated, though it was then hidden by the clouds. I next told him to point in the direction of Mount Goyito; he thought for a moment, and then said he could not. I inquired why, and he scornfully replied that there was no such mountain. Then I asked for Mount Kinangop; but the guide looked still more scornful, and told me I knew nothing about it, for Kinangop was a "gopo," or grazing plain, and not a "doenyo" or mountain. Finally, I told him to guide us along the Guaso Narok as far as it went; he declared that we had already passed its source, and it went no farther. On my saying that it must do so, for it was marked on the map as going for nearly two days farther, the guide flew into a violent rage. He told me I was a fool, for he had been there, and I had not; that I knew nothing at all about it, and was worse than the idiot Jumbe, for I preferred to believe a sheet of paper which had never been in the district, rather than a man who had lived there all his life. As soon as his rage abated, he sat on the ground and demanded tobacco. Unfortunately, as I do not smoke, I had none, and was accordingly denounced as a hard-hearted brute. As he wanted it to chew, it would have been useless to make him some out of brown paper and tea-leaves, a substitute which might otherwise have served. So there was nothing for it but to leave the guide to recover from his sulks, while I went on with the Askari, Mwini, to reconnoitre. We soon reached the depression in the main Laikipia scarp by which Thomson had reached the plateau. We went far enough to make sure there were no mountains between us and the Rift Valley, and that the main mountain mass lay to the south-east. A terraced grass slope led southward to the grazing lands of Kinangop, above Naivasha. We saw the smoke of some kraals in the distance, and as the guide had

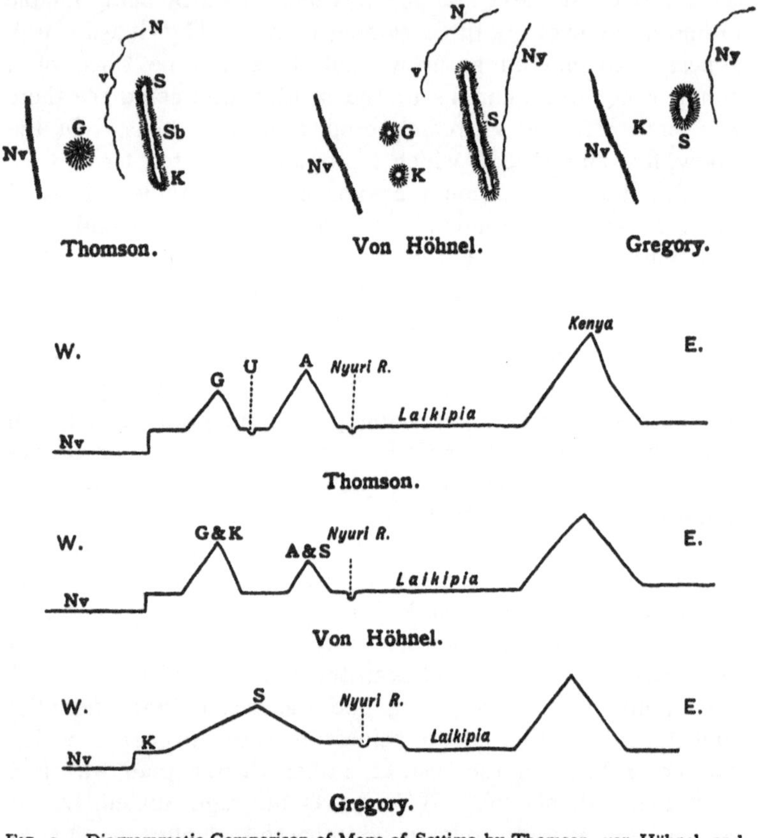

FIG. 3.—Diagrammatic Comparison of Maps of Settima by Thomson, von Höhnel, and the Author. A, Aberdare Mountains; G, Goyito; K, Kinangop; N, Guaso Narok; Nv, Naivasha; Ny, Nyuri; S, Settima; Sb, Subugu; U, Ururi.

told us there were many Masai in this district when he passed through a fortnight before, we returned to the caravan. In the evening the guide came to me and said he wanted to make friends. He plucked some grass, spat on it, and sprinkled it on my head. I returned the compliment, and then he told me that in the afternoon I had been to Rangatan

Busi, and that next day we should camp at Rangatan Ndari. I was delighted to hear these names, as they are two of the few in this district, marked on Ravenstein's map.

The next two marches were among the most interesting in the expedition. They cleared up the perplexities as to our position, threw much light on both the geographical and geological structure of the country, and yielded some interesting additions to the collections, including a couple of black crabs, identified by Prof. Jeffrey Bell as *Thelphusa berardi*, and a cobra, determined by Dr. Gunther as *Naja nigricollis*, Rnhdt. The marching in this district was easy, and the scenery grand, as immediately to the west of us rose the broken eastern slopes of Settima. Another source of interest was meeting some of the natives, called Wanderobbo by the Suahili, by which we obtained information about them (pp. 328-332), which seems to show that some of the people included under this term are really allied to the dwarfs. We arranged, moreover, for some of them to go with us to the Kikuyu country, whereby we became independent of our guide.

Shortly after this, in chasing a rhinoceros, I met with an accident attended with the most unfortunate consequences; I fell and dislocated one toe and severely strained another. So far, however, that was a mere matter of detail. Next day, as ill luck would have it, I slipped when going over some rocks and hurt my foot again, and this led to synovitis. That also was detail, though next morning I could not walk. In defiance of the Zanzibari proverb, "Never mount a donkey that has no saddle," a sack was tied on to the remaining donkey, and I rode. During the day the beast shied, the improvised saddle slipped, and I was thrown and dragged a few paces along the ground. During this the accident happened. In a collision with a basalt boulder my watch sustained a blow which smashed the glass and stopped the works. I did all I knew to start it again; I oiled it, and patted it, and sang hymns to it. But it would not go; and for the rest of the expedition I had to guess the time as best I could.

At length, late one afternoon, we reached the edge of the Settima plateau near a conspicuous boss of lava, called from its shape Narol Gwinia, or the "Man's Head Hill." From this point we looked down into the valley of the Guaso Nyuri or

Nyiro (as the Wanderobbo indifferently called it), and saw in the distance the hills of the northern frontier of the Kikuyu country. We therefore felt that this stage of the journey was almost at an end, and relaxed our supervision of the guide. He repaid this confidence by trying to steal some arrows from our native companions. He was detected, and fearing punishment quietly slipped out of camp. His escape was not discovered till the morning, when I at once started to search for him, accompanied by a porter, who was the only man in camp for whom the guide had ever betrayed the slightest feeling of regard. We hoped to find him sulking under a tree, but the track showed that he had started with no uncertain stride. We followed it for some distance, but it went straight away from camp, and the quest was hopeless. The guide had been a continual source of trouble to us. He had broken every rule in camp, and his behaviour had been most erratic and capricious. We had treated him like a spoilt child, tolerating every irregularity, and humouring every whim, though at the same time we watched him night and day. When we found that he had gone, we all felt genuinely sorry for him ; for we knew that he could never reach Njemps. Even if he got there, he had nothing to gain ; for he had gone without the marked paper that was to be a sign to the chief that he had fulfilled his contract and been properly discharged. He would therefore not get his pay, and would probably be punished as a deserter. But the chances of his arrival at Njemps were so remote that the treatment he might receive there was not worthy of consideration. No doubt, unless the lions found him, the poor guide perished miserably of cold and hunger on the steppes of his native plateau.

Our next camp was at the foot of a hill known as Doenyo Longari, where the Kikuyu dig the ochre with which they decorate their shields. Our last half ration of food had been served out the day before, but during the evening Omari found that a few of the porters had a little flour left. This was collected and found to amount to nine pounds in weight. I bought it at famine price, and it was made into a thin gruel and shared between the men. As the only food I could eat at the time was arrowroot, which in the absence of milk had to be boiled in water, we all fared rather badly.

Next day we entered the Kikuyu forests, which were traversed in every direction by elephant paths. In one or two places we saw elephant tracks that could only have been made that morning; but I had no time for hunting, even if a donkey had been good enough as a mount. We crossed the Guaso Nairobi ("Cold River"), and shortly afterwards emerged from the forests and had an extensive view to the south, across a great tract of undulating country, from which rose numerous clouds of smoke. These signs of habitation and cultivation cheered us, though we remembered the treacherous reputation of the people, and prepared for a possible struggle. We crossed three deep river gorges separated by lava plateaux, and approached the plantations and villages. We saw that some natives had seen us, so we selected a good site for camp, and fired a couple of shots to announce our arrival.

No answer came, so I sent the best Kikuyu interpreter and half a dozen porters to try to find some natives. They returned an hour later with a promise that food should be brought to us. No one came, however, except one or two men who crept through the grass to watch us. During the afternoon, Omari, with characteristic pluck, went off alone to the village and scolded the elders for not having sent us food. I did not know he had gone until shortly before sunset, when the men, who had become seriously alarmed for his safety, reported his absence. We all feared that he had been taken prisoner. I served out 100 rounds of ammunition to each of five reliable men, and started off with them to demand his surrender. To our intense relief, however, we met him a few hundred yards from camp. His report as to the attitude of the natives was not reassuring, so we strengthened our defences, and made half the men go to sleep, while the rest of us kept guard.

Just after ten o'clock we heard the dull thud of footsteps in a gully that ran from the plateau to the river, and saw a party of natives approaching us. They stopped in the mouth of the gully, and two of them came toward us waving torches. Our interpreter met them, and after a few minutes' conversation returned to say that the Kikuyu wanted to know who we were, whence we had come, whither we were going, and how we had dared to enter their country without permission! They were willing to have a shauri, provided that the white man did not take part in it. Omari, Ramathan, and two porters were sent as our delegates, and they were met by an equal number of the Kikuyu. They lighted a fire, equidistant from our camp and the main body of natives, and round it the shauri was held. The Kikuyu declared that we were a bad lot and had smallpox with us, and that we only wanted to enter their country in order to spread it among them. Omari invited them into camp to inspect us, and offered to let them kill any one who was found to be suffering with the dread disease. He refused to discuss the general question that night, for he said that we were starving, and that it is no use arguing with starving people. He declared that the natives must at once sell us sufficient food to last that night, and next day we would discuss the matter further. Omari was made a blood-brother with one of the Kikuyu; they sold us some food, and the conference adjourned till next day.

When the Kikuyu returned next morning we had our shauri. Ramathan opened it by telling the natives that we had crossed Laikipia, and wanted food to enable us to return northward in order to visit the great mountain of Kilinyaga (the Kikuyu name for Kenya). We therefore requested permission to enter their country and purchase provisions. "You shall not come in," replied one gray-haired elder. "Some white men came some few harvests back to our friends away there at Karthuri," he said, pointing to the south-west; "they stormed the villages, they seized what food they wanted, and then burnt the rest. When the elders asked for payment they were shot, while the young men were taken away as slaves into the land of the Masai, and we have heard of them no more." We pleaded against the verdict, and protested against being made responsible for the acts of our predecessors. Ramathan made a long and eloquent speech, in which he expressed on my behalf the utmost horror at their story. Ramathan said that the white men who had done these things were bad, and that I was good; that they were men of war, while I was a man of peace; that they were great fighters, and I was a great medicine man. He promised that I would pay for everything, steal nothing, and harm no one. He warned them that if they refused to sell food I should infer that their friends at Karthuri had done the same, and that therefore the white men were justified in taking it by force, and I should have to do the same myself. He assured them that I should be very sorry if this were necessary, for I liked the people, and thought they were good to work so hard and plant such big shambas, and were clever to grow so many different kinds of grain. I fear Ramathan said many things that I had not authorised, playing on the people's superstitions, and extolling my powers as a great *Mganga* or medicine man. He however assured them that I wished my visit to do them good and not harm, and if the latter happened it would be their own fault.

After a long and, to us, anxious shauri, the suspicions of the Kikuyu were allayed. Omari and Ramathan were made blood-brothers with the sons of the principal chief, and then permission was given to the women to sell us food.

They flocked into camp, carrying gourds of flour, baskets of grain and beans, sheaves of sugar-cane, huge bunches of bananas, bundles of yams, sweet potatoes (*vikwa* and *viazi*), and maize cobs. With the exception of the sugar-cane we bought everything they brought us, for we wanted at least a ton of food, and with people so uncertain as the Kikuyu, a change in their attitude might come at any moment. In the afternoon another party of natives appeared on the scene, and protested against the food being sold to us. There was a noisy quarrel between the two factions, until our friends gave way; then they all withdrew, leaving some warriors to watch the approaches to the camp, and prevent the women bringing us supplies by stealth.

Next morning some ulcerated elders came for medicine; it was a pleasure to dress the wounds of men who stood pain so well. A cretin also visited us, and by his hideous appearance and idiotic ways caused much merriment in camp. But there was no sign of more food. In the afternoon a party of warriors brought us some small bundles. This struck us as suspicious, for trading is usually left to the women; so several men were kept in hiding in my tent ready for emergencies. At a given signal the Kikuyu raised a tremendous shout, and each of them tried to seize something and bolt with it from camp. We were too quick for them, and they gained nothing of any value. The son of the chief was in camp at the time, and did not escape. He was placed under arrest, and told that, in case of treachery, he would be shot.

We saw no more natives that day except an old invalid, who came for medicine, bringing with him a few bananas as a present. His wants were satisfied, and he was sent back with the message that our prisoner would not be released until as much food as we wanted had been brought in. Twice during the night we heard natives creeping through the grass close to camp, but the attack which we all expected was not delivered.

It drizzled with rain all night, but Mwini kept us awake and in good spirits with his merry songs. Next morning the invalids returned to camp. They came alone, so the rule of "no food, no medicine" was rigidly enforced, and they went sadly away. The cretin stayed with us, and I gave him a lump of salt, which he licked and enjoyed immensely; he tried to do some trade with us, offering a rotten potato in exchange for a rifle, and a banana rind for a spade; finally he tried to run away with the thermometer screen. The rest of the day was spent in a three-cornered quarrel between two parties of natives and ourselves. The situation several times looked serious, but the Kikuyu knew that if they attacked us the son of their chief would be the first to suffer.

At last the natives sent a deputation to say they wanted to make friends again. The deputies came to us with the slowness befitting their dignity as messengers of peace; but they returned much more quickly. The temptation of an empty meat tin and some broken bottles was too great; they seized the coveted articles and fled.

A number of women had come down with loads of food, which they intended to sell us, as soon as the quarrel should have been made up. They dropped their goods in terror, and ran away with the men. We collected their loads and carried them into camp, amid the wails of the women, who had stopped on a hillock a few hundred yards away. As soon as they had recovered from their fright they came to beg for their loads of food, as they said they had no right to bring them down. We allowed them to come into camp to identify their bundles, and told them that if they would persuade their friends to sell us what we wanted next day, they should be paid for what we had confiscated. Otherwise we should take it away with us, and they would get nothing.

The plan answered; the women used their influence, and

the necessary supply of food was brought in on the morrow. Lomweri and Iyutha, our two blood-brothers, were told to be ready at dawn next day to guide us through the forests to Ndoro. When the time came to start, the guides insisted on delay, and as it was raining, and was bitterly cold, we did not object. Later on, just as we were starting, a party of warriors, in full war array, emerged from the woods and marched quickly towards us. We prepared to receive them, and I led Lomweri a few paces in front of our line; I levelled my revolver at his head, and signed to him to order the men to stop. He did so, and they obeyed at once. Lomweri sat on the ground, and looked calmly up at me, far less frightened than I should have been had the muzzle of a powerful revolver been almost touching my ear. The leader of the warriors came up, and explained that their women were collecting firewood in the forests, and they were going out to protect them from Masai—a statement received by the porters with a shout of incredulous laughter. I remarked that it was only right for women to be protected during the discharge of their domestic duties, and that we should be delighted to help in this good work. Extra cartridges accordingly were served out with the greatest possible ostentation, and the warriors filed off along a path through the forest. We waited for some time, and then followed them, marching with every precaution, and keeping a strict guard upon the guides. Progress was therefore slow. At three o'clock the two Kikuyu led us to a clearing, where they proposed we should stop for the night. From the point of view of defence, this place was a model of everything a camp ought not to be. So we continued our march; we crossed the deep gorge of the Guaso Thegu, and camped at sunset, on the open plains at the western foot of Mount Kenya.

CHAPTER X

ON THE SNOWFIELDS AND GLACIERS OF KENYA

"The only two essentials to happiness are a sound digestion and a hard heart."

"There shall he see
No enemy,
But winter and rough weather."

As You Like It, ii. 5.

MOUNT KENYA was first seen by a European on the 3rd December 1849, when a break in its veil of clouds enabled Dr. Ludwig Krapf to discern its snow-capped summit, from a hill above the Wa-kamba village of Kitui. Krapf admittedly saw it from a distance of about ninety miles, and though he stayed in the same district for some weeks he only saw it once, and then, but for a few minutes at sunset. European geographers, at this time, were not convinced of the existence of snow on Kilima Njaro, which had been discovered the year before by Rebmann. The evidence in the case of this more accessible mountain was far more definite, and it is therefore not surprising that Krapf's story was discredited, in spite of his description of the appropriate emotions that overcame him. To silence his traducers Krapf returned to the same district in 1851. He reached Kitui, but Kenya[1]—as he called the mountain—was not to be seen. He went forty miles nearer than on the previous occasion, but in vain. His small caravan was dispersed by a raiding party of Wa-kamba on the banks of the Tana, and he had to return to the coast without having

[1] The name Kenya was given to the mountain owing to a misunderstanding. The proper Kikuyu name is Kilinyaga. The Masai call it Doenyo Ebor—the White Mountain; the "Wanderobbo" Doenyo Egeri—the Spotted Mountain; the Wa-kamba "Njalo," a term they also apply to Kilima Njaro; and the Zanzibari "Meru," a name accepted in Europe for the peak west of Kilima Njaro.

No. XII. KENYA FROM THE KAPTE PLAINS WEST OF MACHAKOS. Page 162.

(From a Sketch by Mr. Ainsworth.)

seen the object of his quest. Hildebrandt followed on Krapf's footsteps in 1877, and spent some weeks making botanical collections in Kitui; but he also does not appear to have seen the mountain,[1] and the suspicions as to Krapf's veracity were strengthened. It was not seen, indeed, for the second time by a European until, in 1883, Joseph Thomson saw its western face across the plateau of Laikipia. The doubts as to the existence of the mountain had previously been removed by Wakefield and Denhardt's collections of the routes of various Arab and Suahili traders, to whom it was a familiar landmark. Thomson was the first to give any information about its structure, for he had the good fortune to enjoy several clear views of the mountain, from which, with his usual acumen, he correctly concluded it to be the denuded remnant of an old volcano. This peak, he tells us, "without a doubt represents the column of lava which closed the volcanic life of the mountain. . . . The crater has been gradually washed away."[2]

Four years later Kenya was visited by Count Teleki. He camped at Ndoro, marched through the bamboo forests to the Alpine meadows above, and reached the height of a little over 13,800 feet. Here the failure of his food supply, and the sufferings of his men, compelled him to return. Count Teleki's account of his ascent gives us the first definite information we possess about the mountain. He made a small collection of plants, which proved the occurrence on it of representatives of the Alpine flora of Kilima Njaro and Abyssinia. Unfortunately his conclusions as to its structure are less satisfactory. According to Teleki, Kenya is a well-preserved volcano, having a crater of from 4 to 4½ kilometres in diameter, and from 200 to 300 metres in depth,[3] while the highest point is only a tooth on the northern wall. Moreover, from his collections and descriptions it was concluded that the mountain was a dome of phonolite, and resemblances were detected between it and the phonolite peaks of Central Europe.

This was absolutely different from Thomson's conclusions, but as it was based on an actual visit to the mountain, and not

[1] Vide *Proc. Roy. Geog. Soc.* new ser. vol. iv. (1882), p. 747, footnote.

[2] *Through Masai Land*, p. 224.

[3] Teleki is perfectly right as to the existence of this hollow; his actual observations are all correct as matters of fact; it is only in his interpretation of those facts that I cannot follow him.

on a bird's-eye view of it from a point thirty miles distant, it was generally accepted. The subsequent expeditions sent out by the Imperial British East Africa Company in 1889 and 1891 were unable to contribute anything material to our knowledge of the mountain. Piggott first saw its eastern face early in 1889, and the three Europeans of the Tana expedition—Dundas, Bird Thompson, and Hobley—made a determined effort to ascend it from the south. They failed, however, to penetrate the whole of the forest zone, and had to return from the height of 8600 feet. Dr. Peters passed near Kenya in 1889-90 with the German Emin Pasha Relief Expedition, and his companion, Lieut. von Tiedemann, has given a sketch seen from the south-south-east; but they were neither of them much nearer the mountain than Thomson had been. In 1892 Lieut. von Höhnel, who had previously accompanied Count Teleki, but had been detained by illness in Ndoro during the latter's ascent, returned to the district with Mr. Astor Chanler, who bore the heavy expenses of the expedition. It was hoped they would together work out the topography of the north-east side of Kenya, and possibly gain the summit; they carried out some most useful explorations in the Daicho region to the north-east of Kenya, but did not reach the mountain. Their work was stopped by a series of thrilling adventures; Lieut. von Höhnel was seriously injured by a wounded rhinoceros, and later on the porters mutinied and returned in a body to the coast.

This forms a complete record of previous exploration of Mount Kenya, so that our knowledge of the mountain is very limited when compared with that of its great southern rival, Kilima Njaro. As this is near the coast, and has in fact occasionally been seen from the sea (as *e.g.* off Melindi in May 1893), as it is accessible by an easy road, and as its natives were once friendly, it has been visited by more than a hundred Europeans and carefully explored. Kilima Njaro has been ascended to the summit, it has been geologically mapped, and its flora and fauna described in some detail. In regard to Kenya, however, our knowledge was most rudimentary. It was not certain whether it had a well-preserved crater, or whether it was a volcano in the last stages of decay; and the estimates of its height varied from 18,000 feet (Ravenstein) to 23,000 feet (Peters).

I therefore resolved to attempt to reach Kenya to settle these and other problems connected with it. I assured every one on the coast that I had no intention of completing the ascent; for I guessed from Peters' photograph that it was not likely to yield to any one climbing alone, and none of the Zanzibari could be expected to venture upon the snow.

The preliminary accounts given by Teleki and Hobley of their attempts to ascend the mountain, showed that the first difficulty was the traverse of the forests on the lower slopes. It was advisable, therefore, to select a route which would lead through these at their narrowest part. The full reports of these ascents had not been published when I left England, so I had been careful to engage porters who had accompanied both expeditions. The descriptions and statements of these men showed that the forests are narrower on the west than on the south. They are probably thinnest on the eastern side; but from the direction from which we approached, this was quite inaccessible to us.

On emerging from the forests of western Kikuyu we therefore marched northward through Ndoro, at a sufficient distance from the base of Kenya to get a general view of the mountain. At noon we selected the bay of moorland that seemed to run highest into the forests, and at its head chose a place for camp.

The rest of that day was spent in a bustle of preparations. A "zeriba" or "boma" was built to protect the camp, and a strong shed begun to shelter the stores. Loads were re-sorted, rations served out for ten days, and reserve food packed. Twice Omari and I went through the baggage, to pick out things that could possibly be left behind. My own kit was cut down to the lowest possible limit, though I took all available clothing to lend to the men.

Our camp here was at the foot of the mountain—in fact a few hundred feet up its lowest slope; but we could not see it, for an impenetrable mist obscured everything. Luckily for us, however, the mist dispersed for a few minutes at sunset. I was thus able to take a bearing of the summit, to note the direction of the valleys and ridges that lead to it, and to plan a route by which to traverse the forests to the peaks above.

At eight o'clock next morning (24th June) all was ready. The men paraded, and the twelve chosen for the expedition were called out. Loads were given them, and without allowing any time for discussion or murmurs, Omari led the way out of camp. I kept in the rear to prevent dawdling or desertion ; for the men who had been with Dundas and Hobley, in their attempted ascent from the south, told such pitiful tales of their sufferings, that the porters were loth to enter the dreaded forests. We plunged into these almost immediately, and found they consisted of lofty junipers and *Podocarpus* rising from a matted undergrowth of bush and shrubs. We forced a passage through the jungle and startled a pack of monkeys (*Colobus occidentalis*, Rochb.), whose long black and white fur so closely resembled the Beardmoss (*Usnea*) on the trees, that they were unrecognisable at a very short distance. By watching the men in front of me, and listening to the noise made by the leaders as they lopped off branches or trampled down the undergrowth, I could estimate and guide the direction of our line of march. After passing for two hours through level forest, we came to the foot of the ridge by which I had resolved to make the ascent. It was low and broad and we easily reached its summit, and followed along it to the east-north-east.

In the afternoon we found a swamp in a hollow, on the flank of the ridge. It was fringed by the common English rush (*Juncus effusus*, L.) and reed-mace (*Typha angustifolia*, L.), while dense clumps of tall bamboos were scattered over the hillside above. These clumps rapidly increased both in number and size, till they united into a continuous belt which seemed to bar our further progress. It had been drizzling ever since we entered the forests, and as the rain now increased to a steady downpour, we pitched camp under some lofty *Podocarpus* trees. I spent the rest of the day collecting the snails and slugs which lived in the damp undergrowth. The rain continued all night, and at daybreak next morning a cold raw mist lay around us. We waited for a couple of hours for this to lift, but as there seemed no prospect of it doing so, we resumed our march. At first an elephant track ran past our camp, and we followed it through the jungle until it left the crest of the ridge, and ran down the slope toward the roaring torrent in the valley below. This was out of our way, and we had to cut a path through the jungle. The work from this point was the most trying I have ever experienced. The bamboos are usually about 40 feet in height and from 2 to 3 inches in diameter, and are often packed together so closely that it is impossible to force a way between them. At the height of from 10 to 15 feet from the ground they branch repeatedly, and their long grass-like leaves interlace into a dense canopy of vegetation. Above this the trees give off broad-spreading branches, which are connected by lianas and other climbing plants into a second roof of vegetation. A mist hung over the forests during the whole time we were in them, and kept the vegetation sodden with moisture, and made the mossy soil as saturated as a sponge.

Through this dark and dismal forest we had to force a way. Occasionally an elephant path would run in our direction, and we could then make comparatively rapid progress, delaying only to lop off the lower branches of the bamboos, to cut through fallen stems, or to climb over dead tree trunks. The elephants, however, did not obey the rules of mountaineering, and their tracks soon ran down into the valleys, so that most of the way we had to cut a path step by step. Every blow of our matlocks upon the bamboos shook the sodden canopy overhead, and continual shower-baths of water kept us wet and miserable. My clothes were soon soaked through, while the raw, damp cold chilled the porters to the marrow. We had to stop every hour to light fires to warm them, and even then they found the climate almost unbearable, and one or two cried like children.

We trusted by hard work to traverse the forest belt in two days, for I was anxious to spend only one night within it. But, as we soon saw, this hope was vain. On the evening of the second day we had to pitch camp on a slope, where the bamboos were so dense that we had to clear every foot of ground we wanted, while it was so swampy, that we had to spread out the bamboos as a platform on which to support the tents. I went some distance farther, to try to find a better place, and so knew that we were still in the heart of the bamboo zone; but even then I was not prepared for the revelation made by the boiling-point thermometers in the evening, that, despite all our efforts, we had only risen 1700 feet in two days. Determined not to lose a moment's time next morning, Omari, Fundi, and I went ahead at daybreak to cut the path, leaving the porters to follow as soon as it had become less cold. We made a desperate effort to get out of the forests, but when night fell we were still within them, and the bamboos as thick as ever. We were so exhausted that, when the order to camp was given, we all lay down where we stood; and it was not till some time afterwards that we could rouse ourselves to light fires and prepare food. So far the work had been simply miserable. We had not once seen or felt the sun since we left the meadows of Laikipia. We had never once seen more than 20 yards ahead, and it was only rarely that we could see up to the tree tops. The natural history had also been disappointing. Of vegetation there was enough and to spare; but the species were few, and the plants so sodden with moisture that I could not press them; the bamboos were especially irritating, for I could not find a single flower or fruit, and thus it has been impossible to determine the genus to which they belong. Many animals were not to be expected, and we only saw one pack of monkeys, some red-breasted birds rather larger than robins, many slugs and snails, and a few insects. At night we heard the shrill cry of a cony (*Procavia shoana*, Gigl.), of which I found a skull.

On the morning of the fourth day, however, there came a welcome change. While cutting a way through the bamboos we suddenly stumbled upon a block of lava (andesite). I was delighted to see it, for I had not previously seen as much as a pebble since we left Laikipia. As I examined it, my interest was roused; for its grooved and rounded surface suggested that it had been carried to its present position by ice. The denseness of the jungle, however, prevented any further evidences being obtained here, and I was accordingly all the more delighted when shortly afterwards we entered a clearing, and through the mist dimly discerned a high rocky ridge a little distance above us. With a cheer we hurried forward. The bamboos became smaller and scarcer, and were soon left behind. The forests gave place to scattered clumps of trees, and the rank undergrowth to a firm rich turf; the long monotonous slope broke up into a belt of undulating ground, which, with its numerous swampy, mossy hollows, its irregularly scattered boulders, and its stiff, greasy clay, reminded me of a glacial moraine. The men threw down their loads and basked in the sunshine, while I examined the sections in the stream banks, and collected the flowers on the meadows. Many of these were old friends. There were clover and blackberry, dandelions and bitter-cress (*Cardamine*, sp.) Some silvery gray bushes about 6 feet in height, which clothed the hillside, proved to be an arborescent form of that most typical of Alpine meadow plants, the Lady's Mantle (*Alchemilla*). These plants were associated with others, such as the gladiolus, which I knew only in gardens, and others, such as a tree-lobelia, which were of a type quite new to me. I also caught some specimens of a common English butterfly, the Clouded Yellow (*Colias edusa*, Linn.)

An hour passed all too quickly, and then some rising clouds warned us to hasten on to camp. Immediately after we had started we surprised, and were surprised by, a small antelope (a bush-buck). It made no effort to run away, but stood and watched us, until a bullet from Omari's Snider cut up the ground beside it. The path cutting had so unsteadied my hand, that I preferred to leave the responsibility of the shot to some one else.

Above us rose a steep wall of rock, up the face of which ran an elephant path. While scrambling up this the storm broke upon us, and my men were alarmed by the appearance of hail and sleet. Fortunately some of them had seen hail in Kavirondo, and on the summit of Mau, and they were therefore not as much startled by the sleet as they might otherwise have been.

The men faced the storm most pluckily, but as we ascended the slope it became much more severe, and I went on ahead to select a place for camp. I was recalled by a signal-shot, and found that the porters had collapsed on a half-frozen peat swamp. They were crouching under the lee of boulders and bushes for shelter from the driving sleet. A few of the stoutest put up the tents, and I went on again to find a better place for the permanent camp, as I intended to leave the men here while I went higher up the mountain. On my return I found that one of the porters had not come in. The Askari, who were responsible for his safety, as it was their duty to see that no one lagged behind, and my plucky headman had all tried to go back to the rescue, but had been unable to face the storm. The missing porter was a man named Wadi Sadi, whom I had especially chosen to join the party; for he was a sailor, and thus might have been expected to be a good climber; moreover he was the lightest man in the caravan.

I was therefore disappointed to hear that he was lost, for it showed he was useless for the work for which he had been chosen. I rushed back at once; but as the snow had hidden our trail, I missed it, and had to search for an hour before I found him. He was lying on his load about 300 feet below the level of the camp; he was covered with snow and nearly frozen to death. A little brandy revived him, but he was too weak to stand. As it was still snowing it would have been useless to have returned for help, for the porters were so cowed that they would have refused to move. I recollected that Wadi weighed less than the burdens some of my men had to bear all day long, so I resolved to carry him. He was able to cling to my back, and slowly, and with many halts, I struggled with him up the slope. If the porter had left his load when he first became too weak to carry it, he could no doubt have walked on with the others. I thought his action in staying out in the snow with it simply Quixotic, and, annoyed at the trouble it had given me, I rather brutally told him next morning that he was a fool. It is a point of honour among Zanzibari never to leave their loads, and I shall not forget the man's reproachful look as he asked, "How could I leave my load without my master's orders to do so?"

Another trait in the Zanzibari character was shown at the same camp. In the morning the men came to tell me that the water they had left in their cooking-pots was all bewitched. They said it was white, and would not shake; the adventurous Fundi had even hit it with a stick, which would not go in. They begged me to look at it, and I told them to bring it to me. They declined, however, to touch it, and implored me to go to it. The water of course had frozen solid. I handled the ice and told the men they were silly to be afraid of it, for this change always came over water on the tops of high mountains. I put one of the pots on the fire, and predicted it would soon turn again into water. The men sat round and anxiously watched it; when it had melted they joyfully told me that the demon was expelled, and I told them

No. XIII. VIEW IN THE HÖHNEL VALLEY. Page 171.

(ALT. 13,000 FT. ON KENYA.)

Senecio Kenyensis, Lobelia Gregoriana, and heath occur in the foreground.

they could now use the water; but as soon as my back was turned they poured it away, and refilled their pots from an adjoining brook.

We moved camp next day to a drier and more protected situation, where we left the invalids. My tent was carried to a point 1700 feet higher up. We had to march for some time across a peat bog, over which we made fair progress, until the crust thawed; after this we were for most of the time plunged to the waist in half-frozen peat. Dragging the loads through this was terribly fatiguing; and a depressing sleet, which sometimes passed into rain and sometimes into snow, added to our discomfort and obscured the view. After escaping from the bog, we reached the crest of a ridge. We found there a sheltered nook amid some crags of agglomerate, at a place where a grove of giant groundsels and tree lobelias supplied abundant firewood. The men begged that we might go no farther, and I was only too willing to grant their request. They put up the tent, and then, all but my cook, my personal attendant, Fundi, and a porter who was too weak to return, went back to the lower camp.

From this point I made a series of excursions to work out the topography and geology of the ridges of the mountain.

In order that the account of the excursions among the central peaks of Kenya may be understood, it is necessary to give a short sketch of the topography of this part of the mountain, and of the nomenclature of its peaks, valleys, and lakes. For more detailed information reference must be given to the descriptions in the *Geographical Journal* and *Quarterly Journal* of the Geological Society.[1] As it is impossible to describe the mountain without names, some have had to be invented. I should not, of course, think of applying European terms to places for which native names are already in existence; but in a locality where there are no names, there can be no reasonable objection to proposing them.

Several names are obvious. The valley in which Teleki built his boma, and reached his highest point, it is natural to call the Teleki Valley. The name of this courageous Hungarian

[1] J. W. Gregory, "Contributions to the Physical Geography of British East Africa," *Geog. Journ.* vol. iv. pp. 413-421; "The Glacial Geology of Mount Kenya," *Quart. Journ. Geol. Soc.* vol. i. pp. 515-530.

explorer may also be conveniently applied to the two lakes on the floor of this valley, and to the ridge at its head. To the south of the Teleki Ridge is a peak 15,800 feet in height, which

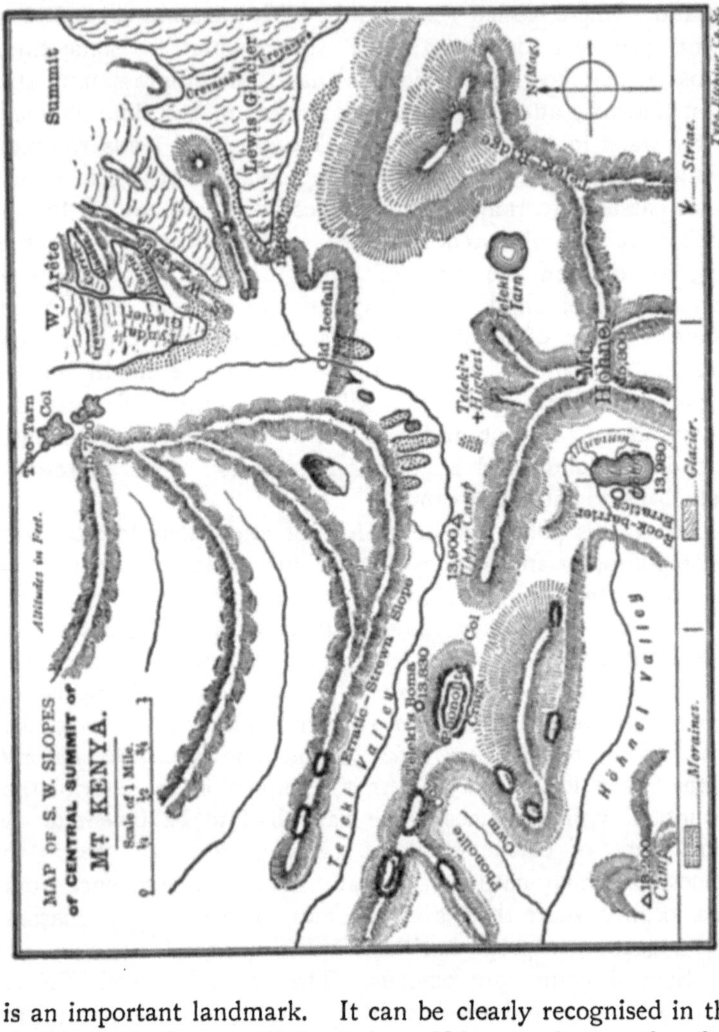

FIG. 4.—The South-Western Quadrant of the Central Part of Kenya.

is an important landmark. It can be clearly recognised in the sketch made by von Höhnel from Ndoro, and may therefore be called Mount Höhnel. The name of this accomplished cartographer may also be applied to a lake in a cirque upon this mountain, and to the valley down which flows the stream which rises in this lake. To the south of the Höhnel Valley is the head stream of the Guaso Mairi, and the valley may therefore be named after this river. For the main valley on the south side of the mountain I propose the name of the Hobley Valley, as it is probably the one which the British East Africa Company's expedition would have entered had it traversed the whole of the forest zone. The tarns upon the floor of this valley I beg to call after Mr. W. Bird Thompson, the caravan leader of that expedition. When we come to the glaciers and the central peak the names are not so obvious. The actual peak consists of five pyramids, viz. the double central peak, two large aiguilles which tower above the Lewis Glacier, and a sharp triangular peak on the western arête. The last, 17,500 feet in height, I named Point Piggott, after the last Administrator of the British East Africa Company. The glaciers are on the south-western quadrant of the mountain. There are none to the south-east, and I could not see any to the north-west, though they may occur there, having been hidden by clouds at the time when I might otherwise have seen them. The principal glacier is on the south side of the mountain, and to it I gave the name of the Lewis Glacier, out of respect to the late Henry Carvell Lewis, whose brilliant researches have thrown so much light on glacial problems in England and America. The other glaciers are the Tyndal Glacier on the west side of the mountain (Pl. XVII.), and the Darwin Glacier in the valley between the other two. The latter was so named as we owe to Darwin the first precise description of a glacial valley in England. The Heim and Forel Glaciers are two small corrie glaciers on the western face, which were named as they proved very useful in surveying. The limits of the three valley glaciers were marked by moraines.

The most interesting excursion made from the camp on the agglomerate ridge was the ascent of Mount Höhnel. I took with me my favourite porter, Fundi Mabruk, who was especially useful, as he had accompanied Teleki during his visit to Kenya and had been with him to the highest point then reached. We crossed the Höhnel Valley, and up a cwm, which from the fine crags of phonolite at its head I called Phonolite Cwm. This led us to the Teleki Valley, which we followed to a col leading south-eastward to Lake Höhnel. From this point we climbed the western ridge; this was at first very easy, but we soon came to a part where the arête was steeper, and broken by a series of lava cliffs. We could easily have avoided these, but they were very easy, and so, to test my companion's climbing powers, I resolved to scale them. Fundi did not like the look of the cliffs, but after a little persuasion he consented to be roped, and I went ahead up a gully. Having taken my position I had another little argument with Fundi, accompanying it by a few gentle pulls on the rope. Thus exhorted he climbed up about 10 feet, but his foothold having once given way, he was useless. I had to lower him to the bottom, and nothing would induce him to try it again. "That is all very well for wajuzi (lizards) and Wazungu (white men), but Zanzibari can't do that," was his verdict. "You'd better come back, master," he cried; "I promised to follow you anywhere, but how can I, when the path stands up on end?" He would not come on, so there was nothing for it but to order "fungua" (unrope), and continue the climb alone. At several places vertical lava walls barred progress on the crest, and I finally gained the summit from the south-eastern arête, which I had reached by a traverse. I rested here a while to examine the ridges round this peak, and to consider the various possible lines of attack upon the central summit. This rose abruptly as a great black pyramid from the snowfields and glaciers at its base. The face opposite me was so steep that the snow only rested on it in cracks and hollows, and down gullies. There was not a cloud upon it, but here and there a streak of wind-borne snow-dust softened the harsh form of the rocks and the dazzling lines of snow. It was so delightful to sit upon the summit of the one peak, estimating the advantages offered by the different lines of attack upon the other, and comparing it with old Alpine friends, that I paid no heed to the clouds creeping up the valleys from the west. I stayed on the summit till they had reached me, and robbed me of the view. Then, in order to obtain a fuller series of exposures of the succession of beds of lavas and volcanic ashes, of which Mount Höhnel is composed, than I had seen during the ascent, I began to descend its western face toward the lake.

The climbing was at first rather difficult, as the face of the

No. XIV. THE CENTRAL SUMMIT OF KENYA FROM THE SUMMIT OF MOUNT HÖHNEL. *Page 174.*

mountain consists of a series of almost vertical lava cliffs, separated by slopes of volcanic ash and talus. The snow on the ledges was in a very unstable condition, and rendered the traverses along them highly interesting. Had there not been good hitching points for the rope, it would have been risky. Owing to the mist, it was impossible to see ahead and select the best way down, and I was often compelled to retrace my steps, and strike out blindly in a fresh direction, throwing stones ahead to obtain warning of my approach to the next cliff of lava. I was, therefore, not sorry when the snow was restricted to mere patches, and the slope became more gradual and passed into the swampy meadows beside the lake.

Some dead lobelias supplied wood for a fire, on which I boiled my thermometers, and warmed some beans. Meanwhile I sketched the outline of the lake, and examined the glaciated boulder-strewn barrier that supports it. Then, refreshed by the food and rest, I crossed the ridge on the southern side of the Höhnel valley, and collected in the valley to the south, till gathering clouds and approaching sunset warned me to return to camp.

The camp on the ridge of agglomerate was too far from the glaciers and the central peak to be of any use as a centre for excursions upon them. Fundi Mabruk, a young porter named Yussuf, and I therefore carried one of the men's small shelter tents and a few stores to the Teleki Valley. We pitched our diminutive camp under the shelter of some groundsel trees, a few hundred yards from Teleki's highest point, and near some old glacial moraines, which ran across the valley like railway embankments. My first excursion from this point was spoiled by a slight attack of mountain sickness. As this occurred at only the elevation of a little under 15,000 feet, I was surprised and disappointed. I did not like to confess this to Fundi, as it seemed safer to keep him in absolute ignorance of the existence of this malady. I therefore became absorbingly interested in the study of a coarse block of lava, one part of which happened to form a comfortable seat. Shortly afterwards the usual afternoon snow-storm broke upon us, and I was glad of this excuse to return to camp. The attack was very slight, I soon recovered, and suffered no more inconvenience, except undue fatigue.

Next day I intended to begin the exploration of the glaciers. I was especially anxious to measure their rate of flow, for which purpose it was necessary to put up a line of stakes across one of them. As it would not be fair to expect any one to attempt snow-work in bare feet, I had brought up a reserve pair of boots and leggings, which I tried to persuade Fundi to wear. He put on the boots—under protest, but absolutely refused to keep them on. As he also declined to allow me to put nails into the soles of his feet (his hide would probably have held them), I had sadly to reconcile myself to the knowledge that whatever snow-work was necessary would have to be done alone. Being anxious to start before daybreak, at sunset we wrapped ourselves up in our blankets and lay down on our improvised mattresses of Alchemilla bush. The snow-storm, instead of stopping at its usual time, increased in severity, and such a furious gale of wind swept down the valley, that I thought several times the tent would be blown away. At about two in the morning we were alarmed by the rush and whistle of slipping snow; with a crash the wall of turf and stone we had built above the camp fell in upon the tent. The doorway was blocked by snow and earth, but a desperate jerk tore up the loosened tent-pegs, and we sprang out into the storm to examine, as well as the blinding snowstorm would permit us, the exact extent of the peril. The slip, however, had stopped, though it had covered our tent ground, and buried the few things and food we had with us. There was no danger, but it was impossible to light a fire or repitch the tent in the darkness and the storm, and there was nothing for it but to jump about, wrapped up in our blankets, to keep ourselves alive until the morning. The two Zanzibari suffered terribly, which, as the thermometer marked 28° below freezing,[1] was not surprising. My boy Yussuf simply sobbed with the cold. The more stolid Fundi, in answer to my exhortations to cheer up, as the sun was coming, when we should all be warm, only replied with his usual fatalism, "Yes, it will come,

[1] It should be pointed out that this was not the proper air temperature. The thermometer was simply slung on my ice axe, and not properly protected from the wind. The instrument was an ordinary bath thermometer that had been given me by Mr. Ainsworth, as none of my thermometers were graduated below 40° F. The instrument was broken on the return journey, and so its index error could not be verified. Meyer, however, in the warm season recorded a temperature of 10.6° F., at an elevation of 14,200 feet, on Kilima Njaro (*Ostafrikanische Gletscherfahrten*, 1890, p. 165).

No. XV. THE LEWIS GLACIER. *Page* 176.

and Inshallah (God pleasing), we shall be alive when it does come."

This little accident put an end to my hopes of starting before daylight. It was not till some time after sunrise that we felt sufficiently warmed to face the cold wind that swept down the valley. Breakfast was impossible, as our food was buried in the snow-drift, but I had long since ceased to regard this meal as a necessity. Leaving Yussuf to look after the camp, Fundi and I hurried along the valley to the foot of the old cliff, which must have marked the site of an icefall in the old glacier. Here the valley bent abruptly to the north. We climbed the cliff, and then, turning again to the east, toiled up a long tiring slope, covered with moraine. Numerous dead lobelia stems lay on this, though there were none then living at this elevation. We crossed some dykes of phonolite, and scrambled up some rocks, over which the stream from the glacier plunged in a series of leaps. Above this we reached a platform, upon which rested the end (or snout) of the Lewis glacier. I was greatly interested to find here a series of five small concentric moraines, the uppermost of which has been broken through by the glacier, so that it is now re-advancing. I rested on the edge of the platform, to sketch the moraines and wait for Fundi. That morning he was weak and ill, but he plodded steadily, though painfully, upward. He had often talked to me about the great white fields he had seen with Dachi-tumbo (Teleki), and how bitterly disappointed he had been at not reaching them. He had taken a keen personal interest in this expedition, and his influence with the men had been most useful. I therefore waited for him to pass me, that he might be the first man to set foot on the glaciers of Kenya. He came up, laid his load upon the ground, kicked off his

FIG. 5.—Terminal Moraines of the Lewis Glacier.

zebra-hide sandals, and mounted upon a boulder. Then, with his hands together before him, he began to pray. I could not understand all he said, but sufficient to know that he thanked Allah for having enabled him to come where neither native nor white man had ever been before, and to stand on the edge of the great white fields he had seen with Dachi-tumbo from afar. He assured Allah that he was now more anxious to return in safety to the coast than he had ever been before, so that he might tell his friends of the wonders he had seen.

After the prayer was over, I told Fundi to go on to the glacier. He went a few steps farther, and then, with a pleading look, said, "No farther, master; it is too white."

There we lighted a fire, and boiled the thermometers, obtaining data which placed the altitude at 15,580 feet. As soon as the instruments had cooled, I prepared to continue the ascent. But Fundi, whose curiosity was now satisfied, begged to be allowed to return. He complained that his head was aching, that his stomach was very bad, that he felt very sick, and that his legs *would* not do what he told them. It was obvious he was suffering from mountain sickness, and it was not fair to take him farther. I therefore added his share of the load of instruments, firewood, and "pitons" (or pegs on which to fasten the rope) to my own, and let him go back. Before doing so, I fear I completely ruined any reputation for sanity I might have had left, by executing a Masai war-dance on the snout of the glacier, and then pelting Fundi with snowballs.

I had hoped to ascend by the right moraine of the glacier, but this was too risky, owing to the falls of snow and rock that thundered down on to it from the face of the cliffs on the north. I was therefore obliged to continue along the left or south moraine, until above the icefall of the glacier. The snowfield here extended to the south, and as it would have taken too much time to skirt it, I risked a traverse across to the main south arête of the mountain. The snow was in fairly good condition; and the few crevasses were marked by "droop," and so could be avoided if not easily crossed. The ascent along the arête was less easy. The ridge rose gently till it reached the point above a cliff, which stands out in the middle of the

No. XVI. FUNDI'S PRAYER. *Page* 178.

glacier. On both sides of this the ice was thin and very rotten. Numerous rocks jutted above its surface, and hitching the rope over these gave me sufficient confidence to proceed. As soon as the arête was clear of the rocks, it became sharper and steeper. It was bare of snow, for it was fully exposed to the west, and was thus windswept. It sloped steeply on the left to the "bergschrund," which had prevented me from reaching the arête on the north side of the rocks. The ice, however, was firm, but intensely hard. My ice-axe was too light for the work, and thus step-cutting was even more fatiguing than it would otherwise have been. About this time I became rather sleepy, and so tired that I could only go a few steps at a time. To lighten my burden I tied most of it into a bundle and fixed it to a peg in the ice, carrying on with me only a little firewood and a couple of instruments without their cases. I took one of the boiling-point thermometers out of its brass tube, and carried it fastened into my pith helmet. An accident to it would have detracted from the value of my long series of altitude determinations, and left me dependent on only one instrument; but though I was quite conscious of this, I somehow felt strangely indifferent to the risk of an accident, which would have distressed me deeply.

As I approached the rocks of the central peak the crest of the ice became broken. An occasional traverse was necessary round pillars or "seracs" of ice, and it became quite clear that I could not go much farther. I persevered, however, for though it was hopeless single-handed to reach the summit, I was anxious to ascend a couple of hundred feet higher, when it would have been possible to make a traverse on to the main eastern arête, and gain a view over the north-eastern part of the mountain. But as I approached the rocks the ice again became snow-covered. The slope to the left was less steep than it had been, but the snow was in too bad a condition to be crossed alone. A few yards farther up a snow cornice overhung the cliff to the east. This might have supported me then, but it possibly would not have done so on my return later in the day. I recalled an unpleasant memory of an April afternoon spent on a ledge in the Cottian Alps, waiting till the cold of night should harden some snow I had safely crossed in the morning. I had no desire to repeat this experience, and there was nothing for it but to return, especially as the afternoon snowstorm was blowing up from the west.[1]

I therefore worked back to the nearest rock, and tried to light a fire to boil the thermometer. But the wood was damp and would not burn, though I kept on trying till I had used up my last match.

The storm broke before I had got off the arête, and as it had buried my footmarks I did not care to risk crossing the glacier again, for the ice in the neighbourhood of the cliffs was much crevassed. I kept along the crest to a col at the height of 16,600 feet, which leads over to the Hobley Valley. I stayed for a few minutes to pile some of the blocks of coarse rhyolite that occur there into a small stone man, under which I left a note of the date of the ascent, and then ran down the Lewis glacier moraine to camp, reaching it, very exhausted, two hours before sunset. The fact that I ordered "fannya jocula tiari" (make food ready), as soon as I was within shouting distance, and settled down at once to as hearty a meal as I dared make, seemed to relieve my men considerably; but I could estimate, from the anxious way they watched me, their deep concern for my mental welfare.

Next day I made another attempt by the western arête, hoping that, as this was all rock work, I should be less handicapped by going alone. I ascended the ridge that forms the north wall of the Teleki Valley, and continued along this until I came to a depression which, from the existence of two lakelets on the summit, I have called Two Tarn Col. From this point I had my last view of the snowfields of Kenya, though I was among them for the rest of that day, and was within range of view of the mountain for a month afterwards. Just before sunset that evening a rift in the clouds gave me for three minutes a view of the double-peaked summit, but with that brief exception I never saw it again. The clouds buried it the whole time, and prevented my obtaining a single bearing by which to help in the determination of its position.

This last view, however, was perhaps the best. At my feet, in a deep valley, lay the snow-covered and crevassed Tyndal Glacier, beyond which rose the steep western face of the final

[1] These storms greatly increased the difficulties of climbing on Kenya. They occurred every afternoon, and spoilt a terribly high proportion of the twelve hours of daylight.

No. XVII. THE WESTERN ARÊTE OF MOUNT KENYA. *Page* 180.

WITH POINT PIGGOTT AND THE TYNDAL GLACIER.

peak. Unlike the southern side this was almost entirely covered by snow and ice, and on it rested two small corrie glaciers (the Heim and Forel) supported upon ledges, the larger of them ending in a cliff of ice from two to three hundred feet in height. Huge masses frequently break off from the ends of both of these, and fall with a crash on to the margin of the Tyndal Glacier below, where they accumulate as fan-shaped piles of ice.

To the north rose the steep pyramidal peak of Point Piggott, separated by a depression from the western arête of the mountain. This ridge ends to the south as a cliff, which forms the northern boundary of the basin of the Tyndal Glacier. Couloirs of snow occur in places along it, and one at least of these might be practicable for a full-sized party, but a yawning bergschrund separated the glacier from the névé fields at the foot of the cliff.

I hoped, however, that the northern face of this ridge would be easier, and resolved to work round the western and northern sides of Point Piggott, and thus gain the depression between it and the central peak.

A rough scramble over a slope, covered with a mixture of snow and talus, took me on to the western edge of Point Piggott; but a mist covered the ridges to the north, and thus I was disappointed of my view in this direction. It also drove me farther than I had intended up Point Piggott, before beginning to work round its northern face. During this traverse I continued to ascend, until I must have been level with the "brèche" or depression which I hoped to reach, but the mist prevented my determining my exact position. I was at length stopped, at a height of probably about 17,200 feet, little below that gained on the southern arête, by some cliffs which it was hopeless to attempt to scale; I tried to turn them by a traverse at a slightly lower level, which the loose snow made rather uncomfortable, though I had always plenty of hitching-points for the rope. Progress was very slow, and I could not see that the daily snowstorm was blowing up earlier than usual. My retreat was too late; the storm broke while I was still on the northern face of the peak, and rendered some rocks I had skipped over in the ascent quite impassable. I had to descend a gully in order to try to reach some less steeply inclined snow lower down; this, however, was so powdery that I could not cross it, and had to continue down the north face, which, but for the rope, I could not have done. In consequence of this I missed some sheets of paper which I had left in the morning in order to guide me on my return, entered the wrong valley, and did not get back to camp till three hours after dark. Here the news was awaiting me that my men at the lower camp were ill, and also a request to descend next morning to attend to them.

The cold at night was intense, for the sky was clear and cloudless, and an icy wind from the snowfields whistled among the crags beside the camp. Personally I revel in cold, so I gaily tripped out of my tent to read the thermometer; but on that occasion it was too much for me, and I was glad to get back again to my blanket-bag, covering my retreat by depreciatory comments on the value of meteorological observations.

The night was followed by a wet day. I sent the men down with the goods to the lower camp, and remained behind for a short time with an Askari to collect in the Höhnel Valley. While doing so it began to snow; and when we began the descent, we entered first a cold drizzling mist, and then heavy rain. The Askari delayed progress, for he walked rather slowly, and had a great dread of being lost. Whenever I got a few yards ahead of him, and he lost sight of me in the mist, he would shout excitedly, "Zhuia! bwana, zhuia!" (Stop! master, stop!) The first time he did so I obeyed at once, thinking that I was rushing headlong over a precipice. His only reply to my startled inquiry, "Kumbé?" (What's the matter?) was a moan, "Ah! bwana! I wish I had never left Zanzibar." The first time he did this, I thanked him for the information. On the second occasion, I reminded him he had told me so before. Next time I tried to soothe him with the remark that I was glad he had, as I had thereby obtained the benefit of his services. On the fourth repetition, however, I heartily agreed with him, and suggested that the quicker he walked the sooner he would be back in his beloved island home. I was, however, always compelled to stop for him, for as soon as I disappeared from sight in the fog his cries became most pathetic; but my remarks on the subject need not be chronicled. Shortly after this I also began to wish myself in Zanzibar. Instead of following round the rocks on the edge of the valley, I resolved on taking a short cut across the peat swamps. The walking was very bad, but the compass led us straight to a huge boulder, on which had been erected a small stone man, as a guide to the position of the camp. In five minutes we ought easily to have reached it, but as we did not see it, we shouted. There was no answer, so we went on again for another five minutes, and then for another, but we could see no sign of the camp. As we knew we were now below the right level, we returned on our tracks. I made excursions on either side, as far as the limit of shouting distance of the Zanzibari, but we got back to the stone man without finding any trace of the camp. By taking care and using the compass, we easily found our way to some heaps of ashes and layers of cut bushes, which marked the site of our encampment. It had clearly been deserted for some time. My companion was in despair, while I was greatly alarmed; for I knew Omari would never have moved camp without permission, except under very extreme circumstances. I told the Askari he could come with me at my pace, or follow by himself at his own, and having found the trail followed as quickly as possible along it. It was not always easy to find, for the spongy peat soon closed up and obliterated the footprints.

At length we reached the upper edge of the forests. We knew that if we once lost the trail, amid the woody shrubs on its border, it would be hopeless to find it again, so we took turns at going on hands and knees to track it beneath the bushes. It was not till close upon six o'clock that we heard the report of a rifle. We ran toward it, expecting to find it had been fired by a search party. To our relief we found it was a signal from camp.

Omari came forward at once to welcome me, and express his regret at the illness of the men. He offered no explanation of the change of camping-ground, so I angrily asked for one. It then transpired that my cook, who had been taught writing in a mission school, had put his knowledge into practice by forging a letter in my name, telling the headman to go back, and that I would follow slowly afterwards. I found when I got back to the coast that the man was an adept in the art, for he had scored some successes, and made one failure, in forging cheques in Mombasa. But for Omari's good sense in stopping at the entrance to the forests, I should certainly have spent a night in the open, foodless, and in the pouring rain.

My first impulse was to keep the men in camp here for a few days, while I returned to continue mapping and collecting in the higher zones. But the men were suffering so much from the exposure, that I did not think this fair to them. They nearly all had chilblains and mountain sickness; some of them were badly frost-bitten, and there were two cases of hæmorrhage from the lungs. This was so severe that the milder remedies proved useless, and it only yielded to hypodermic injections of ergot. I was so much alarmed at the condition of the men, that I felt bound to remove them as quickly as possible to a less Arctic climate. Though very loth to leave the mountain, I felt now justified in doing so, as I had accomplished the four main objects of my visit, which were to collect representatives of the fauna and flora of the different zones, determine the geological structure of the mountain, see if there were any glaciers upon it, and especially if these had had at any former time a greater extension than at present.

Omari had already marked the position of the entrance to the path we had made in the ascent. By means of the cut bamboos and barked trees, we found our way back along it. As we had used up our food, we were very lightly laden, and two rapid marches brought us back again to our reserve camp on the sunny steppes of Laikipia.

We had to rest there to nurse the invalids, for it was advisable that every man should be fit for duty before we entered again into the land of the excitable Kikuyu. I had hoped for an excursion to Settima, but this would have taken nearly a month longer, and it would have involved another march through the bamboo forests. My time was nearly exhausted, and I doubted whether any of the men would follow me, while their recent hardships were so fresh in their memories; so I reluctantly abandoned the plan, and promised the men that we would now return to Mombasa as directly as we could.

In concluding this chapter on Mount Kenya, there are two points in connection with the mountaineering to which it may be useful to refer. The first is the warning that the snow and ice of the tropics are not quite the same as in the Alps. Comparing Kenya and Switzerland, on the former the ice is harder; it is more often arranged in alternate layers, one hard and compact, and one friable and spongy; and the snow is more frequently in the powdery or pulverulent condition. The present training of Swiss guides is not conducive to originality, though it makes them very safe in their own immediate district. If such guides be transferred to equatorial snowfields, formed under very different climatic conditions from their own, and there be allowed to apply unchecked the rule-of-thumb methods favoured by the regulations of the "Corporations of Guides" of the Alpine centres, disasters may easily happen.

In the second place, mountain sickness is so intimately connected with the future of mountaineering, that any additional evidence as to the liability to this malady at high altitudes in the tropics is of interest. Very opposite conclusions are held regarding this malady, and somewhat extreme opinions have been expressed on both sides. According to one school, diminished pressure makes little appreciable difference. Thus Mr. Dent,[1] for example, who is an eminent authority both on mountaineering and on medicine, thinks that it need not prevent men accomplishing the ascent of even the highest mountain on the globe. Whymper,[2] on the other hand, regards it more seriously.

My own experience has been that exposure to lower atmospheric pressure means a diminished capacity for steady work. At comparatively moderate elevations the lightness of the atmosphere has an exhilarating influence, and acts as a decided stimulant. But, as with other stimulants, it is followed by reaction.

It is not unless a stay is made for some time above the level of 10,000 or 12,000 feet, and hard work undertaken there, that the effects of the different atmospheric conditions become manifest. No one now notices any appreciable inconvenience at the height of 8000 feet in the Alps; but Mr. Whymper maintains, from experiments in the Andes, that his powers of walking were less there than at lower levels. I did

[1] C. T. Dent, "Can Mount Everest be Ascended?" *Nineteenth Century*, October 1892, vol. xxxii. pp. 606-613.

[2] E. Whymper, *Travels among Great Andes of the Equator*, p. v.

not attempt any such systematic tests as those of Mr. Whymper; but I have no doubt whatever that at the end of a day's march on Laikipia, at about Whymper's elevation, my rifle weighed heavier than it did at a lower level; I was also sooner stopped by shortness of breath when chasing butterflies, and both symptoms made me anxious as to the effect Kenya would have. Recollecting the experiences of Mr. Whymper and his trained guides at the height of 16,600 feet on Chimborazo, I began the ascent of Kenya thinking that, as I could not stay long enough to get used to the diminished pressure, this alone would prevent my rising much above the level of 15,000 or 16,000 feet. I spent, however, five days above the height of 13,000 feet, several times ascended above the level of 16,000 feet, and twice above that of 17,000 feet. During this time anything like undue haste was followed by an attack of loss of breath, from which I recovered very slowly. I had a slight attack of mountain sickness one afternoon, due to over-fatigue. I began the day by chasing some mountain rats (*Otomys irroratus*), and afterwards, though tired by this exertion, carried what was for me a heavy burden up to the camp in the Teleki Valley. This was necessary, for I had only two men available, and as they were unwell I had to share the loads with them. On attempting to work in the afternoon, I felt bilious and had a singing in my ears. Every step upward required a distinct and painful effort. It was not a mere feeling of lassitude, but as if my legs were paralysed, and my left foot numbed. On the descent I experienced less inconvenience, and next day had fully recovered.

I had no return of either symptom during the rest of my stay. On the contrary, to my surprise most of the time I felt unusually well. Probably the interest in the geology of the mountain, and the absence of the worries of camp work, account for my exuberant spirits. The only inconvenience I felt was great liability to fatigue. This was so marked above the level of 17,000 feet that I had to work in spurts; step-cutting was especially trying, and I could only do this for a few minutes at a time. The desire to lessen the toil of the ascent was so strong, that it was continually tempting to run risks, either personally or to the instruments, to effect some trivial saving in trouble. I felt this especially in my second ascent above the level of 17,000 feet, when I was not so well as on the first. The difference was probably due to food. Then I had had breakfast of porridge and beans, and felt traces of indigestion and flatulence. On the former occasion I had had no solid food at all, for it was buried beneath a snowslip, and I had only taken some spoonfuls of Schweitzer's cocoatina.

The only previous occasion when I had any feeling that could be attributed to mountain sickness was during an ascent of Gray's Peak, which, though the highest summit in the Rocky Mountains, is only 14,341 feet high. Owing to the breakdown of our transport arrangements, my sister and I had a very fatiguing march of twelve hours, and spent twenty-eight hours without food, except some bread and meat which had gone bad. At the height of 13,000 feet my sister fainted, and I had a little trouble in getting her back to a deserted mining hut. After she had recovered I continued the ascent, but during the last 300 feet was so weak that I had to crawl part of the way on my hands and knees, and on the summit was too exhausted to put up my camera.

I could never understand why I should have suffered so severely then, while later on, at elevations only a trifle less (as on Pike's Peak and the Tetons), I was only bothered by a feeling of laziness and depression. But when, six months later, Mr. Whymper's account of his illness on Chimborazo was published, I was so struck by the resemblance between our symptoms and his, that I felt bound to accept the illness from which my sister and I then suffered as true mountain sickness.

In a paper read before the Alpine Club I suggested that its occurrence at this comparatively moderate elevation was due to lack of food. The explanation never quite satisfied me. Professor Roy,[1] however, has recently explained Mr. Whymper's illness as due to the fact that some of his tinned meat had gone bad. The same explanation is applicable in our case. The only food we had been able to get in Georgetown was some steak and bread, which we had packed up in the form of sandwiches. On the way up the meat went bad, and of course also spoilt the bread. We nevertheless ate a little, and our subsequent indisposition was therefore probably due to bad food, instead of to lack of food.

[1] C. S. Roy, "Mountain Sickness," *Sci. Prog.* vol. iii. (1895), p. 92.

The effect of exposure at high altitudes on Zanzibari is more important than that upon Europeans in regard to the prospects of future work upon the mountain. The men left in the camp above the forests suffered from biliousness, but I had no opportunity of watching them. Fundi Mabruk at the height of 15,400 feet showed all the symptoms of genuine mountain sickness, and suffered from a severe headache for days afterwards. He frequently told me during the return march what a funny feeling came over him; but this, with the impossibility of boiling his beans, and the freezing of the water at night, he attributed to the witchcraft which possessed the mountain. The other men, however, did not remain long above 11,200 feet, and any symptoms of mountain sickness were masked by the ordinary results of cold, exposure, and imperfectly cooked food. But if the effects of the low pressure had been very marked, they would probably have been recognisable. It is therefore likely that the men might have reached a much higher elevation if they had been properly clothed and fed; while as far as I personally was concerned, I felt no doubt that I could have ascended at least another two thousand feet before being stopped by the rarefaction of the air.

CHAPTER XI

THE RETURN MARCH

"Kurudi nyuma si kazi"
(Returning is not work).
Zanzibari Proverb.

AFTER the decision to abandon the projected visit to Settima we all felt that we were well on the "home trail," and were not in the mood to brook delay; so, when we reached the borders of the Kikuyu country, our request to enter it was somewhat peremptory. We camped at a place called Karati, beside a swamp, about a couple of miles from our former camp. Our "blood-brothers" soon arrived with presents of food for me. They came at once to show me that their ulcers had entirely healed, and to express their gratitude. They promised food for next day. Enough came in to enable us to march southward on the morrow; but a fresh outbreak of illness delayed us. The return to the damp cold mists of the forests of Kikuyu brought back the old lung troubles. For a couple of days the lives of two of the porters hung in the balance, and the time was wasted to me, as constant attendance on the men prevented collecting. Fortunately they recovered, but the next few marches had to be taken easily.

As soon as we were ready to resume our journey we arranged for permission to cross the country of our friends. They insisted on certain religious rites being performed. They said that if we entered their shambas without blessing our way with blood, their crops would yield only empty grain; but if we went as friends, after the proper ceremonials, then we should be a blessing and not a curse, for the sun would shine upon their shambas, and these would yield abundant harvests. I replied that I should be most delighted to do anything that would secure them a gleam of sunshine, and I only hoped it would come soon. I had never yet seen the sun while in their country, as on both previous occasions the land had been wrapped till midday in a kind of Scotch mist; in the afternoon this changed into a drizzle, or rose to form dense, dark clouds. A determination of latitude and longitude in this country would have been quite impossible, even if I had had the instruments, as not once would the weather have permitted an astronomical observation.

I therefore readily consented to buy a goat, which was to be sacrificed according to Kikuyu rites, and the blood to be sprinkled over the path before us as we entered the country. It was arranged that we should start on Monday 10th July. At daybreak we struck camp, and crossed the swamp to the south; it was an especially vile one, but in compensation it had helped to form a defence of our camp. We crossed two ridges of black lava (andesite), and started on a south-south-east course toward a rounded tree-covered hill named Geitaita. After an hour's march we reached a small brook—the Luiji Reru—which formed the frontier of the inhabited district. Here we were met by the old chief, Nathan Nyuki, and a number of elders, at the head of a formidable crowd of warriors. The chief said we must camp here, so that the rites might be duly observed. This I flatly declined to do. He said they had no goat, and must send to a far distant place to fetch one, as they dared keep none themselves for fear of the Masai. I pointed to some fresh goat-tracks beside the path, and asked if they were made by elephants or crocodiles. He seemed not in the slightest degree abashed at being thus caught in a barefaced lie, but said a goat would soon arrive, and that we must wait till it did. A man in the crowd had a reed-rat or *Aulacodus*, much like those I had seen at Ngatana, but smaller and with softer, shorter hair. I purchased it for an empty meat tin and two cartridge cases, killed, and skinned it. As the goat still did not come, I shot and skinned some birds. At intervals of half an hour I went over to the chief and asked for the goat. It was always "coming soon." At three o'clock the chief sent over to say that it was getting late, and I had better camp. I at once went to Nathan Nyuki, pretending to be in a furious rage.

It was my rule never to show any feeling of excitement or emotion before natives. In the stormiest shauris with the Masai I always affected an air of the most stolid indifference as to the result, and supreme contempt for them. As I could not show self-possession by smoking, I generally wrote or sketched during the conference while others were speaking, and sharpened my pencil or knife while I was speaking myself. On this occasion I made an exception. It was very obvious that in this republic of Kikuyuland there were two parties, one of which was friendly to us, while the other was opposed to our passage through the country. It was necessary to overawe the latter section of the natives. I pretended to fly into a violent rage, paced up and down in front of the chief and his elders, called them liars, told them they were trying to play the fool with us, and warned them that this would not do. I fixed a meat tin with a hole in the bottom on a stick, filled it with sand, and said that when all the sand had run through we should cross the bridge and continue our march to the south. I said I could find my way without their guidance, and would fight if necessary. I added that if they compelled me to do this, I should, of course, burn their villages and raid their cattle. In very few minutes our two blood-brothers, both of them sons of the old chief, returned with a goat. I believe they had been waiting in the valley on the other side of the next ridge.

The goat was killed, the stomach cut out, and filled with blood, which was mixed with the contents of the stomach. Nathan Nyuki and the elders stood on the bridge and sang a kind of chant. The chief then said that he was ready to lead us, while the warriors formed up in two lines, one on each side of the path, on the opposite bank. I explained that I also had a religion, and to this religion I was devotedly attached. The first article in its creed was never to put your head into a lion's mouth; the second was always to get out of the way when an elephant was coming to walk over you; the third was never to place a line of your own men between two lines of spearsmen. I pointed out that according to their arrangements this last article would be infringed, and that my religion must be respected as well as theirs. So the warriors were all sent ahead, the chief and I coming between them and the porters, who kept well together, ready for emergencies. The bridge was of the type known in America as "corduroy," except that sheaves of rushes, bamboos, and banana stems were used instead of logs. These were built into a causeway, the stems being placed parallel to the course of the stream, so that the water percolates through the interspaces. We crossed the bridge, the chief sprinkling the blood and the contents of the stomach upon it, and chanting his benedictions. We climbed the ridge and then descended to another stream, the Guaso Uini. Here we entered the shambas, the extent of which surprised us. Had the natives wished they could have smothered us with food in an hour, instead of keeping us for three days, as they did on our first visit. There were also several herds of goats browsing on the slopes, a fact upon which I made some remarks to Nathan Nyuki. From a plateau to the west of Geitaita the chief pointed out to me the hills of Maranga, the point at which we wished to leave the Kikuyu country. I took their bearings through a prismatic compass, and replied at once that he was a wicked old man to tell such lies, for he knew it was nothing of the sort. I only suspected this from his manner, but he was so startled that he at once pointed out some other hills, and said they *were* Maranga. I took their bearings through the prismatic compass, and verified the information by making another elder tell me the name of the hills. From this point I could see the valley of a broad river —the Ilyaini—flowing south-eastward toward the Tana. I thus felt independent of my Kikuyu guides. We marched on till a little before sunset, when I resolved to camp. The chief objected, and urged me to come on farther. We were now approaching a tract of country covered by the woody shrubs which grow up whenever forests are cleared. Once entangled in the narrow tortuous paths through this shrub jungle, we were at the mercy of the natives. So I chose a suitable place for camp, and explained to Nathan Nyuki that the fourth article in my religion was never to march after dark through jungle, when there were Kikuyu in the neighbourhood. The chief said I must not camp in the place I had chosen, but must move to another site a few hundred yards away. There seemed no especial reason for this, so I flatly declined to obey. The chief and his elders sat in a circle round me and grumbled as we watched the men pitch camp. Then they went away, lighted a fire half a mile from camp, and had a great shauri. My blood-brother Iyutha, who was now a firm friend, said he did not know what would happen, for though his father was friendly, all the neighbouring chiefs were opposed to letting us through their country. He warned us to be careful. He told me that the leader of the malcontents was a chief from a hill named Tuntum, which we should see next day, some distance nearer the sunset than we should pass. After the shauri, Nathan Nyuki and several others came over to say that the people did not want us to go on, and that there was to be another shauri next day. We were, therefore, to stay in our present camp till this was over. I was busy at the time and said I could not be bothered to discuss the matter then, but that they were to return at daybreak.

The natives kept up a tremendous hubbub round their fires all night. Several times some drunken warriors came near us, to execute a wild dance near our fires. This was bad manners. No native ought to approach a camp at night, except in a case of urgency. So when another drunken Kikuyu came near and prepared to dance, I welcomed him with a little dust shot, fired with half a charge of powder. He returned to his friends more quickly than he came, and we had no more dancers that night.

Some of the hostile chiefs came into camp at daybreak, and said we were not to go on to Maranga, and must not move till next day. I said I had no objection to the first, but was not going to obey the second. If they preferred it we would go on to Tuntum, as they grew nice pumpkins and fat sheep there. This was the village of the leader of the opposition, who was rather staggered by the suggestion, while the chiefs of the villages along the proper line of march seemed very pleased. This confirmed me in my opinion that we had nothing to fear from the people, and that they were really afraid we should do mischief to them. It was, therefore, only necessary to be very firm in resisting attempts to delay us, and to let the natives see that we had no intention of robbing them. The elders withdrew for a few minutes' consultation, and returned to say that we must not strike camp without permission. Camp was accordingly struck at once, and we waited to see what would happen. I thought that if we were to be attacked it was best that the attack should be delivered before we were on the march. The interpreter, Ramathan, knelt before me, kissed my hand, and begged and prayed that we would not go on. Omari, however, was as angry as I was, and equally convinced that with such a fickle treacherous tribe as the Kikuyu the safest course was to put on an attitude of determined indifference, and bluff our way through. As the old chief Nathan Nyuki did not come, after we had waited about a quarter of an hour for him, I gave the order to lift up the loads and march. Our two guides protested on the ground that they wanted to go home and get food. I promised to supply them with everything necessary, and said they must stay with us. We walked on to the ridge, above the old lake basin beside which we had camped. We skirted some dense jungle, which I declined to enter at any price, and crossed the shambas. There were several furious discussions, conducted with all the passion of an *odium theologicum*, for it was against the religion of the natives for us to pass through their fields, and it was against mine to go through the shrub jungle. I compensated the elder of the nearest village with ten strings of beads, worth three strings a penny, to cover any damage we might do to his crops. As this sacrifice appeased the native conscience, the protests ceased. Two last efforts were made to stop us. A message came upbraiding me for leaving without bidding farewell to Nathan Nyuki, and saying that as he was an old man he could not get up till late. The only reply was the regret that Nathan Nyuki should be so drunk, and the suggestion that next time a white man honoured the country with his presence, the chief should go to bed in proper time. The final attempt was equally ineffectual, though I was more sorry to have to resist it. At one place on the path they had collected as many of the lame and sick of the district as they could hastily assemble. They begged me to cure them. Most of the fifty invalids had bad ulcers, and to have dressed all these would have taken a day, while a single dressing would have done but little good. I therefore had to break another of my chief rules, which was never to refuse medicine while I had any to spare. This, however, was such a manifest imposition that I declined to look at any of the invalids. I was especially sorry not to be able to help some men who were suffering from inflamed eyes. But I could not wait.

We marched all day along the ridge that formed the water-parting between the rivers that drain the southern slopes of Kenya, and those to the west that form the main source of the Tana. The streams to our left flowed into the Ilyaini, and those to the west into the Thegana. Our course was in the main to the south-south-east. Though the trail ran up as well as down, we descended 400 feet during the day. We left the hill of Tuntum about three or four miles to the west. The leader of the Kikuyu opposition, who lived there, looked very much relieved when he saw we did not take the path that led to it. He drank a deep draught of "pombé," rolled off in a state of drunken jollity, and troubled us no more.

We camped on a piece of open heath country in Kithunguli, at the height of 5440 feet. At the village close by, the people were all more or less drunk. Some natives came in for medicine, amongst them a man ill with smallpox. The moment my men saw him they fled, shouting, "Ndui, ndui" (Smallpox). I seized the man by the neck and ran him out of camp. With the aid of a porter named Stahabu, who knew some Kikuyu, and was so much pockmarked that he was also safe, I explained the difference between in-patients and out-patients. The native was told that he was one of the latter, and that if he tried to return to camp he would be shot; if he kept away, I promised to come out and see him again next morning before we started. The rest of the afternoon was occupied in a similar series of evictions. The Kikuyu were all intoxicated, and more than usually quarrelsome. The warriors came rolling into camp, shouting unintelligible cries, and flourishing their "simés," heavy double-edged, somewhat lance-shaped, swords. If one of the men had laid hands on them, he would probably have been cut down at once. Fortunately they were so much in awe of me that I easily ejected them and ran them out of camp. A noisy crowd of women and children were got rid off with little effort. I went towards them to speak to them, and as I did so took off my pith helmet. Apparently they thought I was taking off part of my head, for the whole crowd fled with a shriek of terror. The natives made several of the usual attempts to snatch goods and bolt with them, but the people were so drunk that they were easily frustrated. Next day, as we got farther from the Masai frontier, the natives became more friendly. We continued over undulating uplands, composed of volcanic rocks. The path led by a pretty waterfall, a pond named Kitui, and then the site of a Suahili camping-place named Chanjega. This was situated beside a cluster of trees which the men called "niamba." They probably belong to a species of dragon-tree or *Dracæna*, having small hard fruits, of the consistency of vegetable ivory.

We camped that night at Thiriati, on the southern frontier of the district of Kornu. Here we said farewell to our two guides. They had reached the southern borders of their own country, and dared go no farther. They said the chief of the next district would do his best to send us back; but they hoped we should get through. I gave them presents and thanked them for their help. We had become fast friends. The elder brother Iyutha was very intelligent; he knew the country well, and gave me much information, which proved to be reliable whenever I could check it. He had had a very trying time while with us, but had never once lost his temper. What with the hostile chiefs on one side, and me on the other, he was rather between the devil and the deep sea, and I am not sure how he would have applied the simile. We each bluffed and abused him in turn; but he sat through it with a quiet smile on his face, as if he knew all would come right in the end. He smiled just as serenely when we were chatting together in the evening over the camp fire, as when my clumsy fingers were dressing the ulcers on his legs, or when I was holding my revolver to his head during a threatened attack. I parted from him with greater regret than I did from any other up-country native. As a tribe the Kikuyu are probably as treacherous and fickle as they are represented to be; but Iyutha never forgot he was my blood-brother, or the duties which that relationship entailed. He acted all through with splendid devotion, which amply atoned for the scores of lies he told when under the influence of the counsel of hostile chiefs.

Next day a deluge of rain made the passage of the river Karati very troublesome. We pressed on, however, to Kithu-Uri, the village of the chief of the district of Maranga. Having heard of our approach, he had sent two men to meet us and to spy out who we were. They could not conceal their surprise and contempt at the smallness of our force. The natives, however, were friendly and interested ; they stood in crowds by the path to watch us pass. They had some cattle, the first we had seen since leaving Lake Naivasha nearly three months before. The abundance of pumpkins, and the use of small stools and poisoned arrows, indicated intercourse with the Wa-kamba. The chief, however, at first refused to see me, and his people informed us that he was very angry, for he had a devil inside him. When I succeeded in gaining an interview, he flatly refused me permission to proceed, as he declared that though white men had faces that smiled like the sky, they were bad inside. As no white man had previously entered his country, though four had passed not far from the frontier, this was a rash generalisation ; but at least the latter part of the verdict was true in my case, as I was then very bad inside. Luckily for us, however, the chief was no better. His "devil" was lodged in a tooth, from which it was exorcised by an injection of cocaine. The chief looked pleased, but went away without the slightest expression of gratitude for the eviction of the agent of the evil one.

Our camp here was on the edge of an escarpment at the height of 3900 feet. At the foot of the steep slope to the south were some meadows with only a stream—the Mothambi—between them and the steppes of the Tana. In the meadows we saw a great number of men engaged in a war-dance, and movements that appeared to be a rather complicated military game. They were too far away for me to see the details ; and as I did not know what the assembly indicated, I did not care to go far from camp.

The hyenas here were very impudent. Two of them came into camp during the night and each went away with a mouthful of donkey. The excitement over this, and an attack of fever from which I was suffering, destroyed my night's rest, and increased my longing for the peace of the uninhabited plains to the south.

When the chief came down in the morning to report that he was better, I urged him to allow us to go at once. He first declined, but when I said that if I had to go back the devil would go back too, he changed his mind. He begged for medicine that would protect his people from the Masai. I said that to do this his people need only use more sand and less pombé. With the sand they could keep their spears and simés sharp, while if they gave up pombé they would be brave and strong, and would not quarrel amongst themselves. The chief's entreaties ceased, not, I fear, because he was satisfied with my advice, but because my interpreter gave him some scraps of paper as a present from me of anti-Masai medicine. I was always very particular to disclaim any magical power, as, although a little simple imposition would have helped me forward, it might have hindered those who followed me. But on this morning the attack of fever had left me so limp, that I could not keep sufficiently alert to prevent an infringement of this rule.

The chief gave me a sheep, which was killed according to the native custom, and blood poured into the ground. After this we were free to depart; we hastened down the hill, and across the meadows. We forded the Mothambi, on the banks of which grew some wild date-palms (*Phœnix*), and we worked round the eastern slope of the ridge of Changabuba. This consists of gneiss, the first non-volcanic rock (excluding gravels, sands, and lake deposits) which we had seen since entering the Kapte plains west of Machakos, three months before.

A ford crossed the Tana, due south of our camp at Kithu-Uri, but it was said to be impracticable owing to the swollen state of the river. We were told of another ford, two days' journey farther east, which would be passable before the other; but we were assured that even this was now so flooded, that no one would go as our guide. We found the ford and resolved to force it. The river was as broad as the Thames at Richmond, and the stream was swift. The water came up to my shoulders, and the current was so powerful that only a few of the men were strong enough to carry loads across. A school of hippopotami was playing in the pool above the ford, and added to the excitement by occasionally charging to within twenty yards of us. A sad accident marred our pleasure at effecting the passage. All the loads were over except two or three, and I was dressing on the bank, when a cry from Omari of "Kretasi, kretasi" (The papers, the papers) called my attention to the fact that the bundle of Kenya plants had been dropped into the river. I sprang in at once, and after half an hour's search Omari found the heavy pressing-case wedged between two rocks in the bed of the river. Half the plants were ruined, but I was lucky to have saved the rest. A load of ammunition met with the same mishap, but now that the Tana flowed between us and the Kikuyu, this was of little importance. We ought to have been thankful for having effected the passage with so little loss of time, since both Peters and Dundas wasted weeks beside the river in this district. Peters marched for a month beside it in the search for an easy passage, and, after vainly trying to build a bridge or a raft, had to abandon the attempt to cross the river. We ought not to have risked the ford with the Tana in flood; but I had no time to lose, for to be back in London at the appointed date, it was necessary to catch a steamer announced to leave Mombasa on the 25th of August.

We spent the rest of the day beside the river, drying the soaked plants, a work which had to be continued all night and late into the following day. As soon as the plants were ready we marched southward across the barra for a jagged hill-range, the name of which we afterwards found to be Voroni. The stream beds were all dry, and we began to fear a waterless camp, when we suddenly came to the bank of a broad rapid river, which ultimately proved to be the Thika-thika. This was recorded by Krapf on native information in 1848, but was not seen by any European till Peters crossed it at its junction with the Tana in 1889. Nothing was known about its course. On Peters' map it is called the Dika, and is marked as flowing due east and west, whereas we found it flowing north and south. I resolved therefore to follow it, in order to settle whence it came, and because I hoped thus to collect some geological evidence, which the troubles with the natives had made impossible when in the Kikuyu country.

The next few days were the most delightful in the whole expedition. There were no natives to worry us, and I could again make afternoon excursions alone. The scenery was charming, the air cool, and the rain fell only at night. Game was abundant and tame; hence, though our store of food was nearly exhausted, I was able to supplement it with zebra meat. The river, moreover, acted not only as guide, but supplied us with both water and food, for fish of good size and fine flavour were abundant in it. This march reminded me of Emerson's lines—

"The water-courses were my guide,
The falling waters led me,
The foodful waters fed me."

The river came due from the south, and I feared that if we crossed to the east side, it might suddenly turn to the west and leave us. To avoid losing it, we therefore kept along its western or left bank. It increased in volume, and finally bent abruptly to the west-north-west round the ridge of Voroni. I climbed this ridge and scrambled along its top to examine the country before us. The main result was a scare from a rhinoceros. On my return in the dusk I had taken the wrong ridge,

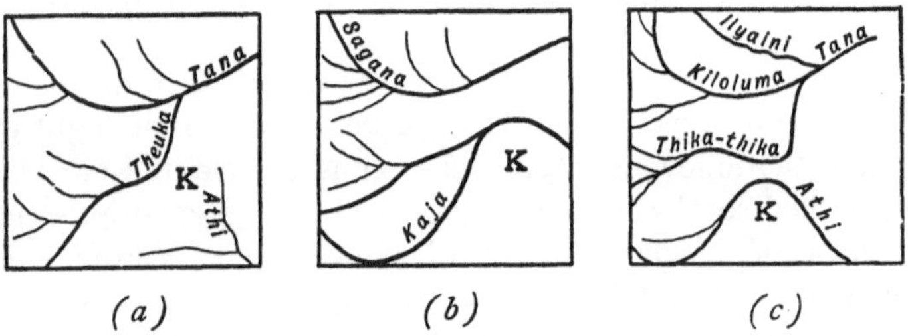

FIG. 6.—Diagrams showing Relations of Head Streams of the Tana and Athi, after (*a*) Ravenstein, (*b*) von Höhnel, and (*c*) the Author.

and when hastening down it to get back to the right one, I almost ran against a rhinoceros. It charged, and I had to climb the rocks again; the result was that I lost myself in the acacia scrub, and did not reach the camp till long after dark.

After passing Voroni the river-bed changed from gneiss to lava, and the country rose in terraces, over which the river plunged in a series of rapids and one grand waterfall about 80 feet high. Then we entered a deep valley, up which we marched until it contracted to a gorge, at the head of which was a series of waterfalls, grander and more picturesque than the lower ones. We climbed the wall to the plateau and found we had now reached the "Kapte" plains, the vast lava prairie that we had crossed on the journey between Machakos and Fort Smith. The hill of Chanjavi rose to the south, and the whole course of the river was now quite plain to me.

We therefore tried to cross it, but found it impracticable. As is so often the case with African rivers, they become larger instead of smaller when traced towards their source. In the barren gneiss desert of its lower course, this river loses so much water by absorption in the sandy soil that, where we first met it, its volume is probably less than a third of that of the point where we tried to cross. Omari, Fundi, and I therefore went ahead to search for a ford. Instead of this we found a party of Kikuyu, or rather they found us, in a way that was not to our credit. They saw us first, and hid in the bush in two parties. We walked straight into the ambush before we saw them. It was gross carelessness, and we felt as much ashamed as we were alarmed. We held up tufts of grass as signs of friendship, and called out "Moratta, moratta" (Friends). When they saw we were alone and meant no mischief, they relaxed their bows, consented to a shauri, and the leader came forward to shake hands. They were a party of Kikuyu out hippopotamus-hunting, as all their crops had been destroyed by the Masai. The great raid, the preparations for which I had seen at Naivasha, had been only too successful. An army of El-Moran had burst upon the district of Igeti, and cleared it like a hurricane. The villages had been burnt, the cattle seized, the crops destroyed, and all the people massacred, except those who escaped to the woods and the Kapte plains. While we were waiting for the porters, the leader of the party told us the sickening story, his men sitting round sobbing. I gave them some wire, and they guided us to a place where some islands had been connected with both banks by some felled trees. While the porters were crossing this, I shot a hippopotamus for the natives, and some birds for a small boy who was trying in vain to kill them with a bow and arrow. I further increased my reputation with these friendly Kikuyu by challenging their best fire-maker to see who could make fire most quickly with a fire-stick. By slipping the head of a lucifer match into the hole in which the stick was twirled, I won easily.

When my men had crossed the river we marched southward, and traversed the almost imperceptible divide between the basins of the Tana and the Sabaki. Dusk found us at the edge of the Athi gorge. Just after sunset we were charged by a rhinoceros; I fired, and the dull thud of the bullet and the animal's grunt of pain told that I had hit it, but in the dark we could not follow it.

Later in the evening, while fishing by firelight in the Athi, I caught something which behaved very differently from any fish that I had ever angled. It was a heavy mass, which yielded gradually when I pulled it. A dark body appeared at the water's edge, and I thought it was a young crocodile. I approached it cautiously, keeping its head covered with my revolver, and found that it was a large water-turtle (*Sternothærus sinuatus*, Smith). The fish-hook was so deeply embedded in its beak that I could not extract it. Yet the animal lived to reach England, where it died a few days after its arrival.

The passage of the Athi next morning was most fatiguing. The ford was broad and deep, and the water bitterly cold. Most of the work fell as usual on Omari, and both of us had renewed attacks of fever. We took huge doses of quinine, drank hot tea, and then started at the head of the men for the mountains of Ukamba. The fever was most unfortunate, for we had used up our last food, and I had to go on ahead to try to shoot some antelope. We saw some game, but my hand was too unsteady; both shots fired missed so badly that I would not try again.

We passed the southern end of Chanjavi, from which we were delighted to see our goal, the mountains of Ukamba, and our old friends the ridge of Lokenya and the sharp cone of Malili. The smoke of some shamba fires cheered us with the signs of the proximity of natives, whom we expected would be friendly, and have abundant food. They did not, however, at first prove very amiable. When we realised, late in the afternoon, that we should not get off the lava plains that day, Fundi and I went ahead to find water. We failed in this, but we discovered a number of natives who were hiding in the long grass. They would not come to us when we called to them, and as they seemed inclined to cut us off from the porters, we thought it advisable to stop till these came up. By the time the caravan arrived it was dark, and we had to camp where we were, though it was very cold, and we had neither firewood nor water. About an hour later a fire broke out to windward, and a long line of flame swept down upon us. By its glare we saw some natives following behind it. We had, of course, taken the precaution of firing the grass before we camped, so that the fire swept past us without any worse consequences than irritated throats and aching eyes. As we had no water to relieve either throat or eye we passed a dismal night. The men were so uncomfortable that at three in the morning they asked if we could not start at once. We did so, and marched until at eight we reached a brook; beside it we rested to quench our thirst, and wash. Soon afterwards we reached a Wa-kamba village, and I asked for a guide. Two men came, but they soon quarrelled over the way, and I had to dismiss them both. We left them fighting, and marched on toward the mouth of the main valley, which was reached at two. Some women came down to sell us pumpkins, and told us that we were in the valley of Kavaluki, and that if we walked on at once we could reach Machakos before dark. The elders came in, and were at first in a great rage, as we ought not to have bought food without first exchanging presents with the chief. This was a fair cause of complaint, so I did my best to apologise, explaining how hungry we were. My soft answer pacified the chief, and he offered to lend me a guide to take me to Machakos. The men were all too tired to go any farther, for they had had no breakfast, and had not my own prospect of a batch of letters to lure them on. So leaving Omari to follow with them next day, with only a couple of Wa-kamba I started over the hills for Machakos.

We crossed the river, ascended the opposite slope, and entered an upland valley that had been converted into a great sugar-cane plantation by an elaborate series of dams and irrigation channels. One of the guides—an old man—would not, or could not, keep up during the ascent, so I sent him back. Just at sunset we reached the summit of the final ridge at the height of 5740 feet, or 1300 feet above our camp in the Kavaluki valley. I could not see the station, and fearing we should not get there that night we started to run. Twice the guide took the wrong path, and we had to return. At last we saw some shepherds sitting round a fire on the hillside, and I asked one of them to act as guide. He pointed to the stars as a sign that it was late, and then led off in a canter. This we kept up till I had twice fallen over the ditches which serve both for irrigation and as boundary marks. After this I said we would walk. At length we reached the valley, and began the ascent of the slope to the plateau on which the station is situated. When we approached it, the guides held back. I cautiously approached, shouting "Mzungu" (European), and keeping sufficiently down the slope to be out of range of fire from the sentry-post. The sentries were startled and puzzled. Presently I called out in Kisuahili that I was a European and wanted Bwana Ainsworth. The excited sentries discussed the unusual situation, the guard was called out, and finally I was told to approach. I walked up into the glare of the firelight, and the guard met me. My guides then came up, and we walked round to the entrance to the fort, where I had a hearty welcome from Mr. Ainsworth. He was not used to visitors arriving alone at night, and my appearance had created a scare.

At Machakos I enjoyed two welcome days of rest, reading letters and newspapers, and having long conversations with Ainsworth about the country and its people. His administration seemed to me more successful than that at any other of the Company's stations. When Ainsworth took over the district there was war between the garrison and the people of the hills, which I had crossed with only a Wa-kamba guide. He succeeded in making peace, and getting into the most friendly relations with the natives. The old chiefs were constant visitors to the fort, and frequently sent in a present of a ton of food. The people were rapidly acquiring a taste for European goods, and learning to work, in order to be able to earn them. The old Suahili garrison was being replaced by Wa-kamba, and Ainsworth was organising a postal service of runners to carry the mails to and from Tzavo, which he expected to prove both quicker and cheaper.

The neighbourhood of Machakos certainly offers better prospects of European colonisation than any other district that I saw in East Africa. In the early days of the history of this station the Wa-kamba acquired the reputation of being hostile, indolent, and treacherous. Ainsworth's administration has shown that, when properly handled, they are friendly, industrious, and faithful. They only need training and protection from the Masai to enable them to make their country one of the richest and most fertile in British East Africa.

At Machakos we were back again on a known road, and my work was done. I resolved, however, on as much variety as possible, and so left the ordinary track and marched southward across the Kapte plains. We entered the Iveti Mountains near the village of Wa-kilome, the residence of the chief of the district of Maka. The chief was a fine old warrior named Kiketi—the Suahili name of a variety of large blue beads. On our way across the Kapte plains we encountered a small raiding party of Masai. As we were in dense bush we met suddenly, and neither party could tell the strength of the other. The one thing Zanzibari apparently cannot do under such circumstances is to stand still. They like to run, and do not seem to care much which way they go. So we charged through the bush and the Masai fled. The party was a very small one, and was apparently trying to sneak into the district of Iuni to steal sheep. So our bloodless victory was nothing to be proud of. However, we could not be sure that the Masai were not the scouts of a larger party, and as Omari and several men were behind with our sheep, we spent an anxious half-hour till their safety was assured.

Two days later we reached the Uganda road again at Nzaoi, and along this we rushed back to Tzavo. The rainless season had now commenced; the swamps through which we had waded on the way up were now dry; water-holes were empty; and we had to march for hours over the hot smoking embers of prairie fires, or rush occasionally through belts of flame.

We were now well on the home trail, and the men needed no encouragement to quicken their pace. We usually marched from midnight or three in the morning till eight, and then from two till five. The stage from Nzaoi to Tzavo occupied only six days, instead of the usual ten.

At Kibwezi I received a warm welcome from Dr. Charters and his colleagues, and a budget of bad news. Mr. Astor Chanler had failed to reach Basso Narok, and was in difficulties. A relief caravan had been sent for, and though we did not then know it, his accomplished companion Lieut. von Höhnel had been seriously injured by a rhinoceros, and was on his way back to Kibwezi for medical treatment. Mr. Dick, with whom I had at one time arranged to travel up to Baringo, had gone off to trade with the Wa-kamba, and had died in a camp beside the Athi, a few days' march away. Edmonds, the last man I had seen when leaving the coast, and who had come with me for two days to nurse me on the way, had also fallen a victim to fever. Watson and Charters had both been so seriously ill at Kibwezi that their faith in its healthiness was shaken. A message had just come up from Mombasa to announce a terrible loss to East Africa—the death of Sir William Mackinnon.

Dr. Charters showed me over the farm, and I was sorry to see that the experiments with the grains and vines had not been very successful. American maize had answered fairly well, but wheat, oats, and barley had failed. The "quick-growing" Russian maize had not grown at all. In a country with such rapid variations between the extremes of excessive rain and drought, the quickly-ripening cereals are the most promising, and the failure of this experiment was especially disappointing.

My kind hosts here were not surprised that I had had trouble at Naivasha, for two Masai boys then living at Kibwezi had told them that an attack had been planned on the powerful caravan of the Railway Survey. The attack was only abandoned when it was found that the expedition included four Europeans and an escort of Sikhs.

After leaving Kibwezi a double march or "telekeza" took us to the Kambu river. We left this at midnight, and at dawn reached a hill known as Ndawi; shortly afterwards we passed our old camp at Mtoto wa Ande. Here a deep well had been sunk; but it was quite dry. It never had contained a drop of water, and it looked as if it never would, except when the stream beside it was flooded. So on the principle of the Englishman in Madrid, who poured some water that was offered him into the Mazanares, as the river needed it more than he did, I emptied my flask into the well as a christening. But the proceeding was not purely ceremonial. My boy had forgotten to boil the water, so, after the few hours' exposure in the sun, it was stinking.

At this point we entered the splendid road then being

made by Mr. George Wilson, at the expense of the late Sir William Mackinnon. The narrow winding path by which we had previously traversed the Motito Wakalia, or "Five Mile Jungle," was replaced by a broad, straight, smooth track. I revelled in the luxury of walking on a good road once more, so went ahead, and marched to Kinani, if Mr. Wilson's mile-posts are to be trusted, at the pace of four and a half miles an hour.

At Kinani I found Major Smith and Mr. Martin with a powerful caravan bound for Uganda. I had a pleasant afternoon with them; they told me more news, and said they had only that day been wondering if I were still in the land of the living. At three next morning they left for the interior and we for the coast. The heat that day was intense. My men collapsed beside the dry water-hole at Ngomeni. I left Omari to look after them, and with two of my best porters pushed on to the Tzavo river for water. I found Wilson there, and he kindly sent some of his men with water for mine. He helped me also with food. I had expected to get some from the station here, but the headman was away and had locked up the store. Mr. Wilson kindly sold me a few loads. All being tired, we rested here for a day. Wilson told me much about his adventures in the country, and of the trouble he had experienced before he could succeed in making the natives cut the road. But he had allayed their suspicions, and a large gang of Wa-taita were now at work upon it. I think Mr. Wilson's road has been one of the most successful industrial missions in East Africa.

At Tzavo I diverged again from the beaten track to follow the Sabaki. Eight days' march through thorn scrub brought us to the freed slave settlement of Makongeni. The country was uninhabited and monotonous in the extreme; some of the men knew it, though at first they denied doing so, and did their best to induce me to return by the ordinary road. The march was rewarded, however, by the discovery of some Permo-Carboniferous beds, and a series of fossils. I found these one evening at sunset, and continued the search far into the night by fire-light.

At Makongeni we entered the Giriama country, a tract of wooded upland, with numerous villages. The population seems, however, to be diminishing fast, and numbers of shambas are going out of cultivation. We passed Kahamisi and Lake Lagobuya, and camped on the eastern slope of the mountain of Mangea, the summit of which I had previously twice seen in the far distance. At Sinikumbe we saw the first cocoa-nut palms and papaws. We passed Sokoki and Poroporo to Fuladoya, and the following day crossed the Mwangudo river, which yielded us some welcome fish, which proved to be a new species (*Chromis spilurus*, Gthr.) We passed Akiluma, and went down the Warabo valley to a deserted mission station at Mwaiba. Here we once more saw the sea. Our last hard march took us to the Methodist Mission station at Ribé, at the edge of the great inland plateau, where I was kindly entertained by the Rev. Mr. Carthew.

We left Ribé next morning for the final march. We descended the hill slope that led to the "temborari" or coastal plain, passed the three round hills known as "Coroa Mombaza," or the Crown of Mombasa, and reached the estuary. As the tide was out we marched along its bed toward Freretown, halting only to collect the cartridges from the men, lest a salute should have undesirable consequences. I had to get money for my men's food allowances, and it being Saturday I hurried on, to try to reach Mombasa before the bank closed. At the beach at Freretown a boat was ready, and two stalwart Suahili rowed me rapidly down the harbour to Mombasa, and landed me at the quay.

A great change seemed to have come over the town during the past five months. The British East Africa Company was in a state of suspended animation, and its condition seemed reflected in that of its chief town. In March Mombasa was a bustling mart, full of life and activity. The harbour was then crowded with dhows; the quay and streets were noisy all day, with people of many nationalities. Native Suahili, Soudanese and Somali soldiers, Beluchi and Arab traders, Hindoo Banyans and their Goanese clerks, were all busy in connection with the safaris that were continually coming or going. But now all was quiet, the harbour was empty, the shops and sheds closed, the quay deserted. As I went through the town to try to find some one, I saw only a few boys. I found Hobley at lunch, and he returned to the Transport Office to receive my luggage and distribute rations, while I went back to the terrace above the harbour, to watch the men unload the dhow by which they had crossed from the mainland.

This was the last act in the expedition, and I watched it with mingled feelings. On the one hand, I was glad of the prospect of rest and peace, and relief from the anxiety lest the collections and note-books should after all be lost. But, on the other hand, there was much that I parted from with regret. The active life, the constant change of scene, the novelty of the experiences, the excitement of exploration, and the interest of the geological problems, had stored the past five months with pleasant memories. But beyond this there was a keener grief. I had now to part with my men. The expedition had offered abundant sources of friction. The men had been overworked, and when Hobley asked Omari how he had got on, the latter expressed the general opinion by his reply: "Vema, illa teli khazi, *teli* khazi" (Very well, but lots of hard work, very hard work). Then the Zanzibari strongly object to rush; one of their favourite proverbs is "Haraka, haraka haina baraka" (Hurry, hurry has no blessing). But our caravan had gone at a record-breaking pace. On my return I found that Dr. Moloney claims the record for a journey of this sort, for the Stairs expedition to Katanga, which travelled 1080 miles in six months all but ten days. We went 1650 miles in two days less than five months. The porters are used to a warm, damp, equable climate, such as that of Zanzibar, where the extreme annual range of temperature is only 10° F.; but on Laikipia they had been exposed to a daily variation of fifty degrees, while the climate of the higher camp on Kenya was of a severity of which the natives had never even dreamed. Then I had been relentless, and insisted on the caravan going on and on, stopping for neither rain nor flooded rivers, hostile tribes nor fear of famine. The men occasionally had complained, and some of them would have been glad to do more. So we had had our little quarrels; the men had grumbled, and my temper had not been of the sweetest. But the memory of these occasional disagreements sank into insignificance, in comparison with the long record of ready obedience, willing self-sacrifice, and personal devotion. Now that the last farewells had come, I realised that the impression stamped most deeply on my mind was not of peril or privation. Recollections of these had been blotted out by regard for the men who had braved danger and hardship, who had injured their health in long marches across the waterless wastes of the Nyika, and by exposure to the blizzards of the Kenyan snowfields, and who had worked all through from a simple instinct of duty, suffering for objects they could never understand.

PART III

EASTERN BRITISH EAST AFRICA

"The Land of Light and Liberty."

Motto of I. B. E. A. Co.

"Nature is cruel, man is sick of blood;
Nature is stubborn, man would fain adore;
Nature is fickle, man hath need of rest."

MATTHEW ARNOLD.

CHAPTER XII

THE PHYSICAL GEOGRAPHY AND THE GEOLOGY OF BRITISH EAST AFRICA

" Hier aber war's ! Plutonisch grimmig Feuer
Aeolischer Dünste knallkraft, umgeheuer,
Durchbrach des flachen Bodens alte Kruste,
Dass neu ein Berg sogleich entstehen musste."

GOETHE.

IN 1852 Sir Roderick Murchison advanced the hypothesis that Africa, south of the Sahara, was a continent of great antiquity and simplicity, which had maintained the form of a great basin ever since the age of the New Red Sandstone. Murchison based his theory on the discoveries of Bain, the pioneer of South African geology ; but he drew support for it from the probability that Lakes Ngami and Tchad, at the two ends of the supposed basin, were connected by others reported by classical traditions and modern traders. Murchison regarded these lakes as the remnants of a series which had existed uninterruptedly throughout two of the three great eras of the earth's history.[1]

This brilliant speculation was reaffirmed after the discoveries of Livingstone,[2] Burton and Speke,[3] and Speke and Grant,[4] and was finally summarised in 1864 in the presidential address to the Geographical Society, and in a paper entitled " On the Antiquity of the Physical Geography of Inner Africa."[5] In these he claimed the country as of interest, because it was

[1] R. I. Murchison, President's Address, *Journ. Roy. Geog. Soc.* vol. xxii. (1852), p. cxxiii.
[2] *Ibid.* vol. xxvii. (1857), p. clxix.
[3] *Ibid.* vol. xxviii. (1858), p. ccviii.
[4] *Ibid.* vol. xxxiii. (1863), p. clxxxi.
[5] *Ibid.* vol. xxxiv. (1864), p. clxxxvii. and pp. 201-205.

"geologically unique in the long conservation of ancient terrestrial conditions. This inference is further supported by the concomitant absence throughout the larger portion of all this vast area, *i.e.* south of the Equator, of any of those volcanic rocks which are so often associated with oscillations of the *terra firma*."

For a while every addition to our knowledge of the geology of the equatorial region seemed to confirm the truth of this hypothesis by showing that it consisted of one solid block of gneiss and schist. These rocks were proved by Speke to form the basis of the central basin; they were found to the east of this by Thornton (1862), Gustav Rose (1863), and Roth (1864); to the north and north-west by Baker, Schweinfurth, Emin, and Junker; to the south by Livingstone, Burton, and Speke; and to the west, in later times, by Wolf and Wissmann.[1]

That part of Murchison's theory, which affirms that Central Africa has never been below the level of the sea, is still in harmony with the known facts, for no deposits of marine origin have as yet been found in the interior. The sedimentary beds, found by Speke and Stanley to the west of the Victoria Nyanza, and recently described by Cornet in the Upper Congo, have yielded no fossils to indicate the conditions under which they were formed. It was hoped, however, that the deposits of the long series of lakes would contain the bones of the land animals, that lived on the plains around them; just as the South American pampas have supplied those of giant sloths, the United States prairies those of horned reptiles, the Karroo of

[1] The five chief contributions to our knowledge of the geology of British East Africa are those of Thornton (1862), Beyrich (1878), and Sadebeck (1879) for the stratigraphy of the coast region; of Mügge (1886) and the joint memoirs of von Höhnel, Rosiwal, Toula, and Suess (1892) for the petrography and physical geography of the interior. Much incidental light is thrown on the country by the works of Blanford (1870) and Baldacci (1891), and of Ebert (1887) and Cornet (1894), of whom the two first describe Abyssinia, and the two last German East Africa and the Upper Congo. The petrographical studies on rocks from Kilima Njaro by Rose (1863), Roth (1864), Bonney (1886), Hyland (1889), and Tenne (1890) must also be noticed. Information as to the range of the various rocks, used in the construction of the map, occurs in the records of most travellers who have written on the country, especially in those of Speke, Baker, Stanley, Stuhlmann, Baumann, and Scott Elliot for the Nyanza region; of Hildebrandt, Thomson, Piggott, Lugard, and Hobley for the eastern half of the country. Gibson's short account (*Proc. Brit. Assoc.* 1893, pp. 758, 759) of his traverse from Mombasa to Uganda, so far as it goes, is admirable. It did not, however, seem advisable to burden the present chapter with references to this literature, for it is only in the case of the Archean rocks and the coast deposits that previous results have been incorporated. Reference to Suess's important memoir is given in the Introduction.

uncouth lizards, the Siberian tundras of buried mammoths, and the Weald of Kent of bird-like saurians. The lake deposits of Equatorial Africa might have been expected to be especially prolific in extinct animals, for the existing fauna is much richer than those of the regions enumerated. Nevertheless, so far they have proved barren. The conclusion has therefore been gradually accepted that Tropical Africa is geologically uninteresting, for as a continent it has no history, and it has no message as to the development of life on the globe.

A quotation in the Introduction (p. 2) shows that the old view of the simplicity of the geology of Tropical Africa has lasted as late as 1891, but in the next year Professor Suess's memoir threw a special interest over the region adjoining the Great Rift Valley. This, however, did not contradict Murchison's brilliant guess as to the antiquity of the continent, and I was not surprised when the four months on the coast-lands and the first month of the march inland showed nothing of any special geological interest. But nearer the Rift Valley the conditions are different, and the region presents a combination of features which elsewhere can be paralleled only, and that imperfectly, in the Western States of America.

Plateau Eruptions.—The first of these special types of structures was seen the day after leaving Machakos, when we reached a tract of country marked on the best existing geological sketch-map as recent alluvium. It was part of the district known as the Kapte Plains, and I expected to find this to be an old lake basin or a desert of wind-borne drift. But on reaching the summit of the last ridge of the Iveti Mountains, we found before us a vast expanse of undulating prairie, stretching away to the western horizon. The rock of which this consisted ended abruptly against the flank of the old gneiss ridge on its margin, but it ran up the valleys and into the hollows of the mountains, just as the water of a lake follows the irregularities of its shore. So much did this view remind me of that across the great Snake River lava fields of Idaho, when seen from the range of the Tetons, that I felt sure at once that this was a plain of lava and not of alluvium. I hastened down to it, and the inference was confirmed. The resemblance between this plain and the American lava sheet became still more apparent as we marched across it, and also when, on the return journey, we traversed it farther to the north. As this plain was therefore lava and not alluvium, it took its place as part of the great volcanic series of Africa, the approximate extent of which is marked on the accompanying sketch-map (Fig. 7). An inset map of Southern Italy on the same scale shows the insignificance of the lava flows of Vesuvius and Etna in comparison with those of Eastern Africa. The latter cover so vast an area, that it is useless to compare them with the flows from ordinary volcanic vents; they must be due to the same method of eruption as that which discharged the rocks of the great lava sheets of America and the Deccan Traps in India. The extent of these is equally enormous; the latter cover an area of 200,000 square miles, and those of the Western States of America have been estimated to occupy a tract of country as large as Great Britain and France combined. Moreover, these vast lava fields have been described as free from the tuff and ashes which form the larger part of the material ejected by volcanoes. Hence, for two reasons, it seemed hopeless to explain their formation by the ordinary type of volcanic action. Baron von Richthofen therefore proposed the theory, which was subsequently supported by Sir Archibald Geikie, that these lava seas were discharged from subterranean reservoirs through fissures, possibly hundreds of miles in length, instead of through simple circular vents. That cracks may form for great distances across a country and be filled by igneous materials is proved by the occurrence of dykes, which cut across northern England and Scotland; that the contents of such fissures occasionally reach the surface is known from observations in the Sandwich Islands. It was obvious at once that these Kapte plains were older than those of North America, and there was accordingly more chance of finding exposed upon them some of the supposed fissures. A ride on the great lava sheets of the Snake River of Idaho in 1891 had temporarily shaken my faith in the fissure-eruption hypothesis, but it had not destroyed it, and I therefore left the Iveti Mountains for the plains expecting these would yield important evidence in support of the theory. The rocks have been extensively denuded; the Athi has cut a deep gorge through the plain, but nowhere could I find any sign of the fissures. In places many small dykes occur, but their existence

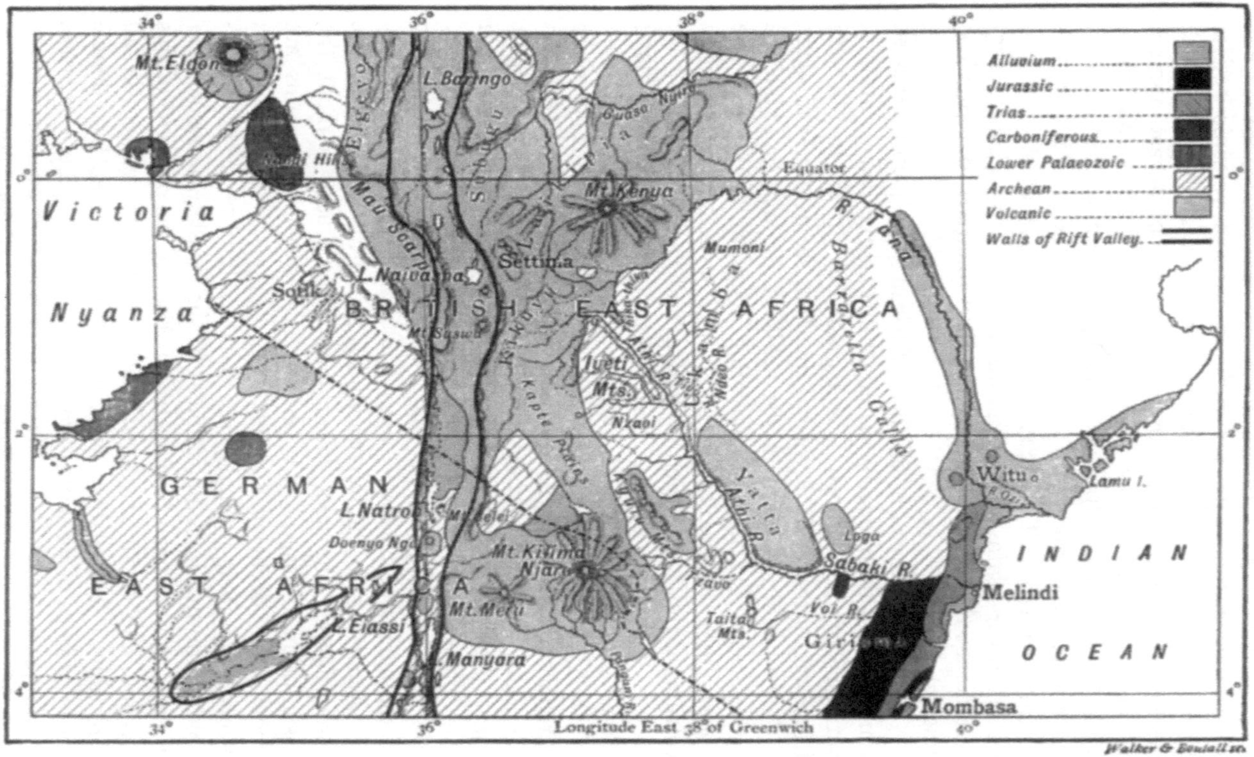

FIG. 7.—Geological Sketch-Map of British East Africa.

is not inconsistent with the ordinary type of volcanic action. Moreover, beds of volcanic tuff and ashes are abundant, and all the evidence points to the former occurrence of a vast number of small craters scattered over the area. Later on, when marching across Laikipia, we passed over another lava plain of more recent date, and there found dozens of the denuded stumps of small craters. That craters may also have existed on the Snake River plains is not improbable, for several facts suggested that these were older than I had inferred from previous descriptions.

Hence, as the study of these African lava plains revealed abundance of fragmentary materials due to explosive eruptions, but none of the fissures, I was gradually driven to abandon Richthofen's theory in favour of one that seems better to agree with the conditions seen in the field. This may be explained by reference to the accompanying diagram (Fig. 8). The upper figure represents an ideal section across an ordinary volcanic crater in a region of folded and contorted rocks. In such a case lines of weakness will traverse the country along the axes of the folds. If beneath the area there be a mass of rock, either molten or ready to melt when the pressure is relieved, then this will force its way to the surface through a vent, and by a series of explosive eruptions build up a volcanic cone. According to the fissure-eruption hypothesis, the strain of this subjacent lava is supposed to tear open a crack or fissure instead of a single short vent; and, owing to the great size of the aperture, the rock quietly wells forth without explosions.

According to the hypothesis suggested in the lower of the three figures, the lava sheets may be produced without the assumption of fissures. Wherever these great lava plains are known, they occur on high plateaux composed of rocks which either retain their originally horizontal position, or are of remarkable uniformity in composition. If such a plateau occurs over a great subterranean lava reservoir, then, as the contents of this expand, the rocks above will be subjected to tension in all directions. The plateau will therefore be traversed by a double series of lines of weakness crossing one another like a network. The intersections of these lines will be points of equal weakness, and when the pressure from below is sufficient to force an opening to the surface by one of these, many others will give way simultaneously. Thus, instead of one big vent, there will be numerous scattered small ones, the flows from which will coalesce into continuous sheets of lava. These may therefore be better described as Plateau Eruptions rather than as Fissure Eruptions.

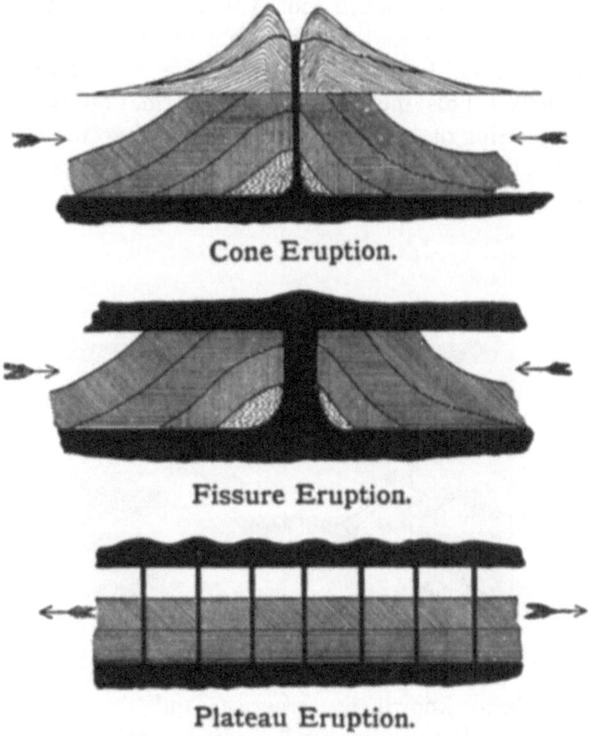

FIG. 8.—Three Types of Volcanic Eruptions.

Rift Valley.—After leaving the Kapte plains we came upon a second interesting type of structure. The valleys we had previously seen were constructed on the same plan as those of England; for their courses were sinuous and their slopes rounded, since they were made by the familiar processes of denudation and erosion. But, on emerging from the Kikuyu forests, we entered one which was straight in direction, and was bounded by parallel and almost vertical sides; its characteristic features were that its lines were straight, and that its angles retained some of their original sharpness, for the direct action of faults and earth-movements still dominated the scenery. An hour after entering this valley, we reached the edge of the Great Rift Valley, which, like the former, must be directly due to earth-movements. Once the plateaux of Mau and Kikuyu were continuous across the site of the Rift Valley; a double series of north and south cut through the plateaux, and allowed the block of material between them to subside. This left a great open Rift Valley (or, to use Prof. Suess's term, a "Graben"). This method of valley formation is illustrated by Fig. 9; strips of country have fallen owing to a series of parallel cracks or "faults," and thus a valley has been formed with precipitous, and sometimes step-like sides. Such valleys

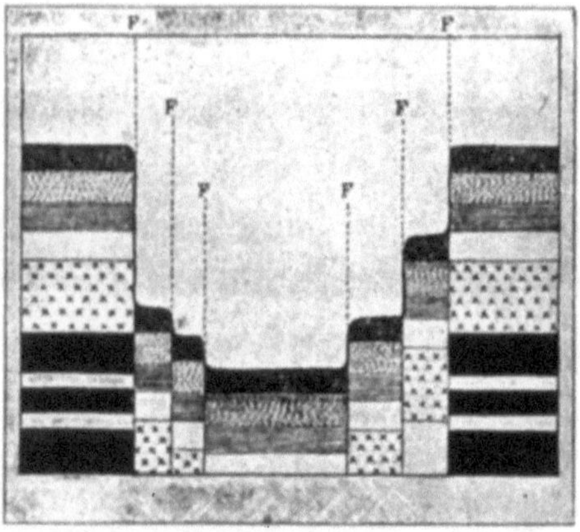

FIG. 9.—Section across Rift Valley. (F = Faults.)

have long been known in America, and the extraordinary steepness of their bounding walls may be seen in photographs of the Yosemite Cañon in California.

Block Mountains.—A third feature in this region of Africa is mountains constructed on a different plan from those typical of Europe; there they are formed by actual elevation owing to the intrusion of igneous rocks or to the folding of beds once laid down as horizontal sheets of sediment; some of the mountains of the Rift Valley, on the other hand, are formed of layers which are still horizontal (as in Fig. 10); each mountain consists of a huge block of material[1] that has been left standing, while the rest around it has fallen to a lower level.

[1] A "Scholl" in the terminology of Professor Suess.

Thus, instead of there being no new type of structure in this region, and the geological facts being of wearisome monotony, it teems with novel problems, and all its conditions seem different from those of Europe. For the valleys are often due to rifts instead of to erosion; the mountains are sometimes formed of blocks instead of by folds; while the lava flows are on a scale that shows the impossibility of measuring the universe by European standards. There was yet a fourth difference which was especially noticeable, as it afforded great aid in the study of the geology of the district. England is an old-fashioned piece of the world; its physical features were determined long ago. Most of the inequalities caused by dislocations and faults have been levelled,

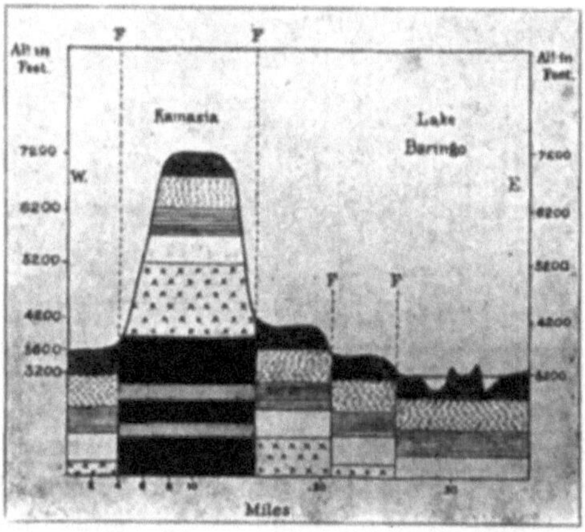

FIG. 10.—Section across a "Block Mountain." (F=Faults.)

and those that remain are mainly due to denudation and erosion. But in Africa it is not so. Great earth-movements have happened so recently that rock scarps, 1000 to 2000 feet in height, still stand bare and precipitous as though formed but yesterday, and straight lines and sharp angles still dominate the scenery. The recent date of such earth-movements has therefore rendered the physical features of the country such a direct expression of its geological structure, that this can be recognised in a hasty traverse.

So intimate is the connection between the physical geography of British East Africa and its geology, that it is convenient to refer to the former before attempting to summarise the latter.

The Geographical Zones of British East Africa.—In the *Geographical Journal* it was pointed out that British East Africa may be considered to consist of seven zones or belts, running north and south approximately parallel to the coast.[1]

The first of these is the coastal plain, known to the Suahili as the "Temborari." In some parts of East Africa this is a wide tract of malarial country, but in most of the British dominions it is narrow, and can be traversed in one or two marches. It consists of a low-lying plain, formed of raised coral reefs and old sea-beaches, and is generally covered by a soil formed of wind-borne sand. On the seaward margin there is a line of dunes, which are sometimes 150 feet in height. This zone is deeply indented by estuaries, which, as they are followed inland, are found to branch repeatedly; some of the

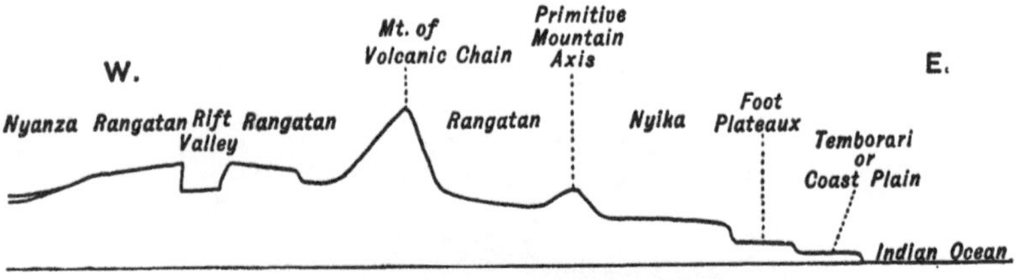

FIG. 11.—Section across British East Africa.

small creeks unite with those of an adjoining estuary, and thus islands are cut off from the mainland. The shores of the estuaries are mud flats, on which grow dense jungles of mangroves. The flats are uncovered at low tide, and the decomposition of the vegetation and dead marine animals left upon them causes the malarial exhalations which are the bane of the coast.

The next zone—the Foot-hills—is also narrow, and is entered to the west of Mombasa at Chamgamwe, at the summit of a steep slope that rises 200 feet above the shore. From this point there extends westward an undulating plateau covered with woods, groves of palms and plantains, orchards of mango and papaw, and fields of dry rice, maize, and dhurra. This district is the granary of Mombasa, as it was in the fourteenth century at the visit of Ibn Batuta. The foot plateau is not

[1] J. W. Gregory, "Contributions to the Physical Geography of British East Africa," *Geog. Journ.* vol. iv. (1894), pp. 293-297.

present throughout the whole of British East Africa. In some places a steep ascent of 800 feet or so leads from the coast plain to the edge of the broadest of the East African zones—the great Nyika.

Nyika (or U-nyika) is the Suahili word for the country of the tribe known as the Wa-nyika, and it appears to mean either "wilderness" or "lifted up."[1] It has been gradually adopted for the whole of the great scrub-covered plains of East Africa, of which the Wa-nyika occupy only an insignificant proportion. The Nyika, in this wider sense, extends southward from Somaliland, opposite Aden, across the Italian, British, and German dominions in East Africa, its continuity being broken only by the valleys of the six principal rivers, the Shebeyli, Juba, Tana, Sabaki, Rufigi, and Rovuma. The peculiar scenery of the Nyika has often been described, and a good idea of its aspect can be obtained from an exquisite photograph published by Paulitschke.[2] The soil of this zone is sandy, and usually stained bright red by oxide of iron. It is porous, and water is scarce except in the rainy season; there is no turf, for the grass is dry and grows in scattered tufts. The higher vegetation (see pp. 286-287) consists of loose acacia scrub, and of plants and trees with succulent stems and spine-shaped leaves, such as the Birthwort and the Spurge.

Rising through the Nyika, or occasionally forming its western boundary, is the next zone, the "primitive mountain axis" of East Africa. This is part of a chain composed of very ancient rocks, which once formed the backbone of the continent. It probably extended from the Drakensberg of Natal to the mountains of Abyssinia, and possibly to the Ababd Mountains of Egypt and the central peaks of Cyprus. Its original continuity has been destroyed, as it was breached by earth-movements and denudation, and it has been in places buried under vast piles of volcanic materials. But the old rocks continually reappear upon this line, as a series of hog's-back-like ridges running from north to south. The Bura, Taita, and Ongalea Mountains, the numerous "Bare Hills" that rise above the Nyika in the Ukamba country, Bwinzau

[1] The etymology of the word is considered in *Geog. Journ.* vol. iv. (1894), p. 296.

[2] Ph. Paulitschke, *Beiträge zur Ethnographie und Anthropologie der Somâl, Galla und Harari* (1886), pl. xiii.

near Kibwezi, the Iveti Mountains, the gneiss ridges of Ithamba and Changabuba, Doenyo lol Daika, and the Loroghi and Matthews Mountains, are the remnants in British East Africa of this ancient mountain axis.

Proceeding still farther to the west, we come to the next zone—broad plains of volcanic rocks, which are the great grazing lands of the nomadic tribes, and for which we may adopt the Masai name of "Rangatan." The largest of these is the "Kapte Plain," which extends from near the base of Kilima Njaro, northward to the upper valley of the Tana. Beyond this is another Rangatan, forming the high plateau of Laikipia; farther to the west is Rangatan Nyuki, or the "Red Grazing Land," probably so called from the colour of the river which flows through it. These Rangatan may play an important part in the future of British East Africa, for, as their soil is fertile and retains moisture well, and as the climate is cool and bracing, they offer the best sites for European colonisation.

The Rangatan are formed of sheets of volcanic material, which occurs also piled into lofty cones. The largest of these run along a line from north to south, forming the "Volcanic Chain." The highest mountain of this series is Kilima Njaro in German East Africa, but the largest is Kenya, 280 miles to the north. Between the two occurs the line of craters of the Kyulu Mountains, while the series is continued to the north by Mounts Loldibo and Kulall, which link it to the lofty volcanic peaks of Abyssinia.

The last zone was originally also a highland, but it is now broken by a series of north and south faults into the "Great Rift Valley."

The Geological Basis of the Seven Zones.—When we turn from the superficial features of the country to its geological structure, we find that the seven zones are all different in composition. The coastal plain consists of raised coral reefs, marine sands, shell beds, and recent alluvium, with an occasional outlier of deposits belonging to the next zone.

The foot plateau consists of shales and sandstones referable to the middle period of geological history, though including some carboniferous beds belonging to the latter part of the previous era.

The Nyika and the primitive mountain axis are both

formed of crystalline rocks (gneisses and schists), which occur in the former as level plains, and in the latter as hog's-back-like ridges, composed of the harder parts of the series. The Rangatan and the volcanic chain are both volcanic, one being formed of sheets of lava due to "plateau eruptions," and the other of piles of ashes and lava as in ordinary craters.

The floor of the Rift Valley is occupied by more varied materials than either of the other zones, for we find upon it ancient and modern lavas of various ages, the alluvium of dried lake basins, recent river gravels, and deserts of loose drifting sand.

The Geological Sequence.—Before attempting to show the relations to one another of the different rocks which form these zones, it is necessary to choose a simple standard of comparison. A convenient method is to refer the African rocks to their position in the sequence observed between Holyhead and London. This begins at the former town with some schists, and next reaches a ridge of "gneiss," which is one of the oldest rocks known to the geologist. Passing thence to the south-east, the traveller continually moves from older to newer deposits. Thus after crossing the Menai Strait the Archean gneisses and Cambrian slates are left behind, and to the south can be seen the summits of the extinct volcanoes of Snowdon. Continuing eastward, still younger rocks belonging to the Silurian period[1] are seen, followed by those of the Flintshire coal-field. Here the systems belonging to the "era[1] of ancient life" (Palæozoic) are left, and a few miles before Chester the railway reaches the New Red Sandstone or Trias, which is traversed till three miles before Rugby. There the sandstones disappear below some clays, and the route crosses an alternation of ridges of limestone and valleys of clay, belonging to the period of the Oolitic Limestones or Jurassic. These in turn are succeeded by the rocks of the Cretaceous period; of these the best known member is the Chalk, which can be seen on both sides of the railway, as it passes along an old river valley cut through the Chiltern Hills. At Watford the Chalk, the latest English representative of the "era of

[1] The terms used for divisions of time in the present chapter have a somewhat technical meaning: thus, an epoch is a subdivision of a period, and this in turn is part of an era.

middle life" (Mesozoic), is seen for the last time, and the rest of the journey is over "London Clay," deposited during the Eocene. The next two periods, the Miocene and Pliocene, are not represented in this district, and we pass at once from the London Clay to the gravels of the "Pleistocene." In some of these, formed by the Thames when it flowed to the north of its present position, we find the flint implements made by early man.

Thus, in a journey from Holyhead to London, we may see a sequence of deposits which includes representatives of most of the great periods in the earth's history. The following table summarises this sequence, marking the rocks which belong to the different divisions in England, and those which, as we shall see, represent them in British East Africa.

Era.	Period.	English Representative.	East African Representative.
Cainozoic or Era of Recent Life	Pleistocene	Neolithic gravels	Gravels with Neolithic Implements.
		Paleolithic ,,	(?)
	Pliocene	(?) Hill gravels on summit of Chilterns	Deposits of Lake Suess, etc.
	Miocene	Unrepresented	Basalts of Plateau Eruptions.
	Eocene	London Clay	Deposits of Lake Kamasia.
Mesozoic or Era of Intermediate Life	Cretaceous	Chalk Hills of Chilterns	Older Plateau Eruptions.
	Jurassic	Limestones of Northamptonshire	Clays with fossils near Mombasa.
	Triassic	New Red Sandstone	Magarini Sandstones.
Palæozoic or Era of Ancient Life	Carboniferous	Coal Measures of Flintshire	Sabaki Shales.
	Devonian	Unrepresented	Karagwe Series.
	Silurian	Sandstones of Denbighshire	
	Ordovician	Volcanic rocks of Snowdon	
	Cambrian		
Archean (Fossils unknown)	...	Gneisses and Schists of Anglesey	Gneisses and Schist

The Geological History of British East Africa.—In British East Africa there is no long succession of fossiliferous deposits, as in England. In 1892 the only fossils recorded from it were a few plants, ammonites, and reptile bones from the hills of Mombasa, and a few sub-fossil shells of living species. The only rocks to which any definite age was assigned were those from which the above fossils came, and the gneiss, which was regarded as Archean.

In spite, however, of the absence of a definite series of fossiliferous deposits, it is possible to determine the relative ages of the rocks in East Africa, and to find among them representatives of most of the principal divisions of the geological sequence.

The earliest era of the earth's history is the Archean, which is represented in Anglesey by a ridge of gneiss and by schists. Rocks of the same kind underlie the whole of British East Africa, and form two-thirds of its surface. Three days after leaving Mombasa the traveller finds these rocks on the plains near Taro, and continues upon them for 230 miles until past Machakos. Here they sink beneath sheets of volcanic rocks, from which they rise to the west, forming the whole of the Uganda plateau and the Nandi Hills.

The rocks of the Archean series may be divided into two groups, the Upper and the Lower. The former occurs in a part of the Iveti Mountains between Zuni and the pass of Kwazome, and has been found by Mr. Scott Elliot on the flanks of Ruwenzori. It is composed of a series of schists—banded rocks which split into flakes like slates. These vary in colour in accordance with the mineral which predominates in their composition. Some are silvery gray, as they consist of mica and quartz ; others are dark green, as they contain hornblende instead of mica. The Lower Archean series consists of gneiss, in which occurs bands of an irregular coarse-grained rock of light colour (pegmatite), and of fine-grained fissile rocks which are dark green or black.

It is not easy to decide as to the exact origin of this Archean series. The gneisses were probably formed from various rocks, some of which were igneous and some sedimentary. They have, however, all been completely altered by burial deep below the surface, whereby they came under the influence of the internal heat of the globe. They were thus rendered crystalline, and their original characters obliterated. The gneisses contain two sets of rocks, which were probably igneous in origin, and were forced into the gneiss in a molten condition. When the main series was altered, these igneous intrusions underwent the same change. One set consisted of so-called "acid" rocks, as they contained much quartz; these have formed "pegmatite" veins. The other set was poor in quartz, but rich in such materials as iron and magnesia, and thus had a more complex chemical composition; they therefore underwent more marked alterations, and were crushed into a series of dark-coloured banded schists in the way so clearly described by Teall in the case of the Scourie Dyke in Sutherland. These rocks, therefore, carry us back to a very early period in the earth's history, for they doubtless date from the Archean era, and are of approximately the same age as the gneiss of Anglesey. Though their interpretation is difficult, they tell us that Africa was originally a country where igneous activity prevailed, and they remind us that all the rocks, now exposed on the surface, were once deeply buried underground.

After leaving the Archean system there is a gap in the geological record, and we have no certain knowledge of the condition of Equatorial Africa during the first four of the five ages of the era characterised by the "ancient forms of life." Deposits formed during this interval are known, such as the Karagwe series, to the west of the Victoria Nyanza, and the Lualaba series of the Upper Congo basin. But they have not yet yielded any fossils; and as the shales and sandstones of which they consisted have been crushed into slates or consolidated into quartzites, they give no certain knowledge of the condition of the country at the time of their formation.

The record does not begin again till the upper part of the Carboniferous period, when England was covered by the swamps and jungles, in which grew the vegetation that has formed our coal-fields.

In the hills to the west of Mombasa there are some sandstones in which occur the famous water-holes of Taro. Thornton found a few plant impressions in these beds, and on this evidence the sandstones have been assigned to the Carboniferous. However, in the Sabaki valley I obtained more definite evidence of the Carboniferous age of some of the deposits, by finding a series of fossils containing some fish scales, and some mollusca determined by Professor Amalitzky of Warsaw and Mr. G. F. Harris as *Palæanodonta fischeri* (Amal.) These show that at this period large freshwater lakes occurred in the valley of the Sabaki.

The next period in the earth's history was the Trias, during which quaint reptiles lived at the Cape in the region of "the baked Karroo," and a great continent extended across the Indian Ocean from the Cape of Good Hope to India. This probably extended northward as far as Western Russia, for one freshwater fauna spread throughout this region. In Eastern Africa a band of brilliant red sands and sandstones occur near the coast behind Mombasa, and thence extend northwards to the hills around the mission station of Ngao on the Tana. These can be especially well seen in the hills above the plantations of Magarini, and I therefore propose to call them the "Magarini Sands."

There is no direct fossil evidence as to the age of these beds, but as they appear to overlie the Carboniferous rocks, and are in turn covered by the next series, they are probably Triassic. They are no doubt a desert sand, and, though here and there the action of rivers and streams can be traced within them, they indicate the existence of an arid climate toward the close of the New Red Sandstone period.

After the formation of the last of the Magarini Sands the land subsided, until part of the coast region was below the sea. The continent over the Indian Ocean was broken up, or at least indented, by gulfs or seas that ran northward from the great Southern Ocean. On the shores beds of shale were formed, some of which are now found at intervals along the African coast from Somaliland to German East Africa. These yield ammonites, which show that the beds from which they come were formed at the same time as the Oxford Clay of our Midland Counties.

In the next great period—that of the Chalk—another change came over the country. In Sofala and Madagascar to the south, and in Somaliland to the north, marine conditions still occurred. The British dominions on the east coast, however, yield no marine deposits of this age. The next fossil from them is Eocene in age, and allied to species found in the Kirthar series (Upper Eocene) of India. The evidence of this fossil is not free from doubt, but as the conclusions it suggests are probable and in accordance with other lines of evidence, and as it fixes a date which otherwise must be left uncertain, it is convenient to admit it.

Accepting, then, that the fossil in question came from East Africa, we know that a new stage in the history of the country began after the close of the Jurassic. Up to this point, ever since the Archean era, Eastern Equatorial Africa had been stable and restful; but then the old volcanic fires broke out anew, and inaugurated a series of events which have given the region its especial interest. In summarising these it is convenient to give names to the principal divisions, and the native terms for the geographical features determined in each of them may be appropriately adopted.

The first of these we may call the Kaptian, as in it were formed the Kapte plains. That this period began later than the Jurassic is probable, for Hobley has shown that the rocks of this age on the coast are cut by dykes. The composition of the rocks suggest that these dykes belong to the first series of eruptions. The volcanic action near the coast was not very powerful, but in the interior innumerable small volcanic cones buried an enormous tract of country under a flood of molten rock.

The earliest lavas ejected by these "plateau eruptions" were of the kind known as "trachytes," and these were followed by others containing less silica, which are named "andesites" owing to their importance in the Andes.

The duration of this period of intense volcanic activity cannot be precisely determined, but it must have been prolonged. When the pressure on the lava sources was relaxed, the volcanoes dwindled, became dormant, and then extinct. They had, however, destroyed the structural stability of the country; for the subterranean reservoirs were empty, while masses of volcanic material had been piled upon the surface. The upper layers of the earth's crust were therefore overweighted above and weakened below, and earth-movements were necessary to restore equilibrium. The first change was probably a subsidence of the country to the east and west of the East African lake-chain, leaving this line as a ridge or arch running north and south from Basso Narok, past Baringo and Naivasha, to the southern end of the volcanic region near Lake Nyasa. Throughout the whole of geological history the dominant lines of weakness in East Africa have run from north to south, and it is therefore only natural that the ridge should have trended in the same direction. As the movements of this epoch raised the line which formed most of the existing watershed between the Atlantic and Indian Oceans, it may be called the "Doenyan," from the Masai name for mountain.

The regions on either side of this central ridge probably continued to subside. The central arch was therefore left unsupported, and parallel cracks opened along its flanks, as they do on a bridge when its buttresses give way. As the summit of the ridge was thus weakened it sank, making the first step in the formation of the Rift Valley.

As we have seen in the Introduction, Suess has shown that the Rift Valley extended from the north of Palestine to Lake Nyasa, but the whole of it was not formed simultaneously. Moreover, each part was probably formed by a series of earth-movements at distant intervals.

The first of the series of north and south faults which formed the Rift Valley happened in the Eocene. This is determined by the occurrence of marine fossils, which must have lived in a gulf that ran into the valley from the Indian Ocean. As the earth-movements of this epoch formed the Nyasa region of the Rift Valley, we may call it the "Nyasan."

After the subsidences of this epoch, volcanic disturbances of the plateau type began again. In the foot-hills of Kamasia there are lake deposits, in which occur beds of pebbles of the Kaptian lavas. Above the lake deposits there are flows of basalt, which have baked the clays into a porcelain as hard and white as the purest china. We may call this second series of plateau eruptions the "Laikipian," as its basalts form the main cap of Laikipia, where they lie on eroded surfaces of the older lavas. After these eruptions had again disturbed the equilibrium of the region by removing the foundations in some places, and piling up accumulations of volcanic materials over others, earth-movements again set in. Probably it was at this time that the plateau which occupied the site of the Victoria Nyanza began to subside. The head streams of the rivers that rose upon it were cut off from their lower courses, and were reversed in direction. They flowed to the centre of the depressed region, where they collected as a great lake. The second of the series of faults which made the main Rift Valley probably happened at the same time, and increased its size and depth, while others enlarged that of the Albert Nyanza. The climate of Africa must then have been less arid than at present. The snowfields of Kenya were certainly larger, and great glaciers flowed from these for several thousand feet down the flanks of the mountain.[1] The heavier rainfall helped the growth of lakes, which extended over places that are now sandy deserts. This period may therefore be called the "Naivashan," from the Masai word for lake. The first of these extinct lakes which we crossed was in the Kedong basin, where we entered the Rift Valley. There on the face of the Kikuyu scarp, which forms the eastern wall of the valley, are the terraces of an ancient lake. This I propose to name Lake Suess, after Prof. Edward Suess of Vienna, whose *Das Antlitz der Erde* I venture to regard as the most original and suggestive geological work that has appeared since Lyell's *Principles*. The terraces of this lake are perfectly visible when seen from a little distance, though often difficult to trace in the jungle that covers the face of the cliff. The terraces were formed by the conservative action of the water, which protected the part of the valley wall below its surface, while that above it was exposed to the wind and rain. The cliff was therefore driven back, while the débris which fell from it accumulated on the floor of the lake. The terraces show that Lake Suess at one time reached the height of 400 feet above the present floor of the valley, along which it extended for a considerable distance.

It is probable that the outlet of the lake was to the north, whence a river flowed along the Rift Valley, through the basins of Baringo and Basso Narok. There it was joined by a river, the Turquell, which, as we shall see in another chapter, probably received the drainage of the basins of the Victoria and Albert Nyanzas, through the valley occupied by the Salisbury or Musaniya lake-chain.[2]

[1] The evidence for this is stated at length in a paper on "The Glacial Geology of Mount Kenya," *Quart. Jour. Geol. Soc.* vol. l. (1894), pp. 515-530.

[2] For topographical information which supports this idea I am indebted to Major Williams, R.A.

Lake Suess must have lasted for a considerable period, gradually dwindling as the climate became more arid, and as earth-movements cut off its extension to north and south. One of these earth-movements formed the ridge crowned by the volcano Longonot, which separates the Kedong and the Naivasha basins. That this ridge is later than the lake is shown by the fact that the terraces on its flanks are uptilted. It is also certain that the lake was earlier than the last of the lava flows from Longonot, for these have spread out over the alluvium, a fact illustrated in Pl. IX. (p. 97). Volcanic action, however, took place in the district while the lake was in existence, for Doenyo Nyuki (the Red Mountain) once rose above its surface as a volcanic island.

In the later stages of the history of Lake Suess volcanic activity of a violent type was renewed, new craters being formed along lines running north and south. Kibo—the highest mountain in Africa—was a member of this series, to which also belong the Kyulu Mountains (a line of craters in Kikumbuliyu to the west of the Uganda road), Elgon and Lekakisera to the north-east of the Victoria Nyanza, and the most recent craters on the floor of the Rift Valley, such as Suswa and Longonot. These volcanic disturbances were followed or accompanied by another series of earth-movements. The chain of volcanoes from Elgon to Lekakisera apparently marks a line of elevation, which cut off the connection between the Nyanza basin and Basso Narok (Lake Rudolf). Parallel to this was a line of depression, which lengthened the Albert Nyanza Rift Valley, and formed the gorge between Wadelai and Lado, through which the waters of the Central Basin of Africa gained an outlet to the Nile.

Mention of this event reminds us that in Equatorial Africa there is not merely one Rift Valley, but a series, which ramifies through the country like the rill systems of the moon. Suess has shown that these "valleys of dislocation" often end by splitting into a series of smaller ones, which he calls a "virgation," from its resemblance to the loose ends of a bundle of twigs. The Great Rift Valley presents a typical case of this in Palestine. In Africa, however, it branches occasionally, and forms an open system. The members of this have not been geologically surveyed, and in many cases the topographical information, from which their existence is inferred, is unsatisfactory. But of the existence of such a series there is now no doubt.

We have thus rapidly summarised the main outlines of the geology of British East Africa. The first striking fact is that, with the exception of a narrow strip upon its margin, the country has never been below the sea. The second point of interest is that the history of the region may be divided into three stages :—

1. The Archean, represented by igneous and schistose rocks.

2. The long uneventful interval between the Archean and the Cretaceous.

3. The subsequent complex series of volcanic eruptions and earth-movements.

This threefold division reminds us at once of the geology of Brazil and of the southern or peninsular half of India, the histories of which agree very closely with that of East Africa. Thus Southern India consists in the main of gneiss and schists ; upon these occur slates and quartzites (the Cuddapah series) much like those of Karagwe. Then comes a long recordless interval, until, in the Cretaceous, plateau eruptions poured out the 200,000 square miles of Deccan Traps, which correspond with the Kaptian series of Africa. After this the scarp of the Western Ghats was probably formed in the same manner as the scarp of the East African plateau.

The main interest in the geology of British East Africa is in the last of the three periods. This is complex, for it has three parallel histories—those of the volcanoes, of the lakes, and of the earth-movements. In the previous sketch the effort has been made to combine the three narratives into a continuous story ; but the following table shows them independently :—

[TABLE

Geological Period.	East African Representatives.	Rift Valley Area.			Nyanza Basin.
		Volcanic Action.	Lakes.	Earth-movements.	
Pleistocene	Upper Series	Longonot, Doenyo Ngai, and Teleki Volcano in eruption	Modern Lakes. Neolithic extension	Last series of faults.	
	Lower Series	Longonot, Kyulu, and existing crater of Kilima Njaro	...	...	Elgon series in eruption. Formation of Nile gorge.
Pliocene	Naivashan	Doenyo Nyuki	Lake Suess	Extension of Kenya glaciers Second series of Rift Valley faults	Depression of area and formation of Semliki Rift Valley.
(?) Miocene	Laikipian	Plateau eruptions of basalt	...	...	Basalt eruptions.
Eocene	Nyasan	...	Lake Kamasia	First series of Rift Valley faults. Sea in southern end.	
	Doenyan	Kenya, Settima, and Mawenzi in eruption	...	Ridge over Rift Valley	Plateau conditions.
Cretaceous	Kaptian	First plateau eruptions.			

In the case of the volcanic history there are probably five main divisions, three of plateau eruptions and two of crater eruptions. The succession of lavas poured forth appears at first to agree with Richthofen's theory of the volcanic rock sequence. The plateau-eruption lavas followed one another in the order —trachyte, andesite, and basalt, and this succession appears to support Richthofen's hypothesis. But if we consider the crater eruptions, no definite succession occurs. In these, trachytes occur after andesites as well as before them, and basalts at the end of the series as well as at the beginning.

After the completion of the system of Rift Valleys, the study of East Africa becomes the work of the archæologist and historian, instead of the geologist. The stone implements of the old terraces of Lake Baringo (see pp. 323, 324) show that man entered the region while yet the lakes were larger than they are to-day. Volcanic eruptions still took place and earth-movements continued, for some of the fault-scarps are so bare and sharp that they must be of very recent date. This continuation of earth-movements into the human period is one of the most striking features of the district. Whereas, according to the old view, British East Africa was supposed to have acquired a condition of stable equilibrium at a very early age, it has, on the contrary, been in a continual state of change since the time of the formation of our Oolitic limestones. During the eras between those of the Archean and of the Chalk, the country may have enjoyed comparative rest; but in the age of the latter, there began one of the two greatest of the series of volcanic outbursts known in the world's history. This and the resultant series of earth-movements have kept the region ever since in a condition of disorder and unrest. One region has been raised and another depressed; in one place a fiord has been opened from the sea, and then separated from it; elsewhere a line of movement has reversed the direction of rivers, and transferred lakes from one river system to another; while differences in elevation have caused variations in climate and rainfall. The evidence of these changes is apparent on every hand. Scars of great earth-movements, extinct volcanic craters, dried lake basins and old river beds, show the extent and recent date of these events, and the structural instability of the region of the Great Rift Valley.

In later chapters we shall see that this structural instability has had a most important influence on all branches of the natural history of British East Africa, for its results have affected the development of both animals and plants, and helped to mould the character of the people.

CHAPTER XIII

PROBLEMS OF THE DISTRIBUTION OF THE EAST AFRICAN FLORA AND FAUNA

"Kisauni kutamea mvinde?"
"Mvita kutamea mgomba?"

(Will Kisauni grow the she-oak?
Will Mombasa grow the banana tree?)

Zanzibari Proverbs.

THE problems of the distribution of animals and plants in a country are riddles, the difficulty of which varies with the complexity of its history. In regions of great stability they are simpler than where important geographical and climatic changes have taken place in the past. Thus, on the old view of the geological uniformity of the continent of Africa, these problems might have been expected to be comparatively simple, whereas they have always proved exceptionally confused and intricate. After making a preliminary collection in East Africa, I compared notes with those of the residents, such as Mr. Ainsworth of Machakos, the late Mr. Bell Smith of Melindi, and the late Dr. Charters of Kibwezi, who had had experience on the west coast. The result sorely puzzled me, by bringing out apparently glaring contradictions in the facts of distribution. Thus certain groups run across Africa from east to west, while others extend from north to south. The commoner beetles, butterflies, and birds seemed to belong to a fauna that spread across the continent from the Atlantic to the Indian Ocean. On the other hand, some less important groups of animals, and some of the more striking of the plants, have their nearest affinities with those of

Abyssinia and the Cape. That different groups of living creatures have different geographical distributions is a well-known fact. As shown by Dr. Blanford, in his remarkable presidential address to the Geological Society in 1890, it can be easily explained by the assumption, that the distribution of land and water has varied greatly at different periods in geological history. A group of animals, therefore, that made its appearance on the earth at one period, was able to spread along very different lines from those followed by a later group, when old land-masses had been broken up, and seas once connected had become separate. Thus, if we compare the distribution of different groups of animals, we find a gradually increasing specialisation as we pass from the oldest to the youngest groups. This is shown by the accompanying four sketch-maps. The oldest of the five classes of vertebrate animals is that of the fish, of which only the fresh-water forms are of any value in this connection; when they were introduced, they were apparently able to spread in any direction, and thus their present distribution appears to be determined mainly by temperature, for the faunas range round the world in three bands. By the time the tortoises appeared in the period of the Trias, or New Red Sandstone, the land of the southern hemisphere had apparently been broken up, and it was only to the north of the equator that the animals were able to range round the globe. According to the tortoises, therefore, North America, Europe, and Asia are all part of the same province; but before the introduction of the lizards, remains of which first occur in the Purbeck limestones, great geographical changes had occurred. These reptiles could not spread westward into North America, but they made up for this restriction by extending southward throughout Africa. After another great lapse of time, snakes appeared upon the scene in the age of the London Clay (a part of the Lower Eocene); the European species were now cut off from Africa, and in Asia were limited to the western half of the continent. The passerine birds, on which Dr. Sclater's classification is mainly based, soon followed the snakes; they were cut off from America and Tropical Africa, but their powers of flight enabled them to spread over the whole of Europe and Asia, though they did not succeed in entering India, Siam, and the Malay Peninsula.

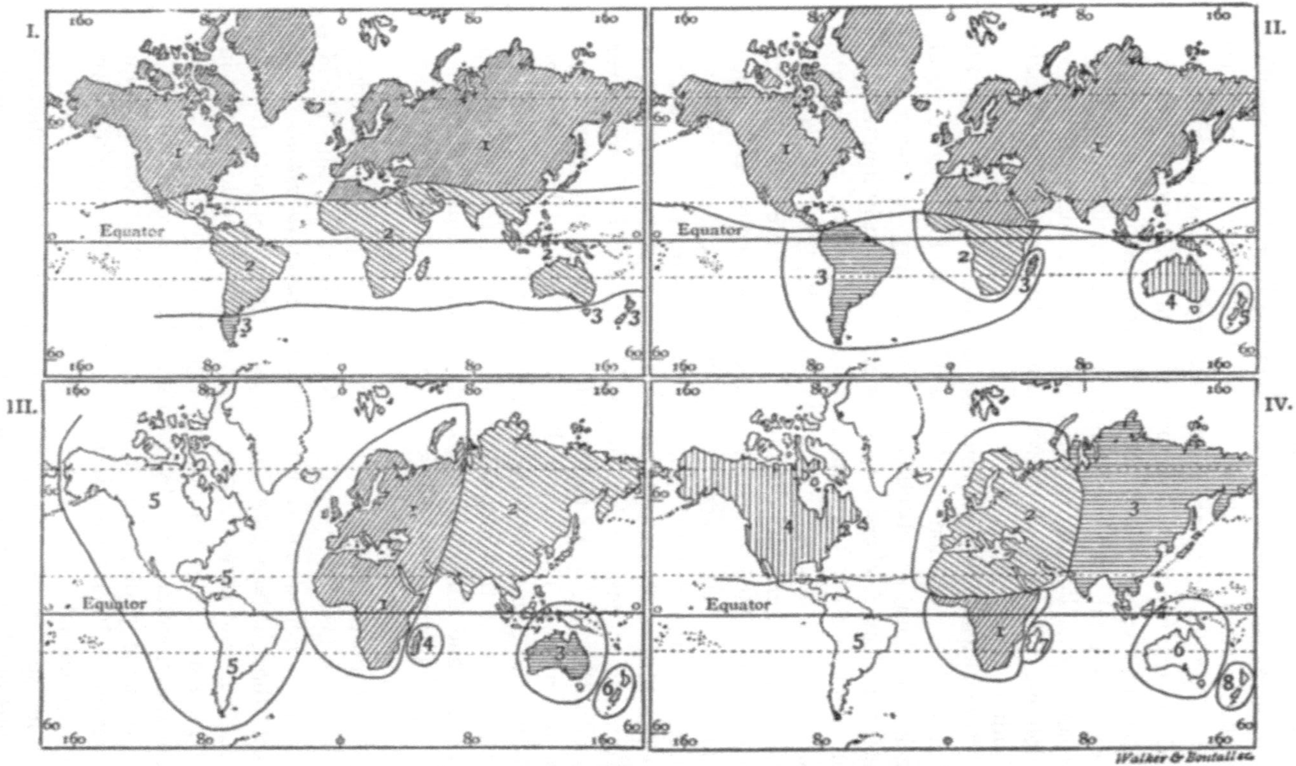

FIG. 12.—Maps of Zoological Distribution. (According to Günther.)
I. The Distribution of Fresh-Water Fish. II. The Distribution of Tortoises.
III. The Distribution of Lizards. IV. The Distribution of Snakes.

The application of this simple explanation to the anomalies in the internal distribution of plants and animals in Africa seemed prohibited by the general assumption that this continent had always maintained its present form. Moreover, this explanation does not fully account for the distribution of some groups in small local patches; and as soon as the scientific exploration of Equatorial Africa began, numerous small outliers belonging to one province were found to occur in the middle of others.

Thus, when Baron von der Decken returned from his memorable expedition to Kilima Njaro in 1862, he brought back with him a collection of plants, many of which were determined by Ascherson to be species previously known from the mountains of Abyssinia, such as *Helichrysum abyssinicum*, Sch. Bip., *Spilanthes abyssinica*, Sch. Bip., and *Achyrocline Hochstetteri*, Sch. Bip. Others, such as the tree lobelias (*Tupa*), are allied to those of the same region, and others belonging to genera such as the Wormwood (*Artemisia*) are typical of the north temperate zone. During New's daring visit to Kilima Njaro in 1871 he obtained specimens of twenty plants, which enabled Sir Joseph Hooker and Professor Oliver to add the *Bartsia* to the list of northern genera growing on the mountain. The later collections of Teleki, Johnston, and Volkens have fully confirmed the fact of a flora existing on the higher part of this mountain, unlike that of the surrounding lowlands, and allied with those of the mountains of Abyssinia and the Cameroons, and to a less extent with those of the Cape and the Mediterranean basin.

The most striking of these plants was a giant groundsel, which was first discovered on Kilima Njaro by Mr. H. H. Johnston, after whom it was named *Senecio Johnstoni*. This plant, though belonging to the same genus as our English groundsels and ragworts, grows as a tree from 20 to 30 feet in height, resembling members of this genus previously known from the mountains of Abyssinia and the Cameroons.

In some photographs taken by Gedge at a corresponding elevation on Mount Elgon, similar arborescent groundsels form a conspicuous feature in the scenery. Count Teleki observed another species on Kenya, and a view published by Stuhlmann demonstrated the existence of a similar form on Ruwenzori.

Moreover, on each of these mountains, the giant groundsels are associated with tree heaths, tree lobelias, and other plants allied to those of Kilima Njaro and the temperate zones. Teleki's observations did not, however, show whether the Alpine flora of Kenya was more closely allied to that of Kilima Njaro, of Abyssinia, or of the Cameroons, and the determination of this point was one of the main objects of my visit to the mountain. More than half of my collections from Kenya were lost in the Tana, and of the remainder, only two out of three botanical groups have been as yet described. These show that the flora is allied to that of Kilima Njaro, though the species are in many cases distinct.

The Alpine birds of Central Africa have the same remarkable distribution as the plants, for upon the highest mountains there is an avifauna unrepresented on the adjoining lowlands. Dr. Bowdler Sharpe has grouped together these isolated colonies into a special sub-region, which he calls the "Cameroonian." On his map, illustrating the geographical distribution of birds,[1] he shows this "Cameroonian sub-region" as a series of areas scattered over Equatorial Africa.

The most simple possible explanation of these facts is, that the seeds have been carried by the wind, or by birds, from one mountain summit to another. It is well known that the wind can carry some seeds for enormous distances; and if the resemblance of the floras of Abyssinia, Kenya, and the Cameroons, were due solely to plants with seeds so small as those of the orchids, or provided with appliances especially adapted for floating in the air, such as the pappus of the groundsels, this explanation might suffice. But seeds so large as those of the Gladiolus, and so hard and heavy as those of *Podocarpus*, can hardly have been carried far by the wind; nor are the birds sufficient to account for the distribution, as the sun-birds, which alone occur at these altitudes, are not migratory.

A second explanation is suggested by the fact that similar anomalies in distribution occur in Europe, where Alpine and Scandinavian plants, such as the little *Luzula arcuata* of the Grampians, live on the summits of mountains, far from the

[1] R. Bowdler Sharpe, "On the Zoo-Geographical Areas of the World, illustrating the Distribution of Birds," *Natural Science*, vol. iii. (1893), pp. 100-108.

present homes of the species. In Europe this admits of very easy explanation, for the former extension of the glaciers in the great Ice age drove the northern plants farther south; as the climate became milder and the ice receded, the plants returned northward, and in the south survived only on the summits of the higher mountains. Analogous cases have been found in North America, and admit of the same explanation. Thus on the White Mountains in New Hampshire there is a group of butterflies of Arctic types, which have been left as a legacy from glacial times.

The occurrence of these isolated patches of an Alpine flora in Equatorial Africa therefore suggests that, in times geologically recent, a change has come over the climate. This is confirmed by other lines of evidence; instances are known of the survival of plants in situations now ill-suited to them. For example, at the southern end of Basso Narok, von Höhnel found a patch of the Hyphæne palm. His photograph shows that the trees are dwarfed and unbranched, and are clearly growing under unfavourable conditions. They are probably survivals from a rich growth of these palms that flourished around Basso Narok at a period when a great river flowed into its southern end, and its waters were less alkaline than they are to-day. The geographical evidence is equally conclusive. Livingstone[1] has graphically described the existence of river gorges in Bechuanaland far larger than could have been cut by the rivers that now flow through them; and the same fact has been reported in the Sahara by Zittel, Rohlfs, and Weld Blundell. Old beaches and terraces indicate the former existence of lakes, whose waters have long since been lost by evaporation into the air and absorption by the soil.

Hence the evidence of the plants, river gorges, and old lake basins, together demonstrate that some change has taken place in the African climate. The two latter could be explained by a mere decrease in the rainfall, but this would not account for the first; and the facts afforded by the Cape, Kilima Njaro, and the Andes seemed to forbid an appeal to the glacial agencies, which explain the analogous cases in Europe and America.

[1] D. Livingstone, *Missionary Travels and Researches in South Africa* (1857), p. 111.

It is true that, in the Himalaya, ice action once operated several thousand feet below its present limit, but there is no proof that it reached the plains;[1] moreover, this occurred in the midst of a continent and 30 degrees from the Equator.

Some evidence has indeed been adduced from the Cape of a recent glaciation, but so far it is inconclusive, and the facts opposed to it are very weighty. For example, one of the most easily recognisable rocks in South Africa is the volcanic "Pipe Amygdaloid" of the Stormberg series. If glacial action had occurred since this was formed, boulders of it must have been distributed across the country. Their absence seems incompatible with the existence of an ice sheet in Cape Colony since the Cretaceous period. It is therefore less surprising that by the Europeans who have visited Kilima Njaro (more than a hundred in all) no extension of its glaciers has been recorded. It may be objected that negative evidence in the case of Equatorial Africa is not reliable; but that of the Andes shows conclusively that there has not been any universal extension of the glaciers in the tropics. The fact that the glaciers at the Andes are now at their maximum extension is well established: D'Orbigny indeed argued from it that the Andes are still undergoing elevation. Mr. Whymper has informed me that only twice in Ecuador did he see any trace of glaciation below the level of the existing glaciers. In the more important case he found some decayed *roches moutonnées* below his second camp on Chimborazo; but they were so little below the "Glacier de Débris," that a mere local variation in wind would account for the slight extension of the ice that made them.

Thus the negative evidence of Equatorial and Southern Africa and the analogy of the Andes seemed to forbid any appeal to glacial theories for an explanation of Central African anomalies in biological distribution. When, therefore, above the forests of Kenya, I found an old moraine several thousand feet below the level of the existing glaciers, I felt that I had found not only a valuable clue to the causes of the spread of the Alpine flora, but a chance of determining the date of its extinction on the lowlands and of the change in the African climate. This evidence has been considered at some length in

[1] C. A. M'Mahon, *Records Geol. Surv. India*, vol. xiv. (1881), p. 310; vol. xv. (1882), p. 49. See also *Man. Geol. India*, ed. 2, p. 484.

a paper on "The Glacial Geology of Mount Kenya,"[1] which deals with both the causes of the greater glaciation and some of its results. The former has been referred to in a previous chapter, but the latter must be considered in relation to the changes in the flora.

In the paper quoted it is shown that the glaciers once extended for over 5400 feet below their present limits, and it is suggested that this was due to the level of the whole country having been much higher than at present. Whether this were the cause or not, there can be no doubt that the Alpine flora, which now descends on Kenya to the level of 10,400 feet, must have been driven as far as the glaciers advanced. It would thus have reached the contour line, which now stands 5000 feet above the sea. Other effects would have tended in the same direction. The greater extent of the snowfields must have increased the rainfall on the surrounding country. Moreover, on the mountains there does not appear to be any division of the year into wet and dry periods, as there is on the plains; the heavier rainfall would therefore have been all the more effective, as it would have been distributed throughout the year. The rainfall on the plains would also have been increased in another way. In all mountainous countries, more rain falls on one particular zone than either above or below it. At present, in British East Africa, this "zone of maximum rainfall" occurs between the levels of 7000 and 11,000 feet, and thus only includes parts of the highest mountains, and not the plains. But during the period of greater elevation of the country, the maximum rainfall would have occurred at a relatively lower level, and probably most of the surface of the high plateaux was included in this zone. In much of the country in which the rainfall is now scanty and uncertain, it would therefore have been regular in distribution and considerable in amount.

The results on the vegetation of the district must have been very great. The air being damper, it would no longer be necessary for the plants to guard themselves against the drain of moisture caused by the process of transpiration. The desert scrub, which now covers the country, would have been replaced by a less specialised and more normal type of vegetation; the

[1] *Quart. Journ. Geol. Soc.* vol. l. (1894), pp. 515-530.

possession of narrow, spiny, or needle-shaped leaves and succulent leafless stems would not then have been an advantage; the foliage would have been more luxuriant and better adapted for animal food; forests that now occur only as belts beside the rivers would have been spread far and wide across the country. The scrub would have been replaced by woodland, and districts that now, as barren sandy deserts, present barriers to animal migration, would have been fertile, well-watered prairie. All the conditions, in fact, that govern the distribution of animal and plant life would have been different from those that obtain at the present day.

At the time, therefore, of the maximum size of the Kenyan glaciers, the Alpine flora would have spread at least throughout the areas marked by dots on the accompanying map (Fig. 13). No doubt it extended lower, as traces of it have been found in the mountains of Usambara, but the precise lower limit cannot be determined from the data available at present. When the climate changed with the decrease of the glaciers, the Alpine flora crept up the mountain sides, and became isolated from its allies in the temperate regions to north and south.

We have assumed, however, that this Alpine flora has received contributions not only from the north but from the south, and this compels us to consider the theory that all faunas and floras have originated in the northern hemisphere, and thence have worked their way to the south. The occurrence of isolated patches of the flora of one region in a more southern zone has been mentioned as not unknown in Europe and North America. Similar evidence occurs in Asia. On the mountains of Ceylon and the plateaux of Southern India there occur plants such as *Rhododendron arboreum*, Sm., and mammals such as the Indian Marten (*Martes flavigula*, Bodd.), which do not live on the surrounding lowlands, but reappear to the north on the Himalaya. Similarly, the Himalayan fauna and flora can be traced along the mountains of Assam and through Burmah into the Malay Peninsula. An outlier occurs even on the hills of Java.[1] Russell Wallace[2] and Haacke[3] have collected a good deal of evidence of the same character, and

[1] Blanford, Medlicott, and Oldham, *Man. Geol. India*, ed. 2 (1894), p. 14.
[2] A. Russell Wallace, *Island Life* (1880), chap. xxiii.
[3] W. Haacke, "Der Nordpol als Schöpfungszentrum der Landfauna," *Biol. Centralbl.* Bd. vi. (1886), pp. 363-370.

upon it advanced the theory that the movement of faunas is always to the south.

This theory of the northern origin of life is unquestionably

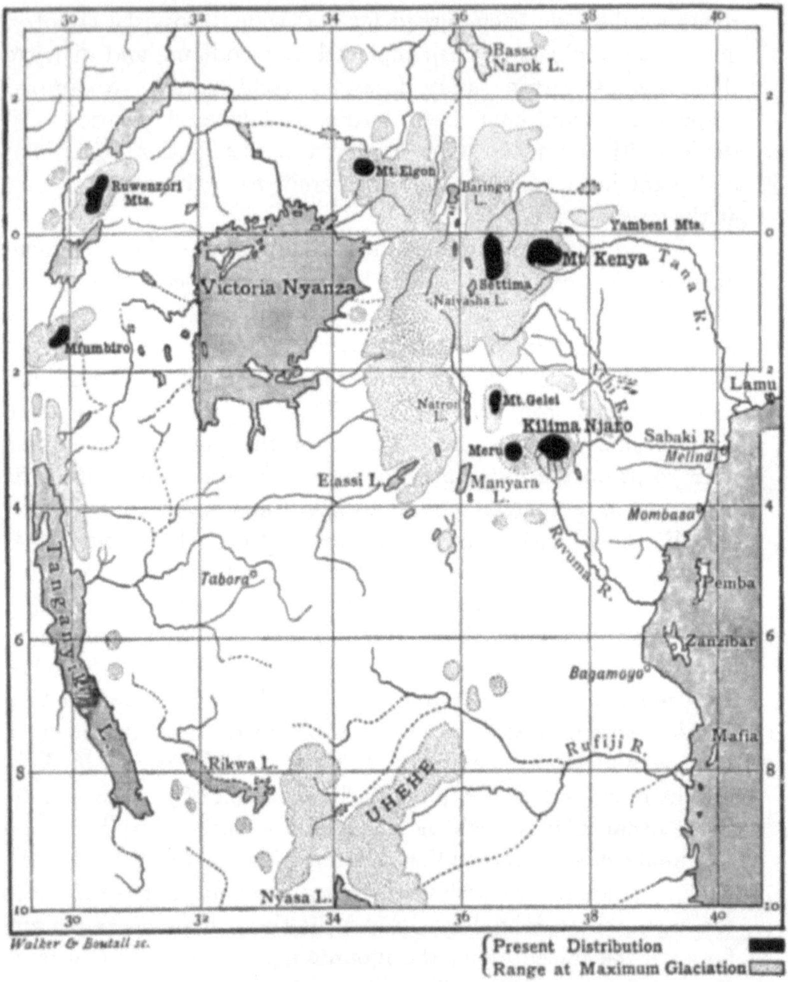

FIG. 13.—Map of Present and Former Range of Alpine Flora.

very suggestive, as it is supported by many different lines of evidençe, explains many difficult problems, and is itself readily explained. Nevertheless, I venture to doubt its universal truth. Colonies occasionally have migrated into Europe and

North America from the south. Thus a South American fauna entered the United States in the Upper Miocene.[1]

Plants so typical of the Cape flora as *Calodendrum capense*, Thunb., and orchids of the genus *Disa*, occur on the plateaux of Equatorial Africa associated with the Mediterranean and north temperate species. This certainly suggests that in Africa plants have been introduced into the equatorial regions from the south, as well as from the north. That the rule that plants and animals have usually travelled from north to south is not so universal as to prohibit this, is shown by the fauna of the Andes. The theory of a migration from north to south during a cooler period has been applied there, but H. W. Bates, in discussing the results obtained by Mr. Whymper's collection, pointed out that the facts give no support to it. He says: "It seems to me a fair deduction from the facts here set forth, that no distinct traces of a migration during the life-time of existing species, from north to south, or *vice versâ*, along the Andes, have as yet been discovered, or are now likely to be discovered."[2] On the contrary, he claims that there are no traces in the Andes of Ecuador of temperate forms of either beetles or butterflies, and the fauna consists of modified representatives of the genera of the neighbouring lowlands.

Mountain floras, therefore, are of two types—those which have been left as relics of an older or foreign flora, and those which have developed *in situ* by the adaptation of local, lowland species. In British East Africa there are representatives of both types. Many of the mountains are not sufficiently high to reach the levels at which alone the temperate plants can live. In such cases they have been killed off, and the summits are occupied by dwarfed, or otherwise modified, representatives of the species of the adjoining lowlands. The occurrence, however, of species of such genera as *Clematis*, *Polygala*, etc., on the summits of the Taita Mountains, shows that in some localities representatives of both groups occur together.

[1] W. B. Scott, "The Mammalia of the Deep River Beds," *Trans. Amer. Phil. Soc.* vol. xvii. (1894), p. 62. Many similar cases could be quoted. A book has, indeed, recently been issued by C. Dixon, *The Migration of British Birds* (1895), which maintains, as the fundamental law of distribution, that species never migrate to the south.

[2] E. Whymper, *Travels amongst the Great Andes of the Equator*, vol. ii., Supplementary Appendix (1891), p. 4.

As has been already suggested, a great change in the character of the flora of a country must inevitably influence the distribution of its animals by altering the food supply. Thus insects will vary with the flowers, and birds with the insects; small mammals will disappear as the vegetation becomes sparser; earthworms will be restricted to the more fertile regions, and termites or "white ants" replace them, as the soil becomes harder and more barren. In the case of the fresh-water fauna, the range of possible variation in the environment is more limited. The current of a river may become slower, or the quantity of water in a pool may lessen; but the conditions do not vary fundamentally, unless the water dries up entirely, and such a change as this is at once fatal to animal life. Nevertheless, the fresh-water fish of Equatorial Africa present as many anomalies in distribution as the land animals and plants. When Günther[1] in 1869 described the collection made by Petherick in the Upper Nile, he pointed out that the fauna of which it is a part has a very remarkable distribution, being more nearly related to the faunas of the Palestine and West African rivers than to those of Lake Nyasa and the Zambesi. Later collections have fully confirmed the truth of Dr. Günther's view. Though the fish of the Zambesi, the Upper Congo, and their connected lakes, such as Nyasa and Tanganyika, are in the main identical in genera with those of the Upper Nile, they differ from them in species; while the Upper Nile contains some of the very same species as the Jordan and the Sea of Galilee. My collections prove that the rivers of British East Africa, such as the Tana, Athi, and Sabaki, and the rivers and lakes of the Rift Valley have the same generic fauna. The Tana, however, is intermediate between the Nile and the Zambesi, for it has two species from each, viz. *Clarias lazera*, C. V., and *Clarotes laticeps*, Rüpp., from the former, and *Eutropius depressirostris*, Ptrs., and *Synodontis zambezensis*, Ptrs., from the latter, as well as one (*Barbus intermedius*, Rüpp.) from Abyssinia.[2]

The fact that renders the evidence of fresh-water fish so instructive is that the distribution of the species is often extremely local. Thus many of the small British and Irish lakes have

[1] In J. and B. H. Petherick's *Travels in Central Africa*, Appendix C, vol. ii. pp. 197-268.

[2] A. Günther, "Report on the Collection of Reptiles and Fishes made by Dr. J. W. Gregory during his Expedition to Mount Kenya," *Proc. Zool. Soc.* (1894), pp. 84-91.

forms peculiar to them, such as the Llanberris Charr (*Salmo perisi*, Gthr.) found in the two Llyns at Llanberris, the *Salmo colii*, Gthr., of Lough Eske, and the *Salmo grayi*, Gthr., of Lough Melvin.[1] The fact, then, that the same species occur both in the Upper Nile and in the Jordan, suggests that there must once have been some connection between them. If the fish which occurred in both rivers were eels, able to wriggle their way for some distance across land, or catfish (*Clarias*), able to live for months embedded in the dried mud on the floor of a desiccated pool, the facts would have little weight. But the fish in question have no such remarkable powers of vitality, and are killed by a few minutes' exposure to the air, or to water of a different composition from that to which they are accustomed.

Dr. Günther suggested that this remarkable fish fauna must have originated in the lakes of the central plateau of Africa, and thence spread in every direction.

That such changes in the river systems of Africa, as are necessary for Dr. Günther's theory, actually occur, is shown by the controversy as to the relations of the Congo and Lake Tanganyika. When Stanley in 1871 proved that the river Ruzizi at the northern end of Tanganyika was an inlet, instead of an outlet, there seemed no point left for the escape of its surplus waters. Cameron accordingly in 1877 circumnavigated the lake, and then found that the Lukuga had a slow current flowing westward from the lake. Stanley returned in 1876, and found no outlet at this point. Hore early in 1879, and Thomson at its close, and other explorers later on, have found a powerful stream flowing from the lake toward the Congo. Tanganyika, therefore, is sometimes connected with the Congo, and at other times it is not.

If Tanganyika had not been visited till twenty years hence, it is quite possible that its connection with the Congo might never have been seen, and the presence in it of so many Congo species would have been difficult to explain.

The dispersal of the fish fauna of the rivers and lakes in the Victoria Nyanza region may have been brought about by

[1] F. Day has denied the validity of these species (*Fishes of Great Britain and Ireland*, vol. ii. (1884), pp. 112-114); but Günther's faith in them is unshaken, and Day admits them as distinct varieties. They are therefore certainly distinct forms.

a change, similar to that which is even now taking place in the case of Tanganyika. Probably at one period a wide plateau of gneiss and schists extended over the Victoria Nyanza basin. As the surface of the plateau was at a great elevation, the rainfall must have been heavy, and the hollows occupied by numerous lakes, with rivers flowing from them in every direction. It is well known that the direction of some of the head streams of the Mississippi is occasionally reversed, and they then flow northward into the Red River, and thus to Hudson Bay. The same thing must have happened on this central African plateau. The different rivers must have had their highest sources in the same swamps; dams would have been formed across the streams by the growth of vegetation, the accumulation of shoals, and the falling in of the banks. Occasionally a slight earth-movement would block the outlet of a lake, and cause it to open at a spot whence its waters passed to another river system. Thus floods, dams, and earth-movements would continually give the fish in one stream an opportunity to work their way into a stream of another river system. Hence all the rivers and lakes of such a plateau would be tenanted by a common fauna.

But at length came a change; the centre of this plateau subsiding, formed a basin without an outlet. The rivers were therefore "beheaded"; instead of rising in the centre of the plateau, their sources were now on the outer slope. The streams, which originally flowed outward from the highest part of the plateau, reversed their direction and flowed inward, so as to help to form a lake in the centre of the depression. Later on the separation between the waters of the central area and the rivers that originally rose on it was rendered more effective by the formation of the two Rift Valleys, when strips of land subsided on either side of the central Nyanza depression. There is therefore no difficulty in understanding why rivers which flow into the Atlantic and into the Indian Oceans, and are now widely separated by an area of internal drainage, are inhabited by the same forms of fish.

So far Dr. Günther's suggestion accounts for all the facts. But it does not remove the real difficulty, which is, that some members of this fish fauna are absent from the Lower Nile and yet are present in the rivers and lakes of Palestine.

As Dr. Günther says, "The system of the Jordan presents so many African types that it has to be included in a description of the African region."[1] He adds, moreover, that "this infusion of African forms cannot be accounted for by any accidental means of dispersal," and thus it appears to afford conclusive proof of an original connection between the rivers of Palestine and Central Africa.

At first it seems to be very easy to account for this connection, by assuming that a river from the Jordan basin was a tributary of the Nile, at the time when the eastern end of the Mediterranean was dry land. We know that the Levant was land in Pliocene times, for thus only is it possible to account for the almost complete difference in the faunas of the two seas on either side of the narrow isthmus of Suez. This difference is perhaps the most impressive fact in the whole range of zoological distribution. Thus in the case of the Sea-Urchins, not one species was common to the two seas at the opening of the Suez Canal. The evidence of the Mollusca is equally conclusive, for according to Mr. Edgar Smith,[2] who has issued the last authoritative statement on the subject, only eight species live on both shores of the isthmus. This number is quite insignificant in proportion to the enormous faunas of the two seas, while the presence of the eight species in both seas can be easily explained. A couple of them (*Chiton siculus*, Gray, and *C. discrepans*, Br.) live attached to seaweeds, and could easily have been blown across the narrow isthmus; the other six occur also in the Atlantic, and could thus have reached the Mediterranean through the Straits of Gibraltar. The evidence of the Mollusca is therefore as conclusive as that of the Sea-Urchins, in disproving any recent connection between the Mediterranean and the Red Sea. But the separation between the two seas cannot have been always effected by the present isthmus of Suez; for the geological evidence shows that the Red Sea extended farther north in Pleistocene times, and then actually occupied part of the present site of the Mediterranean. It is now generally believed that the passage of the Red Sea by the Children of Israel must have been effected at some

[1] A. Günther, *The Study of Fishes*, 1880, p. 227.

[2] E. A. Smith, "On a Collection of Marine Shells from Aden," *Proc. Zool. Soc.* (1891), p. 398.

point farther north than the present end of the Gulf of Suez. The account of the destruction of Pharaoh's host may possibly be based on some catastrophe that happened owing to a change in the relative levels of land and water in this district. But the conclusion does not depend on the uncertainties of tradition. Raised beaches containing marine shells of living species are claimed to occur at a height three times as great as that of the existing watershed, and must have been formed at a time when the sea-level was higher than it is to-day. We are therefore confronted by the probability that the two seas must have been in communication in times geologically quite recent. So clear is this evidence, indeed, that Professor Hull has issued a map showing the two seas connected by a wide strait, at a time for which he accepts the name of the Pluvial period. He accounts for the divergence of the faunas by the assumption that they could not have crossed owing to the shallowness and narrowness of the water. But as a colony of Red Sea species, including the Sea-Urchin (*Heterocentrotus mammillatus*, Brdt.), has worked its way through the Suez Canal, which is only one-sixth of the depth and one four-hundredth of the width of Professor Hull's strait, this explanation seems quite insufficient. In order to reconcile the apparent geological proof of the connection of the two seas, and the zoological proof of their separation, we are bound to accept Professor Suess's suggestion that, when the Red Sea extended to the north, the Mediterranean lay much farther to the west. The Levant must then have been a plain, over which roamed herds of antelope and rhinoceros, and across which flowed rivers in which lived the hippopotami, whose remains have long been known in the island of Crete. Suess suggests that the great scaly dragon slain by one of the Knights of St. John of Malta on the island of Rhodes may have been a crocodile, for these reptiles still live in the Jordan and in the Zerka (or "Crocodile River") on the coast of Palestine. That such traditions are of geological value has been shown in the case of Samos; for the stories of the occurrence of great monsters there led Dr. Forsyth Major to visit the island, and to make the famous collection, which finally proved the former extension of the African fauna across the Levantine area.

It is therefore practically certain that the Nile must have

been continued farther northward and westward than its present mouth, and it seems natural to conclude that the connection between the Jordan and the African river systems was established by a river, which flowed from Palestine into the Nile. Professor Hull[1] has not only accepted this view, but endeavoured to show that the connection was established along the course of the brook Kishon, for, near the head of its valley, a pass only 300 feet in height leads over into the Jordan basin. It is known that in Pliocene times the Jordan valley was occupied by a deep fresh-water lake, and Professor Hull suggests that the drainage from it escaped by this "Esdraelon Gap" into the Levantine basin.

There does not, however, appear to be any sufficient evidence in support of this view. No traces of gravels or any river deposits, such as would probably have been formed by so large a river, have been found upon the pass. The theory that the Jordan discharged by a river flowing along the valley of the Jalud and over the Esdraelon Gap into the valley of the Kishon, assumes that this gap was then but little higher than it is to-day, and that immediately to the west the ground sloped down toward the Mediterranean. But the whole arrangement of the river system in the district renders it more probable that the Esdraelon Gap was formed by a river flowing eastward instead of westward, and that at the time of its formation the Jordan valley was even more effectually separated from the Mediterranean than it is at present. This gives a very different origin for the Esdraelon Gap, which can best be explained by reference to a parallel case near London (see Fig. 14). All travellers to Brighton may notice that on leaving Croydon the railway gradually ascends the broad, dry "Golden Valley." Six miles south of Croydon the railway passes through a tunnel and emerges into a valley, which runs at right angles to the former, and is occupied by a tributary of the Mole. Looking backward, the Golden Valley is seen as a gap on the hill face above. If the traveller has time to walk up to this, he may find upon its floor some beds of gravel containing pebbles of a rock (Lower Greensand Chert), derived from the hills to the south. This material can only have been brought into its

[1] E. Hull, "On the Physical Conditions of the Mediterranean Basin," *Trans. Vict. Instit.* 1895, 10 pp.

present position when a stream from the south could flow directly into the Golden Valley.

This, however, has now been diverted owing to the occurrence of a band of clay east and west across its former course; the Mole has cut its way eastward along the clay, as it is soft and easily removed by water, and thus has absorbed into its own basin the stream that originally flowed northward down the Golden Valley.

The Esdraelon Gap was probably formed in exactly the same way. It occurs on the face of the plateau of Endor and Nain, which presents a steep face to the west, and a long

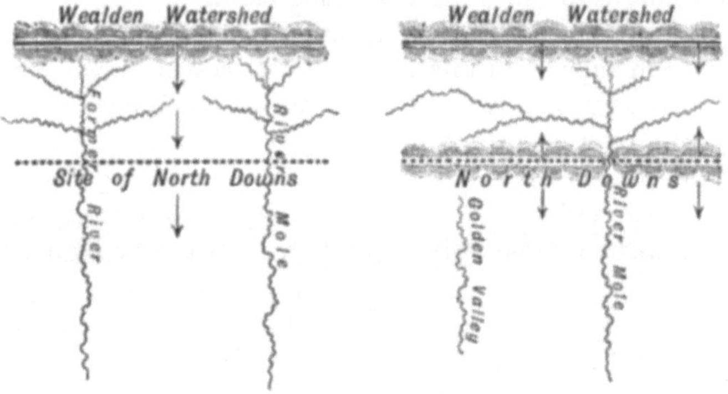

Original Course of Golden Valley Stream and the Mole.

Present Arrangement: the Golden Valley beheaded by T-shaped Course of the Mole.

FIG. 14.—Diagrams illustrating the Beheading of River Valleys.

gradual slope to the east. Originally the plateau must have extended farther to the west, continuing to rise in that direction, so that the rain that fell on it flowed into the Jordan. The Wadi el Muwali, a tributary of the Kishon however, now flows at the foot of the steep western face of the hills. It has cut its way into the plateau, and drained the streams that once rose farther to the west, just as the Mole has diverted the Golden Valley stream. The Wadi el Muwali has therefore "beheaded" the Jalud and cut away the upper part of its basin. What now appears as a notch in the western edge of the plateau is therefore only a section across the upper part of the valley eroded by the Jalud, when this stream rose much farther to the west.

It is, therefore, quite possible to account for the existence of the Esdraelon Gap without assuming that it always marked the watershed between the Mediterranean and the Jordan (for this, indeed, probably lay as far west as Cyprus); while a connection between the Jordan and the Nile would in no way help to explain the difficulty for which it is proposed. Such a connection is directly disproved by the fact that the Mediterranean fish fauna is absent from the Jordan; only one or two species, and those are Blennies, belong to both. If there had been any direct connection between the Jordan and the

FIG. 15.—The Neighbourhood of the Esdraelon Gap.

rivers of the Mediterranean basin, it is inconceivable that the fish of the latter should not have entered the former. Moreover, any such connection would be useless to our present problem, for it ignores the real difficulty. If we follow the Nile downward from its upper basin, genera of fish such as *Clarotes*, *Lepidosiren*, *Gymnarchus*, and *Haplochilus* disappear, and others not known in the Upper Nile take their place. Thus *Mochococus* replaces its close ally *Rhinoglanis*. In the Lower Nile there appear certain fish, such as five species of Mullet (*Mugil*), which have entered from the Mediterranean, and another group, which appears to have entered from the equatorial region, at the time when the Zambesi and the southern lakes received their contribution from this source.

The African fish in the Jordan, however, appear to have entered that river at a later time, and they are not found in the Lower Nile.

As Dr. Günther says, "*Hemichromis* is not represented in the north-eastern part of Africa proper, but chiefly on the west coast and in the Central African lakes." Similarly the *Clarias* of the Jordan is not the species (*C. anguillaris*, Linn.) of the Lower Nile, but that (*C. macracanthus*, Gthr.) of the Upper Nile.[1] It is no use, therefore, to assume the existence of a connection between the Lower Nile and the Jordan, to account for the existence in the latter of fish which do not occur in the former.

As, therefore, the route proposed for the migration of the fish across the Esdraelon Gap is opposed to the evidence, and useless as an explanation, we are compelled to turn to some other line of connection between the Jordan and the rivers of Equatorial Africa.

It has been mentioned as certain that in Pliocene times a large area in Palestine was occupied by a lake, the terraces of which may still be seen on the banks of the Jordan and around the shores of the Dead Sea. The fossils from the terraces indicate that the water was fresh, so the lake probably had an outlet. This must have been either westward across Palestine, or southward by the valley of the Wadi Arabah, which leads to the Gulf of Akabah. The only locality suggested for the former is the Esdraelon Gap, which, as we have already seen, is improbable ; but the latter is supported by several arguments. It is true that the Wadi Arabah depression is crossed by a ridge known as El Saté, which rises to the height of 787 feet above the sea, or nearly 500 feet higher than the Esdraelon pass. But the evidence already quoted shows that the latter has been reduced in elevation by denudation since Pliocene times, while there is reason to think that the height of the Wadi Arabah ridge has been increased by elevation. The equatorial part of the Rift Valley is divided into a series of basins by transverse ridges, similar to that of El Saté. Some of these are certainly due to elevation, for the old lake terraces upon their flanks have been tilted from their originally horizontal position (see Pl. VIII. p. 94). Similar

[1] Günther, *op. cit.* p. 228.

terraces extend for some distance up the Wadi Arabah valley from the southern end of the Dead Sea, and attain the height of 1300 feet above its present level. Hull[1] states that the alluvial deposits slope down to the north, and this supports the view that the ridge has undergone elevation. Anderson,[2] the geologist with the United States Expedition to the Dead Sea, indeed, came to the conclusion that the ridge across the Arabah had been raised. Moreover, in the geological map prepared by Professor Hull,[3] some old lake deposits are marked on the south side of the watershed and only a few feet below it. The account of these given in the text is not very precise, and we cannot decide whether or not these deposits are connected with the former discharge from the Jordan lake; but their existence is an important link in establishing a fresh-water connection along this line.

The opinion that the Dead Sea has always been an isolated basin was held by Lartet;[4] and, as far as concerns any marine connection between it and the Red Sea, his arguments are unanswerable. His conclusion that the rivers on the northern side of the ridge of El Saté have always flowed from south to north is, however, not so well established. The evidence he adduces shows that the rivers flowed thus in the latter stages of the history of the Dead Sea; but this is scarcely open to doubt. It is quite compatible with a discharge to the south at an earlier period, and such has not yet been conclusively disproved.

In order that the fish from Equatorial Africa might have entered Palestine by the route along the Wadi Arabah, it is not necessary to assume that a river flowed the whole way from the Jordan to the northern end of the Red Sea. If fish from the south reached the lake proved by Hull to have existed on the summit of the El Saté ridge, an occasional flood or a slight earth-movement would have enabled them to enter the streams that flowed northward into

[1] E. Hull, *Memoir on the Geology and Geography of Arabia Petræa, Palestine, and Adjoining Districts*, p. 81.

[2] H. J. Anderson, "Geological Reconnaissance of Part of the Holy Land," *Offic. Rep. U.S. Exped. to Dead Sea and Jordan*, part v. (Baltimore, 1852), p. 206.

[3] E. Hull, *op. cit.* map ver. p. 138; text p. 87.

[4] L. Lartet, "Note sur la formation du bassin de la Mer Morte," *Bull. Soc. Géol. France*, sér. 2, t. xxii. (1865), pp. 442-448; "Essai sur la Géologie de la Palestine" (1869), pp. 236-237.

the Jordan basin. An incomplete connection such as this would allow the fish to pass by the Wadi Arabah route; but the evidence, geological and geographical, renders an actual outlet from the Jordan over this pass so probable, that it is simpler to accept it. We may therefore assume that a river flowed southward from Palestine, and along that part of the Rift Valley which is now occupied by the Red Sea, and entered the Indian Ocean somewhere near Aden. Near its mouth this Erythrean River (for it may conveniently be called after the ancient name of the Red Sea) probably united with another from the highlands of Central Africa, which flowed along the Rift Valley from the northern end of Basso Narok (Lake Rudolf),[1] following the course of the Omo and Hawash, and across the depressed basin of Afar. This river would have received the drainage of the southern slopes of Abyssinia, the Rift Valley between Naivasha, Baringo, and Basso Narok, and also of the Victoria and Albert Nyanzas, which were not then connected with the Nile. At this time the Nile itself probably rose on the northern slopes of the mountains of Latuk on the east bank, and of Kakuak on the west, and the drainage of the two Nyanzas passed to the south of this mountain range, along the valley of the Musanyi, through the Salisbury lake-chain, and down the Turquell into Basso Narok.

At this time, then, we may assume that the equatorial lakes were in direct river communication with the Jordan, and their fish could reach Palestine without entering

[1] Dr. Donaldson Smith's very successful expedition across Somaliland to Basso Narok has thrown further light on the possible connection between the Omo and that lake. Dr. Smith, in a paper read before the Geographical Society on 6th January 1896, has expressed his doubts as to this connection; he regards the Omo as one of the head-streams of the Juba, and thinks that the Basso Narok depression ends off in the mountains in southern Abyssinia. If that is the case, there cannot have been any former outlet from Basso Narok to the north-east. But Borelli traced the Omo to a level (3450 feet) which renders its connection with the Juba almost impossible, and all the information given him by natives was in favour of the Omo reaching a great lake. The altitudes given by Dr. Smith to Lake Abbaja render it just possible that the Omo reaches this lake instead of Basso Narok. On Dr. Smith's preliminary map (issued 6th January) two scarps with a depression between them are indicated. The western scarp is only shown to the west of the head of the Nianam; but the occurrence of an apparent continuation of this to the north-east has been shown by previous explorers. The eastern scarp is marked from Fakes to Amara, and the altitudes given render its continuation past Basso Ebor to Basso Narok most probable. This leaves the Zuai lake-chain, the Omo valley, and the country between Borelli's southernmost point on the Omo and Dr. Smith's most northern point on the Nianam, on the floor of the depression. Dr. Smith's exploration therefore appears to me to confirm the idea that the Omo and Lake Rudolf are connected by a line of depression.

the Nile. Later on this river system was broken up by earth-movements and change of climate. Owing to the latter cause the glaciers and snowfields of Lebanon and Central Africa shrank in area, the rainfall became less, the lake levels were lowered, and the rivers diminished in volume. At the same time the elevation of the ridge of El Saté across the Arabah portion of the Rift Valley cut off the southward discharge of the Jordan, a change more readily produced owing to its lessened size. The continued depression of the region to the south brought it below the level of the Indian Ocean, and thus the Red Sea took the place of the Erythrean River. An elevation of the land across the valley of the Omo separated the basins of Basso Narok and Afar, and along the chain of volcanoes of Elgon and Chibchagnani severed the connection of the Victoria Nyanza with the eastern Rift Valley. A parallel depression to the west continued the western Rift Valley farther to the north, and formed the gorge between Wadelai and Lado by which the great lakes are connected with the Nile.

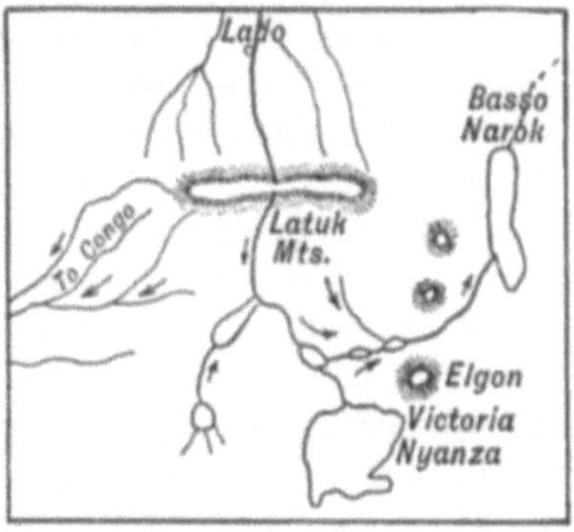

FIG. 16.—Diagram of possible former Source of the Nile and Outlet from the great Equatorial Lakes.

This view is put forward only as the theory which seems to me to best harmonise the known facts. For proof or disproof we must wait, until the river-courses and the gravels of the valleys between the Nyanza, Southern Abyssinia, and Somaliland have been investigated. The theory seems, however, to explain the three main difficulties :—

1. The occurrence of the equatorial fish in the Jordan ; for it shows the possibility of direct river connection between Equatorial Africa and the north of Palestine.

2. The absence of some species of fish which occur in the Jordan from the Lower Nile, by assuming the comparatively late date of the connection between the Nile and these rivers.

3. The complete difference of the faunas of the Medi-

terranean and Red Sea, owing to their being separated in Pliocene times by the whole length of the latter and by most of the Levant.

The theory has, moreover, the recommendation of throwing considerable light upon the relations of the land and fresh-water faunas of Abyssinia and the adjoining region. The evidence furnished by the fish, which alone we have considered, is fully supported by that obtained from other groups of animals, and also from plants.

We may take, first, the land and fresh-water shells. The fauna in Abyssinia has long been known to consist of a remarkable combination. The typical African genera, such as *Ætheria*, *Iridina*, *Achatina*, *Galatea* and *Lanistes* are absent, and the fauna is composed of three distinct elements, all foreign. One group has come from the Cape, another from India, and a third from the north, probably from Palestine. The last group is especially noteworthy. It consists of a group of species which are either still living in Europe, such as the common English pond shells, *Limnæa pereger*, Müll., *L. truncatula*, Müll., and *Succinea debilis*, Mor. ; or species that have only just become extinct there, such as *Corbicula fluminalis*, Müll., which is one of the commonest fossils in the gravels of the Thames valley. This species is still living in the Sea of Galilee, in the Tigris, and also in the Lower Nile. Aubry and Hamon discovered it in 1883 in some old lake beds in the Afar region of Abyssinia, through which the assumed Erythrean River would flow. The shell was found farther south by von Höhnel, and, in the course of a tributary of this river, I found masses forming a shell bed a little to the south of Lake Baringo. There it was associated with *Melania tuberculata*, Müll., which is another common species in the raised beaches of the Dead Sea. It therefore seems probable that the fauna, which is represented by this *Corbicula fluminalis*, worked its way into Abyssinia from Palestine, and entered Abyssinia by following the Erythrean River.

The evidence afforded by the birds and land animals is less complete, because, in consequence of the direct connection with Egypt, the fauna is mainly Mediterranean. But even among the birds there are many species which support the view of the direct connection between Palestine and Eastern Africa. Thus in the oases at the northern end of the Red Sea and above the gorge of the Jordan there is a Sun-Bird (*Cinnyris oseæ*, Bonap.), belonging to a family not found in Africa north of the Sahara, though to the south it ranges east and west through the tropical regions of Africa and Asia.[1] The African Darter (*Plotus levaillanti*, Licht.) is common in Southern Africa, on the Zambesi, along the East African coast, and on the Niger; it is absent from Egypt, Nubia, and the whole of the north-east of Africa, but reappears in Syria, in the Lake of Antioch, and no doubt once extended along the Jordan.[2] Of the three Turtle-Doves of Palestine one (*Turtur communis*, Selby) is European, another (*T. risorius*, Linn.) Indian, and the other (*T. senegalensis*, Linn.) Ethiopian; the last occurs throughout Africa, except on the north of the Atlas, and in Palestine is restricted to the Jordan.[3] Tristram's Grakle (*Amydrus tristrami*, Sclater) and the Fan-Tail Raven (*Corvus affinis*, Rüpp.) have the same remarkable range, being confined to Palestine and Eastern Tropical Africa; the former lives only in Palestine and Equatorial Africa, and the latter only among the cliffs of the Dead Sea and the highlands of Abyssinia and Kordofan.[4]

The literature of any group could be quoted as further evidence for the former connection of Palestine and Equatorial Africa along a line which avoided Egypt and the Lower Nile. The mammals give us the example of the Coney (*Procavia syriaca*, Hemp. and Ehr.), which belongs to a typically East African genus, but does not there occur farther north than Abyssinia. The only Rhizopod quoted by Tristram from the Dead Sea is the *Grammostomum capreolus*, P. and J., an Indian Ocean species.

Among butterflies, Hart's small collection on the Dead Sea yielded two species of the typically Ethiopian genus *Teracolus*, one of which (*T. phisadia*, God.) was known in Aden and Abyssinia, and the other (*T. chrysonome*, Klug.) in Somaliland. The evidence of the plants points in the same direction. Among Mediterranean and European species there are many known elsewhere only in Equatorial Africa and Asia, and—which are absent from Lower Egypt.

[1] See H. B. Tristram, *The Fauna and Flora of Palestine* (London, 1884), pp. 63-64.
[2] *Op. cit.* pp. 108-109.
[3] *Op. cit.* p. 121.
[4] *Op. cit.* pp. 74-76.

The same fact is clearly shown by Hart in his careful analysis of the Flora of Sinai.[1] He has published a list of species found in the Dead Sea basin, which are typically tropical in their range. Thus the mallow *Abutilon fruticosum*, G. and P., ranges from Senegal through Abyssinia to Beluchistan and Scinde, but is not reported from Egypt. *Loranthus acaciæ*, Zucc., is a common parasite on the trees of Nubia, Abyssinia, and the Wadi Arabah. The grass *Cyperus eleusinoides*, Kunth., has an enormous range to the south, but misses North-Eastern Africa. *Calotropis procera*, Willd., is common in Somaliland and the East African Nyika, but its Palestine habitat is separated from these by the whole length of the Red Sea.

It would be easy to quote further evidence, showing that one of the constituents of the complex fauna of Palestine entered that country from the south, but was excluded from Lower Egypt and North-Eastern Africa.

As Canon Tristram concludes—"To sum up our deductions, a review of the botany as well as the zoology of the Dead Sea basin reveals to us the interesting fact that we find in this isolated spot, comprising but a very few square miles, a series of forms of life differing decidedly from the species of the surrounding region, to which they never extend, and bearing a strong affinity to the Ethiopian region, with a trace of Indian admixture."[2] As the species which serve as the most striking illustrations of this fact live either in or beside fresh water, a river connection is the most natural agency by which to account for it; and as these species are absent from the Lower Nile valley and from Egypt, the river connection must have been established along the eastern side of the range of highlands, which separates the Nile from the Red Sea. Thus it is the aim of the present chapter to show that the apparently irreconcilable contradictions of the facts of plant and animal distribution become intelligible when the former geological evidence is taken into account; for the recognition of the existence and history of the Rift Valley affords a simple explanation of the occurrence in Palestine of a colony of the inhabitants of Equatorial Africa.

[1] H. C. Hart, *Some Account of the Fauna and Flora of Sinai, Petra, and Wady Arabah* (London, 1891), pp. 123-172.

[2] H. B. Tristram, *The Fauna and Flora of Palestine* (London, 1884), p. xvi.

CHAPTER XIV

NOTES ON THE FAUNA OF BRITISH EAST AFRICA

"*Lep.* What manner o' thing is your crocodile?

Ant. It is shaped, sir, like itself; and it is as broad as it hath breadth: it is just so high as it is, and moves with its own organs; it lives by that which nourisheth it; and the elements once out of it, it transmigrates.

Lep. What colour is it of?

Ant. Of its own colour too.

Lep. 'Tis a strange serpent.

Ant. 'Tis so. And the tears of it are wet.

Cæs. Will this description satisfy him?"—*Ant. and Cleop.* ii. 7.

"Nasals entering the nasal aperture; splenial elements not entering the mandibular symphysis, which does not extend beyond the eighth tooth. No bony nasal septum."
—The Crocodile according to Boulenger (*Catalogue of Chelonia*).

NECESSITY for rapid marching interferes more with the study of the habits of animals than with any other branch of a naturalist's work. While traversing a country at the rate of fifteen miles a day, there are few opportunities for the quiet observation of animals or of their habits. Besides, in the present condition of our knowledge of the African fauna, it is generally more profitable to collect than to observe; so to avoid risking the loss of an animal, it is often necessary to shoot or capture it at once, instead of quietly watching it at work or at play. Unfortunately, too, the only groups of animals of which I have any special knowledge are marine. For these reasons the present chapter contains only a series of somewhat disjointed notes and observations.

Mammalian life is naturally the first object of interest. In the African fauna it holds the most conspicuous position, although perhaps not the most instructive. The mind at once recalls the giraffe, elephant, lion, hippopotamus, rhinoceros, zebra, as well as the numerous species of antelope; and few scenes leave a deeper impression on the imagination of the traveller than the great herds of game, gambolling and feeding on the vast plains of the interior.

The relative importance of the great mammals is increased by the scarcity of the small ones. This reminds us at once of Darwin's famous comparison of the mammalian faunas of Africa and South America, when he pointed out that the barren steppes of the former are inhabited by vast herds of enormous animals, whereas the luxuriant vegetation of the latter supports only a few small forms. Darwin's statement of the facts is undoubtedly correct, and his conclusions just and instructive, but he gave no explanation of the anomaly. This may be suggested by describing Africa as the land of migratory mammals, instead of that of large mammals. The rains fall at two special periods of the year, which differ in different parts of the country, and at other times the steppes are burnt and foodless. In the dry season the soil is baked into ferruginous brick-like loam, so intensely hard that small mammals cannot burrow into it. The result is that the only animals that can live are those which have sufficiently rapid powers of locomotion to follow the rains, or are sufficiently strong to hold their own in the fight for water round the pools, or to survive through long periods without drinking. Unless animals are followed, one cannot realise how far they wander. Harris and I kept on the track of a herd of elephant on the Tana, and followed it all the way from Merifano till close to the coast. Several times we found tracks that could not have been made more than an hour previously, and once we heard them trumpeting and playing in a swamp beside our camp. But we never caught sight of them. The distance we followed them, however—only thirty miles—is little compared with that which hunters sometimes have to chase a herd. M. Foa, for example, in his recent book,[1] tells us how he tracked a herd of elephant for seventeen days, going as quickly as he could, before he finally came up with it. Other animals, such as the buffalo, travel in the same way. They walk slowly, feeding as they go, but often journeying twenty or thirty miles as the crow flies (or is supposed to fly), from one night's resting-place to the next. Only animals of

[1] Edouard Foa, *Mes grandes Chasses dans l'Afrique Centrale* (1894), p. 285.

considerable size can travel such distances, and since these journeys are necessary in Africa, the continent has gained the name of "the home of large mammals."

The vast herds of game, however, which once roamed over the steppes, are being rapidly reduced in size and number. Plains which in the days of Andrew Smith, Oswell, and Gordon Cumming were thronged with antelope, are now tenantless, and many of the species seem destined soon to follow the quagga and white rhinoceros into extinction. Man no doubt has played a leading part in the annihilation of the enormous herds that once thronged Cape Colony. The fact that, during the last few years, the game has retreated from the Somali coast into the interior, shows how easily it can be driven from a district. Nevertheless I doubt the justice of charging sportsmen with the main responsibility for the destruction of the big game.

In South America a mammalian fauna, much richer than that of Africa (for it included no less than fifty-eight genera of animals larger than a big dog), has been destroyed since a time which, though before the date of the human occupation of the continent, was geologically recent. Man has no doubt helped to exterminate some species, but his influence has probably been insignificant compared with that of natural agencies.

Lions are abundant on all the game-fields, and Jackson[1] and Mackinnon once saw twenty-three in a single herd on the Kapte plains. The number of animals such a herd must destroy every year is enormous, and disease is probably even more effective in the process of destruction. When Jackson[2] returned from Uganda in July 1890 he saw, between Baringo and Naivasha, herds varying in size from 100 to 600 buffalo six times in a single day; and Teleki,[3] while at Njemps in January 1888, shot no less than fifty-three individuals in the month.

In the same district in 1893 I did not see a single buffalo. Five years before the buffalo was almost the commonest of the big game in British East Africa. The whole number I saw was four—a herd of three in the Tana valley, near Ngatana, and a single bull in the valley of the Thika-thika. The explanation

[1] F. J. Jackson in *Big Game Shooting*, Badminton Library, vol. i. (1894), p. 245.
[2] *Op. cit.* p. 217.
[3] Höhnel, *Zum Rudolf-See* (1892), pp. 823-824

has been supplied by Gedge, who followed Jackson a few months later. Several times a day his caravan had to diverge from its path, to avoid the stench from a rotting carcase—in fact he saw fifteen in one day; but he did not see a single living buffalo.[1] Cattle disease had swept through the country, and destroyed them all.

The gnu and the giraffe have suffered almost as badly; I only saw one of the former and one herd of the latter, both on the Kapte plains; but in the valley of the Thika-thika I found giraffe bones nearly every day, and once saw the remains of six skeletons on a single march. Giraffe and gnu are both subject to the same disease as the buffalo, and thus in British East Africa they have almost shared its fate.

Another mammal that is now almost extinct is the Square-mouthed or Burchell's Rhinoceros (*Rhinoceros simus*), which differs from the ordinary *Rhinoceros bicornis* in having the mouth square and adapted for browsing, instead of prehensile and fitted to feed on leaves and shrubs. This species is only certainly known south of the Zambesi, and even there it is almost extinct. In marching across Laikipia I came one evening upon three rhinoceros together, browsing on the steppes. I was attracted by their light gray colour, and stalked them. As I approached, I found they differed from the ordinary rhinoceros, not only in colour, but in the bluntness of the head and in the shape of the horn. I had only a Martini rifle with me, and had by this time come to the conclusion that, in open ground, it was advisable not to attack more than one rhinoceros at a time. But I was so interested in these three, that I resolved to risk the attempt to secure one. I got within about sixty yards, when the birds resting on the backs of the animals saw me; their fluttering and cries disturbed the rhinoceros, which fled. I sprang up and sent a bullet into the hindmost, but the animal went on. It was late in the afternoon, and I was far from camp, so that I could not continue the chase. I reported this as soon as I returned to England, and before having read von Höhnel's statement[2] that Count Teleki had killed a white rhinoceros a little to the north-east of Lake Baringo; nor did I then know that a horn (now in

[1] *Big Game Shooting*, Badminton Library, vol. i. (1894), p. 217.
[2] *Zum Rudolf-See*, p. 542.

the British Museum), similar to those of the rhinoceros I saw in Laikipia, had been brought by a native caravan from this district, and described by Dr. Sclater as a new species, under the name of *R. holmwoodi.* Reports of a similar animal have been received from Loango, and it is probable that these rhinoceros all belong to a type once widely distributed over Africa, but now dying out. The square-mouthed rhinoceros is the nearest living ally of the extinct Tichorine or Woolly Rhinoceros (*Rhinoceros antiquitatis*), which lived in England when this was in a steppe condition at the close of the Glacial period. It is interesting, therefore, to find the existing representatives of this

FIG. 17.—A Bushman Rock-Painting of Burchell's Rhinoceros.

species persisting in the southern region of Africa, and on the high plateaux near the glaciers of Mount Kenya.

The square-mouthed rhinoceros is often called the "White Rhinoceros," as it is identified with that so described by the Dutch hunters at the Cape. The two specimens recently shot by Mr. Coryndon and sent to England, as also that obtained by Mr. A. Eyre in 1895, and now in the Museum at Cape Town, are as dark as the common species. In a discussion at the Zoological Society Dr. Günther made the very plausible suggestion that the White Rhinoceros of the Dutch may not be quite the same as the *Rhinoceros simus*, for the old Dutch farmers were not in the habit of calling black animals white. The colour of *Rhinoceros simus* probably varies within the same limits as does the common species. It is therefore possible that the old "white rhinoceros" is a different and extinct species, belonging to the square-mouthed group.

Though disease unquestionably aids in the work of extermination, it can scarcely cause the destruction of whole faunas, for one malady only seems to attack a few species. We must look to some other cause to account for the vast accumulations of bones belonging to animals of different species and of different habits, from which most of the remains of fossil mammalia have been derived. These huge piles of bones have always been a puzzle to geologists, for, as Sir Henry Howorth remarks—"Nor would any causes we know to be operating now account for the caches or heaps of incongruous beasts found in precisely the same fresh condition, and yet piled together in confused masses. This mixture of animals of different habits and habitats—of carnivores, and pachyderms and herbivores—is most puzzling, especially when the remains show so often a common freshness and an unworn and ungnawed appearance. Death certainly has no favourites, and is singularly neutral in its methods; but it does not, in its normal moods at all events, collect great mylodons and thickly-hided megatheres, nimble opossums and safely-cuirassed glyptodons, cavies and mastodons, and kill them together and bury them together."[1]

This singular association of bones is one of the arguments on which Sir Henry Howorth bases his theory of the destruction of the great extinct mammalia by a deluge. On the march across Laikipia, however, a different explanation of the phenomenon impressed itself forcibly, and even painfully, on my mind. The plateau had been described to me as one of the richest game-fields in Africa, and I trusted to it to supplement our scanty food supply.

Here and there around a water-hole we found acres of ground white with the bones of rhinoceros and zebra, gazelle and antelope, jackal and hyena, and among them we once observed the remains of a lion. All the bones of the skeletons were there, and they were fresh and ungnawed. The explanation is simple. The year before there had been a drought, which had cleared both game and people from the district. Those which did not migrate crowded round the dwindling pools, and fought for the last drop of water. These accumu-

[1] H. H. Howorth, *The Mammoth and the Flood* (1887), pp. 345-346.

lations of bones were therefore due to a drought, and not to a deluge.

My remark in reference to the encounter with the three gray-coloured rhinoceros—that I had learnt to leave these animals alone when there were more than one—may suggest that I regard the rhinoceros as a dangerous beast. Such, however, is not the case. I had been told on the coast many stories about the savageness of this animal, and the fury with which it charges caravans. But Jackson, the most reliable authority on East African big game, gives it a different character, and my own limited experience entirely confirms his. The rhinoceros is really so short-sighted and stupid that with a powerful smashing rifle it is the easiest of game to kill, and it will always run away if it has time to think the matter over. It is only dangerous in dense bush, for it then smells an opponent without seeing him, and dashes wildly about in a fit of fright and fury. I especially remember a valley to the south of Lake Losuguta, where, as we cut our way through the thorn jungle, we roused ten rhinoceros in one day. Some of them dashed away at once, but others charged the caravan, broke through the line, and disappeared into the bush on the other side, while a hail of Snider bullets shivered into dust upon their skins. Some of the brutes were uncertain of our exact position and rushed about, crashing wildly through the scrub, and snorting with rage. At these onsets some of the men would fling down their loads in terror, and rush for the nearest acacia bush to climb it; others would stand with rifles ready, though knowing that at any moment the rhinoceros might charge any one of us, and that as we could not see him till he was within a couple of yards, it would then be too late to dodge or stop it. This affected the men in very different ways; some of the pluckiest men in the caravan were the most alarmed, and some of the most stupid were then the steadiest. The former recognised most clearly their absolute helplessness, while their livelier imaginations gave them a more vivid picture of how they would look when the rhinoceros had finished with them.

That in such positions the rhinoceros is really dangerous was shown by the experiences of Mr. Astor Chanler's caravan, which at this very same time was trying to work northward to Basso Narok, about 100 miles to the east of us. Several of his porters were killed by rhinoceros, and Lieut. von Höhnel was so seriously wounded that he had to be carried back to the coast.

In open country and in thin scrub the rhinoceros is comparatively harmless, but even there he may be mischievous, as I found in my first attempt to hunt this animal. On the day of our entry on to the Kapte plains, we saw a rhinoceros feeding about a mile from the path. Telling the boy to follow with a Snider, I started to stalk it. We found that it was a cow with a young one. I succeeded in reaching a fairly good position, and was just preparing to fire, when some birds sitting upon the back of the animal detected me; they sprang into the air, flapping their wings and uttering a shrill hiss. This alarmed the rhinoceros, and it suddenly turned and faced me. In this position it was impossible to fire with any effect. So I crawled a few yards to the right in the hope of getting a side shot. The rhinoceros, however, moved too, and fidgeted about between us and its young one. The latter gave a short cry, and as the cow turned to look at it, I fired at its backbone. The report of the rifle and the animal's grunt of pain roused the bull, which was sleeping in a hollow close by. It trotted slowly into a position between us and its mate, and stood there until the female and the young one had fled to a safe distance up the hill. I waited to see what would happen, and kept ready for a side shot at the shoulder when the animal turned to follow its family, as I anticipated it would do. Instead of this, to my surprise, it suddenly charged furiously at us. Its head was held forward and was jerking up and down as it ran, so that I could not get a satisfactory shot. If I had aimed for the brain, the bullet would probably have been deflected by the horns. All I could do was to fire at the left cheek, which made the rhinoceros swerve to the right and then stop. I turned to get the Snider from my boy and give him the Martini to reload, when, to my consternation, I found that he had bolted with the second rifle and all the cartridges, except one, which I had in my hand. My spectacles were misty with perspiration, so I took them off and cleaned them. Just as I was replacing them the animal, without the slightest warning, charged again, and there was nothing for it but to run. I thought I could easily escape, for the rhinoceros looked such a slow and ungainly beast. I soon found, however, that he could go more quickly than I could, and double like a hare, so I dodged behind a white-ant-hill into some long grass. My pursuer fortunately missed me, caught the scent of my boy, and followed him. He soon lost the scent, and then sweeping round to the north galloped wildly away after its mate.

The only animal in the extermination of which man is playing the leading part is the elephant. The date of its extinction, however, is far distant, for in some districts it is still so numerous as to be a serious plague to the inhabitants. On the borders of the Kikuyu country elephants occur in such abundance and do such serious damage to the plantations, that an elephant-hunter would be welcomed as warmly as if he were a mediæval knight-errant come to do battle with the dragon. Lions also are numerous, and do terrible damage to the herds. Their tracks occur everywhere; and though I only saw them thrice, I heard them very frequently.

Lions as a rule appear to be timid, and three of them withdrew from the body of a hippopotamus, on which they were feeding, rather than allow me to put a bullet into one of them. But on another occasion I had an unpleasant experience of their audacity when hungry; for, as described on p. 150, they charged the camp at night, and killed two donkeys. The power of lions, however, has been much exaggerated. Those in Algeria have been reported to leap into cattle kraals, seize buffaloes bigger than themselves, and then, with their prey in their mouths, leap over ten-foot palings, and run away at full speed for miles. Neither of the lions I saw could have performed such a feat. Yet their slouching style of movement gives them an aspect of immense muscular strength; and the catalogue of accidents in lion-shooting shows that its dangers are not to be lightly estimated. It has been denied that lions can kill animals as large as the rhinoceros or hippopotamus; but that they can do so we found on the Thika-thika. Three lions had surprised a hippopotamus in some long grass about thirty yards from the river; there had been a desperate fight, in which the grass had been trampled down for yards around, but the hippopotamus had finally succumbed to loss of blood; its skin was terribly scratched by claws and teeth, and the lower part of the neck had been torn away. The hippopotamus was about two-thirds grown, and its skull (which I brought back to London) measured 19 inches in length.

Most of the larger mammals in British East Africa do not seem to adopt any especially protective colouring, but trust for safety to their speed, and to a good outlook against their foes. Many of the antelope when feeding have some of the herd posted as sentries, to watch for the approach of danger. These outposts may often be seen standing on an ant-hill; they do their duty so efficiently that it is generally difficult to stalk the herd. In some cases, however, a coloration, which at first sight would appear to render the animals extremely conspicuous, is found to be really protective. Thus the monkey *Colobus occidentalis* is covered with a long silky fur arranged in alternate stripes of black and white, so handsome that the skin is much prized by the Masai for making head ornaments. The contrast of black and white is so marked, that at first sight it would seem to preclude concealment, but its value is at once evident when the animal is seen at home. This monkey lives in the high forests of Abyssinia, Kenya, Kilima Njaro, and Settima, where the trees have black trunks and branches, draped with long gray masses of beardmoss or lichen. As the monkeys hang from the branches they so closely resemble the lichen, that I found it impossible to recognise them when but a short distance away.

The ornamentation of the zebra was also a puzzle to me till I saw them at home. The ordinary explanation of striped animals, such as the tiger, is that the stripes resemble bands of light seen through tall grasses and jungle. But this is not applicable to the zebra, which lives in open plains. Watch the zebra on these, however, and the value of the coloration is apparent. At a distance of from 250 to 300 yards the stripes of the East African species (*Equus boehmi*) cease to be visible, and the animal appears of a dull gray colour, like that with which warships are painted to render them inconspicuous. In dull cloudy weather, and especially at dawn and sunset, which are the most dangerous times for game, the zebra is practically invisible at a distance of over 500 yards. In bright sunshine, in the middle of the day, I have seen a herd of them at a distance of over three miles; but that is not the time when their enemies prey upon them, and then their visibility involves little risk.

It is, however, among the insects that cases of protective resemblance are most abundant. Insects shaped like seed-pods and leaf-buds, like lichen and dried sticks, or coloured like leaves and rocks, were met with nearly every day. They are too well known to need description, but two cases may be mentioned.

One case was met with during our return march from the first expedition. Strolling one evening out of the camp at Kurawa, I was startled by a hissing noise like that of a snake coming from a clump of grass. As I was wearing knickerbockers and tennis shoes, I sprang back and pelted the grass with handfuls of sand. As this did not drive out the supposed snake, I cautiously approached, peering into the clump. I could just detect a small green head among the stalks, and behind this appeared, whenever the noise was repeated, an expansion like the hood of a cobra. I tried to kill the animal by a few sharp blows with my stick behind the head, and one of these knocked it over. I then found that I had been frightened by a big grasshopper, which, by puffing out its wings, assumed a resemblance to the shape of the head of a hooded snake; while its noise was a good imitation of the dull, jerky hiss of some species of snakes.

The second case was more remarkable. I was working through the woods beside the Kibwezi river with Mr. Watson, one of the missionaries at the station there, when my attention was attracted by a large brightly-coloured flower, like a foxglove or a *Tinnæa* (see *Frontispiece*). It had been raining heavily, and the vegetation was so sodden with moisture that collecting was useless; I should have passed the specimen by had not I noticed some small, white, fluffy patches below the flower. These appeared to be lichen, of a kind which does not usually grow on flower stems; I therefore pushed my stick through the bush to pull the flower towards me; as soon as my stick touched it, to my great surprise, the flowers and buds jumped off in all directions.

There were several similar clusters close by, and when Mr. Watson came up I pointed one out to him and asked him if he had determined to what genus it belonged. He said he had not done so, but that he had seen it before, growing in these woods. He attempted to pick it, and was as surprised as I had been at the result.

The arrangement of the colony, with the green bud-like form at the top of the stem, and the pink flower-like insects below, looked so much like an inflorescence that it deceived both of us, although Mr. Watson is an enthusiastic botanist.

Whether the insects can resume this arrangement on the stem if they are once disturbed I cannot tell. Though we sat and watched beside them for an hour, they made no attempt to return to the stem. The insects were very sluggish, and simply clung to the leaves on which they first alighted. As a rule the members of this genus can fly well, but these seemed only able to hop for a few inches at a time, and would not move if they could help it. It may be that the insects were only rendered sluggish by the cold and rain; but it appears not unlikely that the members of this species have very limited powers of flight, and secured protection from birds by this ingenious mimicry of a cluster of flowers. The colony resembled very closely some species of *Tinnæa*, a genus which grows in this district; and Dr. Rudolf Schlechter tells me that it is even more like a Transvaal plant, which he names *Sesamopteris pentaphylla*, DC.

At the time I was much puzzled, for mistaking the lichen-like form for the larva of a beetle, I thought the pseudo-inflorescence was formed by three distinct species. My colleague Mr. C. J. Gahan, however, identified the insects as members of the hemipterous genus *Flata*, and on examining specimens in the British Museum we found that the larva belonged to the same genus.[1]

The arrangement is therefore intelligible: a female has

[1] The mimicry of the larva is now well known, as a colony of the larvæ of an allied species (*Flata limbata*) has been placed by Sir William Flower in the mimicry cases in the Natural History Museum, to illustrate their resemblance to lichens. Mr. W. F. Kirby has kindly called my attention to a translation of a paper by M. C. Piepers, "On the Habits of East Indian Insects, especially Lepidoptera," in his *Handbook to the Order Lepidoptera* (Nat. Libr. vol i., 1894, pp. lxii.-lxxiv.) In this Piepers describes how two rings of the orange butterfly *Callidryas scylla*, L., surrounded five white species of the genus *Pieris*, giving rise to a group, which was at first mistaken for an orange-coloured flower.

laid her eggs on a stalk, gradually walking up it as she did so. The eggs at the same height on the stem are thus of about the same age; the lowest eggs are hatched first, and the insects there are accordingly adult while those above are still immature. This does not account for the occurrence of the larvæ at the base, which is inexplicable to me, unless they belong to another generation.

This assumes that the green bud-like and the pink flower-like forms belong to the same species. Mr. Gahan has kindly sorted the specimens into male and female, and found that the colour difference is not sexual. Thus on the *Frontispiece*, Fig. 2*b* shows the male, and Figs. 2*a* and 3 the female form; these may be distinguished by the greater length of the transverse mark on the hinder margin of the fore-wings of the male. The same difference occurs between the sexes of the green form.

In every respect, except colour, the two forms are identical, and it therefore seems most probable that they are dimorphic forms of one species, which is *Flata nigrocincta* (Walker).

Another point which interested me in reference to insect coloration was the influence of the different capacities for the absorption of heat possessed by different colours. A black object becomes more heated than a white one, when both are exposed under the same conditions. An insect has so much surface in proportion to its bulk, that dark-coloured species are heavily handicapped when exposed to the intense sun of the tropics. This is the simple explanation of the fact, which impressed itself upon me as soon as I began to collect butterflies—that the light-coloured species fly in the daytime, and the dark ones in early morning and at dusk. I made considerable collections at Ngatana, at all hours of the day, to test this point. Thus on 30th January, I began collecting at 5.45 A.M., and found only species which are mainly of dark brown colour, such as *Hypolimnas misippus*, Linn., and *Junonia clelia*, Cram. At 6.30 a reddish brown species (*Limnas klugi*, Butl.) began to appear, and this was the only species caught during the next half-hour, though this was abundant. A little before half-past seven a light brown species, *Acræa cæcilia*, Fabr., made their appearance, followed immediately by numbers of light-coloured butterflies, such as *Teracolus syrtinus*, Butl., which is all white except for a red tip to the wings, and *Catopsile pyrene*, Swains., which is wholly of a light creamy white. The dark-brown forms disappeared from the open steppes before seven, and they were followed into obscurity by the light brown *Limnas*, of which not a single specimen could be found during the heat of the day. Then the open "barra" was tenanted only by white and light-coloured species.

This rule, however, is not universal, for other factors modify it. Thus in dull, cloudy weather the dark-coloured forms fly abroad all day; while some species of rapid flight habitually do so, such as many of the swallow-tail butterflies. *Papilio demoleus*, Linn., for example, a common species in the Sabaki and Tana valleys, was met with at all times of day; but it lived mainly under trees, darting out across open places from one shady place to another.

In connection with birds, the most interesting point observed was the presence of some of the Kilima Njaro species on the meadows around the glaciers of Kenya; these have been identified by Dr. Bowdler Sharpe as *Nectarinia johnstoni*, Shell., and *Pinarochroa hypospodia*, Shell.

The birds on the Lower Tana were only typical examples of the East African coast forms, such as the Grossbeck (*Melanobucco melanopterus*), and numerous Weaver birds of the genus *Hypantornis;* and on the march inland, neither time nor cartridges could be spared for them. Of the latter I hoped to make better use on Laikipia, where the drought, however, had cleared the district, and a vulture was the only bird seen until we approached the Kikuyu frontier. But elsewhere birds were abundant; wild duck swarmed on the fresh-water lakes; enormous flocks of flamingoes (*Phœnicopterus roseus*, Pall.) lived on the salt-lakes of the Rift Valley; guinea-fowl and pigeons were common in the woods. An interesting adaptation in form is met with on the open plains, where many of the birds have long tail feathers, which flutter in the wind like the tail of a kite, and aid in the balance needed for flight against the powerful winds that sweep across the steppes.

As reptiles can be preserved without the expenditure of time involved in skinning birds, I lost no opportunity of collecting them. The members of this group of most general interest are the crocodiles, and the same feeling was roused in me by Bird Thompson's thrilling stories. He had seen a good deal of these reptiles during his journeys on the Tana, which he described as simply paved with them. I had been taught in England to regard crocodiles as harmless bogies. In some places this estimate is no doubt true, but in others, facts are not in accord with it. In the Tana the crocodiles are of great size, and are certainly not to be despised. In some parts of the Ozi, if I leant my arm on the side of the canoe, the Pokomo canoe men left off paddling, declaring that a crocodile would seize me and capsize the canoe. That this was no idle fancy of the natives I learnt afterwards from Thompson. He was once coming down the Tana with a Pokomo boy, whom he had adopted, when a crocodile suddenly caught hold of the boy and dragged him overboard. The whole thing was over in an instant. No one saw the crocodile until its snout suddenly jerked out of the water, and at once the crocodile had disappeared and the boy was gone. There was a ripple on the water, but neither boy nor beast was ever seen again. During the ascent of the Tana by the stern-wheel steamer *Kenia* in 1891, Thompson had many unpleasant experiences of crocodiles. Several times when porters were going to the river for water, one was seized by the leg and dragged into the river.

The Tana Wa-pokomo regard the crocodiles as their most deadly foe, and strive to wage a war of extermination against them. To encourage this, no man is allowed to marry until he has killed a crocodile. The animals are surprised when asleep on the bank, and killed with spears ; but the work is rather dangerous, and inexperienced men are frequently knocked over by a blow from the reptile's tail, and dragged into the river. On land crocodiles are generally timid ; they lie close beside the water and slip into it at the slightest suspicion of danger ; but when hungry they are more daring, and a child, asleep in the verandah of the mission station of Jilore, was recently carried away by one. In the water they are very courageous. I was once fishing in the river at Ngatana, from a bank about six feet above it, when the chief came and warned me not to sit so near the water, as a crocodile might knock me into it by a blow with its tail. From politeness I moved a yard away from the edge of the bank until the chief was out of sight; but the danger seemed so fanciful, that I thought my old friend merely wished to speak to me, and could think of nothing else to say. Later on, however, I found that the natives of other Pokomo villages attribute the same power to the crocodile, and the German missionaries at Ngao knew of cases where people had been thus swept into the river and killed. The natives on the Nile told Sir Samuel Baker the same story, and it is hardly likely that it would have been independently invented in two such distant localities, and by such different tribes, if it had no basis in fact.

Reports as to the size of crocodiles vary greatly. Lieut. Giraud[1] shows a figure of one, which, on the most moderate computation, must have been 4 feet high and 30 feet long, while Werner[2] states that he saw specimens on the Congo 50 feet in length. I saw no such fine creatures as these. The normal length of the adults on the Tana was from 10 to 12 feet, but a few were perhaps 3 feet longer. In the rivers of the Rift Valley they were even smaller, and certainly less dangerous. When we had to cross a river, if there were crocodile tracks on the bank, we fired a few shots into the water, shouted, and splashed about, and then plunged in freely. Further precautions were only once or twice taken, where the river was swift, muddy, and deep, and the signs of the crocodile exceptionally abundant.

In the shallow jungle-filled swamps of Baringo, crocodiles, varying from 3 to 6 feet in length, are so numerous as to remind me of Bates's statement that the caymans in some of the upper tributaries of the Amazon are as abundant as tadpoles in an English ditch. These small Baringo crocodiles live on water-fowl, and seem afraid of anything larger. We soon learned their harmlessness and timidity, and went into the swamps and pulled out the ducks we had shot without hesitation. Yet even in this district the natives have an ingrained terror of crocodiles; our local guide would never go into the water if he could help it, and then only after taking elaborate precautions. Hence it is probable that the deeper pools of the rivers of the Baringo basin are haunted by

[1] V. Giraud, *Les Lacs de l'Afrique Équatoriale* (1890), p. 441.
[2] J. R. Werner, *River Life on the Congo* (1889), pp. 184-185.

crocodiles of a larger size, which levy a death tribute from the unhappy natives. The cumulative effect of repeated fatalities has apparently led to a hereditary fear, which the visitor who only hears of casual accidents finds it difficult to understand.

CHAPTER XV

THE FLORA OF BRITISH EAST AFRICA [1]

"Even in thy desert, what is like to thee?
Thy very weeds are beautiful."—BYRON.

IN Chap. XIII. reference has been made to the interest of the Alpine flora of Equatorial Africa, in its bearing on the problems of plant distribution, and the past climatic history of that region. The floras of the coast-lands and of the interior

[1] *Note on the Literature.*—As the exploration of Eastern Equatorial Africa began in districts which are now included in the German sphere, the knowledge of the botany of British East Africa has lagged far behind that of the regions to the south.

The three first important collections were made by Grant (1860-63) in the Victoria Nyanza basin; by von der Decken in 1859-61 (described by Ascherson in 1879) in the coast regions and the Tana delta; and by Hildebrandt (in 1878) in the Taita Mts. and Ukamba. The next additions were made in the Rift Valley and on its adjacent plateaux by Fischer (1882-83) and Thomson (1883-84). The collections of the latter were described by Sir J. D. Hooker in a remarkable paper, in which, from comparatively limited data, he drew conclusions as to the affinities of the flora, which have been confirmed by all later work. Extensive additions to the flora of the Rift Valley and Laikipia were made by Teleki and von Höhnel in 1887-88 (described by Schweinfurth, C. Müller, J. Müller, and Stephani). Important collections from the great central basin have been made by Stuhlmann (1890-91), Baumann (1891-93), and Scott-Elliot (1894); the small but instructive collection made by Stairs in 1889 may also be mentioned. Rev. W. E. Taylor has added greatly to the knowledge of the plants of Mombasa Island and Giriama. South of the Anglo-German frontier much more work has been done. Sir John Kirk early collected in Zanzibar and in adjoining parts of the mainland. Decken, New, Thomson, etc., made small collections from Kilima Njaro, from which adequate collections were first made by H. H. Johnston in 1884 (List by Oliver and Baker, 1886). Since the country passed into the possession of Germany, our knowledge of its flora has made vast strides by a number of botanists organised by Engler. His *Ueber die Hochgebirgsflora des tropischen Afrika* 1893 (1894), and *Ueber die Gliederung der Vegetation von Usambara und der angrenzenden Gebiete* 1894 (1895); the *Die Pflanzenwelt Ost-Afrikas und der Nachbargebiete* 1895, and the series of papers in Engler's *Botanische Jahrbücher*, "Beiträge zur Flora von Afrika" (Pts. i.-xi., 1891-95), are unquestionably the most thorough and scientific works yet carried out in Equatorial Africa. For the floras of Abyssinia and British Central Africa, which help to link the Alpine flora of Kenya with those to the north and south, reference may be made to the great monograph of Richard (1847) for the former, and the memoir by Britten, Baker, and Rendle on the flora of Milanji (1894).

plateau are less important in this respect than that of the meadows beside the glaciers and around the snow-fields of the higher peaks; nevertheless they yield interesting evidence as to the factors that govern distribution, and striking examples of the adaptation of plants to conditions of rainfall, soil, and climate.

Taken as a whole, the flora of British East Africa at first sight seems poor and commonplace—very different from the usual conception of tropical vegetation. There is no vast expanse of dense forest, in which the trees are buried below in a matted jungle of undergrowth, and laced together above by twining creepers and lianas, while the trunks are adorned by epiphytic orchids, bearing flowers of exquisite form or powerful odour.

Instead of this wild luxuriance, the vegetation is sparse, the flowers are small and insignificant, wild edible fruits are rare, and the flora appears disappointing, dull, and monotonous. Of course there are exceptions. Belts of dense forest occur on the alluvial plains bordering the rivers, and in the zones of perennial rain on the mountains of the interior, while the swamps on the coast are covered with acres of blue-flowered lotus-water-lily, or Yungi-yungi, and tiny yellow bladderwort (*Utricularia*). Neither is the country wholly lacking in plants of quaint form or interesting habit, for there are the trailing rubber vines (*Landolphia*) and rope-like lianas; the screw-pines (*Pandanus*) with huge spiral rosettes of leaves, and trunks balanced on the apex of a cone of aerial roots; baobabs (*Adansonia digitata*, Linn.) with massive soft trunks and irregular knobbed branches; silk-cotton trees (*Bombax* sp.) with graceful stems and rectangular branches, and forests of branching dum palms (*Hyphæne thebaica*, Mart.) These, however, are exceptions, and, as a rule, the vegetation is commonplace in aspect and ordinary in structure.

Most of the plants above mentioned are confined to the coast zone, and there they are associated with others which, as on Mombasa island, form a vegetation of exquisite beauty and more typical of the tropics. But a few miles inland there is a sudden change in the flora. The coast plants, the palms, mangoes, and fruit orchards disappear; they are replaced by huge candelabra-shaped euphorbias with thick succulent stems, by sharp-spiked aloes, saw-leaved sansevieras, and a wide expanse of thorny acacia scrub. But in spite of its unfamiliar appearance, the wide range and monotony of this flora soon render it tiresome. Here and there, it is true, there is a break; a belt of forest occurs beside a river; a jungle of shrubs crowns the summit of a ridge, or a tract of wood and turf covers a lava sheet. But the contrast of these oases only intensifies the dreary dulness of the surrounding plains, where day after day, week after week, there is nothing to be seen but the same few species.

At length there comes another change, as striking as the first, but more pleasing. As the land rises to a higher level, where the rainfall is heavier and more evenly distributed throughout the year, the spiny-leaved scrub is replaced by shrubs with luxuriant green foliage, the tufts of grass of the desert thicken to a fine rich turf, meadow flowers of European genera cheer the traveller with home associations, and forests of fine timber trees give him welcome shade from the tropical sun.

Throughout the whole country, however, the general aspect of the vegetation, taken in mass, is remarkably similar to that of temperate regions. Both the scrub of the plains and the shrubs of the plateaux, when seen in mass from a little distance, appears much like the bushes and pollards of our woodlands. The grasses and the tussocks of rush and sedge resemble those of England in form and habit, and in the interior many, such as the reed-mace (*Typha angustifolia*, L.) and the common rush (*Juncus effusus*, L.), are the familiar British species.

A superficial examination of the East African flora is sufficient to show that it occurs in zones which, as a rule, run parallel to the coast. These zones correspond with those based on geological and geographical considerations (see p. 222), except among the higher mountains, where the cold and rainfall have induced a local, zonal arrangement on a somewhat different plan. The three lower divisions, viz. the coast plain or "temborari," the first inland plateau or Nyika—including the primitive mountain axis as a subzone,—and the high grass plateaux or Rangatan of the interior, are as well marked botanically as geographically. The forest belt of the lower slopes of Kenya, of the uplands of Kikuyu, and of the summit of the scarp of Mau, belong to a fourth zone, corresponding less exactly to the geographical division, which includes the chain of volcanic mountains. Above this there are four other zones, the limits of which are due in the main to temperature and moisture. The botanical zones are therefore as follows :—

Zone.	Characterised by
1. Coastal Plain . .	Palms, mangroves, etc.
2. Foot-hills . . .	Fruit-trees and herbs of species common round shores of Indian Ocean.
3. Nyika	Acacia scrub, succulent trees and herbs.
4. Rangatan . . .	Flowers and shrubs allied to those of Abyssinia and the Cape.
5. Mountain Forest . .	*Podocarpus*, junipers, blackberries, stinging nettles, etc.
6. Bamboo . . .	Dense thickets of bamboo.
7. Lower Alpine . .	Tree heaths, *Gladiolus*, *Alchemilla*, and a tree lobelia (*L. dechenii*).
8. Upper Alpine . .	Tree groundsels, tree lobelias, and heaths.
9. Snow-fields . . .	A few *Helichrysum* and lichens.[1]

1. *The Coast Zone.*—The Coast Zone contains most novelties to the European visitor. Its most characteristic plants are the palms and the mangroves. The prevalent species of the former on the sandhills of the coast, and in the plantations of the Arab merchants, is the Coco-Nut Palm (*Cocos nucifera*, Linn.), which is to the East African what the reindeer is to the Lapp. He feeds on its nut and leaf-buds, drinks its milk and sap, and lives in huts made from its timber and leaves. He weaves its fibres into cloth, or twists it into rope, and makes the shells of the nuts into spoons and ladles; while he derives his main income from the export of the dried kernel or "kopra."

[1] Since this chapter was written, Engler has published a detailed memoir ("Ueber die Gliederung der Vegetation von Usambara und der angrenzenden Gebiete," *Phys. Abh. k. Akad. Wiss.* Berlin 1894 (1895), pp. 1-86) on the distribution of the plants in part of German East Africa, which he divides into eight zones. The correspondence of these zones to those here adopted is shown by the following table :—

1. Coast Land	Coastal Plain.
2. Creek Zone	Coastal Plain.
3. Bushland of the Jurassic Rocks .	Foot-hills.
4. Nyika	Nyika (zones 5 and 6 being regarded as fertile parts of the great Nyika plains).
5. Bushy Steppes of the Vorland . .	Nyika (zones 5 and 6 being regarded as fertile parts of the great Nyika plains).
6. Tropical Mountain Forest . .	Nyika (zones 5 and 6 being regarded as fertile parts of the great Nyika plains).
7. Parts of higher uplands with few or no trees	Rangatan.
8. High Mountain Forests . . .	Mountain Forest Zone.

The coco-nut palm, however, is more restricted in distribution than in utility. The natives have a proverb that it will not grow out of sight of the sea; this is practically correct, although we found a grove at Charra on the Tana, a day's march from the coast, and the missionaries have also planted some trees a day farther inland at Borabini, where they yield, however, only a scanty crop of fruit. The coco-nut palm is replaced in the interior by the branching Dum Palm (*Hyphæne thebaica*, Mart.), known to the Suahili as the "mlala." It forms dense forests a few miles from the sea, and extends for a considerable distance inland along the banks of the larger rivers; in British East Africa, for example, it occurs at Tzavo, 130 miles up the course of the Sabaki, and there is a small clump of dwarfed forms at the southern end of Basso Narok.[1] Emin[2] reports them in the Nile Valley as far south as Lado in latitude 5°·20 N. In full-grown specimens of this palm the stem bifurcates, and each branch may divide again once or twice, so that it appears very different in form from the familiar coco-nut or date palms. The fruit consists of a cluster of hard brown nuts (more correctly drupes), which in colour and taste resemble ginger-bread full of chopped-up horse-hair. People who have been accustomed to this as food seem able to thrive on it, but to others it is poison. Our men on the Tana would at first insist on eating it, but as it led to severe attacks of dysentery we had to prohibit its use.

The third palm common on the coast zone is the most graceful of the three—the *Borassus flabellifer*, Linn., the "mvuma" of the Suahili, the "doleb" of the Nile, and generally known as the Palmyra Palm. Its most striking feature is the spindle-shaped stem, which is thickest at about half its height from the ground, and tapers both to root and crown. In Eastern British East Africa it does not appear to extend far inland; I did not see it farther from the sea than a little above our camp at Vuju on the Tana. It occurs, however, on the Nile farther south than the dum palm, for Emin[3] found it at Dufilé, a short distance to the north of Mwutan Nzige (the Albert Nyanza).

The only palms that seem to thrive in the interior are a

[1] Shown in Höhnel's view, *Zum Rudolf-See*, p. 705.

[2] *Emin Pasha in Central Africa* (1888), p. 401.

[3] *Ibid.* p. 11.

species of wild date (*Phœnix* sp.), known to the Zanzibari as the "mkindu" (from which the river to the north of Kibwezi takes its name), and a water palm (*Raffia* sp.), both of which occur along the rivers of Kikuyu and Ukamba.

After the palms, the mangroves (mainly belonging to the species *Rhizophora mucronata*, Lam., and *Bruguiera gymnorrhiza*, L.) form the most striking feature in the coast flora. They grow in dense, jungly masses, bordering the estuaries and tidal creeks. The trees usually grow in the water, but the whole of the trunk is raised above it by a series of adventitious roots; it therefore appears as if the tree were supported on a many-legged stand. The most interesting point about the plant is the method by which it prevents its seeds falling by the way-side, and restricts them within the narrow belt in which alone they can grow. If the seeds were scattered on the surface of the estuary, the currents would either cast them ashore or wash them into water too deep for them to root in. The seeds therefore germinate while attached to the tree. The radicle grows into a thick solid spike, eight or ten inches in length, which, when released from its attachment to the parent, falls with sufficient force to drive the spike firmly into the mud beneath. The seeds therefore secure a suitable soil, and grow into dense thickets along the shore between the tide lines.

A third conspicuous feature in the coast flora came rather as a surprise. In front of the Residency at Lamu there are some specimens of that typically Australian tree, the She-Oak or *Casuarina*. These had obviously been planted artificially, so I thought no more of the matter. But on returning to the coast at Marereni, there we found on the headland a clump of these graceful feathery trees. Subsequently I noticed them in similar positions at all the points on the coast where I landed, as far south as Natal. The Suahili proverb, quoted on p. 237, shows that the natives have noticed the very restricted belt in which the she-oak can live, for Kisauni is on an estuary about a mile from the sea: their cones must have been carried from Australia by the West Australian and the "Equatorial Drift" currents, and washed upon the shore. They thus form an impressive lesson on the action of the sea as an agent in plant distribution.

2. *The Foot-hills.*—Accepting the palms and the man-

groves as the typical forms of the coast zone, it is necessary to make a separate zone for the "Foot-hills" that occur in most places between the coast plains and the inland plateau. As, however, most of the shrubs and herbs on these hills are the same as those on the lower zone, except when directly due to difference of soil, this belt is not very important. The most striking feature in this zone is the abundance of fruit-trees, most of which, however, have been introduced. Thus the mango (*Mangifera indica*, Linn.) and the betel-nut palm (*Areca catechu*, Linn.) have been imported from India; the cashew-nut tree (*Anacardium occidentale*, Linn.) and the papaw (*Carica papaya*, Linn.) from South America. The Jack-fruit (*Artocarpus integrifolia*, Linn.) represents the bread-fruit tree of the South Sea Islands; while the orange, lime, lemon, custard-apple, pomegranate, pine-apple, and guava are included among the debts which East Africa owes to the Arab settlers.

Among the indigenous plants there is a considerable proportion of Indian species, for the flora of both the lower zones contains many plants, common round the shores of the Indian Ocean. The specific names of some of the commonest, such as *Cephalandra indica*, Naud., *Abutilon asiaticum*, G. Don., *Glycine javanica*, Linn., and *Centrosema virginianum*, Benth., indicate the affinities of this flora.

3. *The Nyika.*—The third botanical zone commences at the summit of the slope that terminates the coast plains. Thence it extends inland for a great distance, occupying the whole of the "Nyika" and the gneiss mountains that rise above it. The differences between the flora of this zone and that of the coast-lands force themselves upon the notice of the most casual observer. Fruit-trees disappear, palms become scarce, the shady mango trees and sycamores are replaced by thin loose scrub. The main characteristics of this zone are the prevalence of spine-leaved acacias, and plants with massive succulent stems or leaves, such as the aloes, euphorbias, *Sanseviera*, and *Aristolochia*. The acacia scrub extends across the country for hundreds of miles. The trees as a rule are low (usually from 10 to 20 feet in height), flat-topped, and often umbrella-shaped. At times they grow so close together that a path has to be cut through them; as the scrub is thorny, and the undergrowth consists of hideously-barbed briers and of

"nkonge" (aloes and *Sanseviera*) with leaves pointed like bayonets or edged like saws, the work is severe. Elsewhere the scrub is thinner, and the traveller marches for days over a sandy plain, with here and there a thorn bush, and here and there tufts of hard dry grass. In the more fertile regions a few baobabs raise their knobby leafless branches above the scrub; beside the streams some large acacias spread out their flat cedar-like branches; but these are so exceptional that camps where they occur are generally named after them—*Mbuyuni* ("At the Baobabs"), or *Mkuyuni* ("At the Acacias"). Occasionally the vegetation is richer—where the decomposition of lava sheets has formed a fertile soil, where hills rise above the plain and collect a larger proportion of the rain than is their due, or where the springs maintain permanent water-pools. On the hills are dense growths of woody flowering shrubs, yellow marigolds and *Acanthaceæ*, feathery asparagus and wild bananas; around the pools are dense belts of reeds, rushes, and sedges, and upon them float the light-green, cabbage-shaped rosettes of *Pistia stratiotes*, and filamentous growths of *Lagarosiphon*. Such oases, however, break the monotony of the Nyika but rarely, and most of this zone is occupied by thin thorn scrub and sandy plains, and from the summits of hillocks the eye ranges over a vast expanse of flat-topped trees.

During the rainy season, however, a change comes for a while over the Nyika. The whole country is then sodden with moisture; the paths, which have generally been worn into hollows, are occupied by streams; the valleys are converted into swamps. The vegetation suddenly appears to wake up; the baobabs burst into flower and then into leaf; the grass becomes green; creepers climb over the acacias and cover them with a mass of large white flowers, among which the convolvulus is especially conspicuous. But as soon as the rains cease, the Nyika reverts to its normal condition. The grass withers, the undergrowth dies and disappears, prairie fires break out and sweep across the country, and the traveller has to march for days over charred, blackened wastes, which a month before had been green with turf and gorgeous with flowers.

The most remarkable feature in the flora of the "Nyika" is its specialisation to resist desiccation and death during long periods of drought. It is modified to enable it to survive the hot dry season in four different ways, each based on the principle of reducing the loss of moisture in "transpiration." The simplest contrivance is that adopted by the baobab (*Adansonia*), the leaves of which remain on the tree only during the rainy season. A second plan is used by many of the acacias, in which the leaves are reduced to mere spines or needles, the fleshy tissue being lost, and only the veins or "vascular bundles" left. A third arrangement for the same end is the reduction of the surface of the leaf in proportion to its mass; thus the leaf becomes thick and succulent, and the number of "stomata" (or pores through which moisture can escape) lessened, as in the fibre-yielding plants, such as the aloe and *Sanseviera*. The last and extreme method is the entire disappearance of the leaves, which are represented only by thorns and spines, while respiration is effected by the green succulent stem. The loss of moisture from the plant is therefore greatly reduced, for the surface on which it can take place is small in comparison with that exposed on a leaf-bearing tree. The plant secures, in fact, a minimum of surface with a maximum of mass. The thick succulent stems, moreover, contain special stores of moisture, and reservoirs of milky juice or "latex," which, being confined in special vessels or elongated cells (tracheides), can only escape by the slow process of exosmosis. The plants in which this adaptation has been developed are the most remarkable-looking in East Africa. Such are the species of *Aristolochia*, a genus which has one representative—the birthwort (*A. clematitis*, Linn.)—established in England; the forms in the Nyika are huge spherical bulbs, sometimes 3 feet in diameter, from which long trailing branches, armed with thick spines, spread over the ground. The bulb is full of juice; this is reported to be very poisonous, and the only animals that attack the plant are the ants. The Spurges (*Euphorbiaceæ*) offer a still better illustration, for they occur in two very different types. In the grass plains, as on Laikipia, this family is represented by small herbs, with leaves and structure like the common spurges of our woods and fields; but in the Nyika, the species (such as *Euphorbia Nyikæ*, Pax.) are lofty candelabra-shaped trees, from 30 to 60 feet in height, with thick succulent stems like the cactus. In other places, as in the Baringo basin, a closely allied genus forms hedges and thickets, which can only be traversed at some risk; for the spines are sharp and brittle, and if they run into the flesh they break off, and deposit an acrid juice which causes ulceration. The plant is useful, however, medicinally; the Zanzibari, who know it by the name of "mtepu," as well as the native Njempsians, chew the young shoots, in order to benefit by the gently purgative properties of the latex.

The mountains that rise above the Nyika, such as those of Taita and Iveti, at first sight appear to have a very different flora, owing to the prevalence of woody flowering shrubs, which form a dense jungle from 8 to 10 feet in height. But the smaller flowering plants, which, from the point of view of geographical distribution, are the more important, contain so large a proportion of dwarf representatives of the plants of the valleys and the plains, that it appears best to leave these mountains as a subzone of the Nyika. An occasional screw-pine (*Pandanus*) or a dragon-tree (*Dracæna*) even remind the visitor of the vegetation of the coast. The fact that the peaks and the valleys have the same flora instead of a totally different one, as in the case of Kilima Njaro or Kenya, shows that these mountains were probably never covered by an ice-cap.

4. *The Rangatan.*—The fourth zone is that of the "Rangatan," the prairies of the high plateau. This zone consists mainly of vast tracts of undulating moorland, covered with grasses. These grow to the height of from 2 to 4 feet; and as the ears ripen and turn yellow, they give the country a certain resemblance to the great cornfields of Dakota. Others, such as *Tricholæna rosea*, Nees., are tinged with a delicate brownish pink, which, when seen on a distant hill-slope, reminds one of a clover field on a Surrey down. Some of the most conspicuous grasses on the Rangatan are—*Setaria glauca*, Beauv., *Andropogon finitimus*, Hochst., *Themeda Forskalii*, Hack., *Sporobolus indicus*, R. Br., and *Eragrostis Brownei*, Nees.

The great charm of this region consists, however, in the belt of country on its borders. There, clusters of trees and clumps of flowering shrubs are scattered over velvety turf, forming park-like scenery of exquisite beauty. The resemblance to an English park is increased by the number of plants belonging to the same genera, or even species, as those of

Europe. Thus we find our common water-plantain (*Alisma Plantago*, Linn.), a British reed-mace (*Typha angustifolia*, Linn.), and rushes such as the Mediterranean species *Juncus Fontanesii*, Gay. The St. John's Wort is represented by a tree (*Hypericum Schimperi*, Hochst.), the mallows by *Hibiscus gossypinus*, Thunb., and the pond weeds by *Aponogeton abyssinicum*, Hochst.

The flora, therefore, presents resemblances to that of more temperate regions, such as the Cape, the highlands of Abyssinia, and the shores of the Mediterranean.

5. *The Mountain Forest Zone.*—The woods on the border of the prairies gradually thicken, and pass imperceptibly into the "Mountain Forest Zone," which occupies the region of heavy rainfall on the summit of the scarp of Mau and the lower slopes of the main peaks. The forest consists of great Junipers and *Podocarpus*, from 80 to 150 feet in height, which have been determined by Schweinfurth and Rendle as *Juniperus procera*, Hochst., and *Podocarpus elongata*, Thunb., *P.* aff. *Mannii*, Hook. The trees of this forest zone are draped with long pendent masses of gray, beard-like lichens, which Mr. Gepp identifies as *Usnea ceratina*, Ach. A dense canopy of vegetation formed of creepers and lianas spreads from tree to tree, while the ground below is carpeted with a rank growth of plants, among which species of European genera are very abundant. Chief among these are the blackberry (*Rubus*), the meadow-rue (*Thalictrum*), and the stinging-nettle (*Urtica*).

6. *The Bamboo Zone.*—The Mountain Forest Zone passes upward into that of the Bamboo, which on Kenya occupies the slopes from the height of 8000 to 9800 feet. It occurs also on Settima, Mau, and Elgon. The bamboos are packed together so closely that, except where the elephants have forced a way through them, a path has to be cut step by step. The bamboos rise to the height of from over 40 feet, and the stems are often from 3 to 4 inches in diameter at the base. Above they branch repeatedly, and the foliage interlocks into an imperfect thatch, which is always sodden with moisture. In spite of the abundance of the bamboos I was never able to find them in flower, and the foliage is insufficient for the determination of the genus; but Mr. Rendle tells me that they do not belong to *Bambusa*, the ordinary bamboo.

This zone of Kenya did not yield very much of botanical interest, no doubt because during our traverse through it there was no time to collect. The bamboos as a rule leave no space for anything except mosses, but in places the soil was covered by Iceland moss (*Selaginella*) and the maidenhair fern, of a species which Mr. Carruthers identifies with that living on our southern coasts.

7. *The Lower Alpine Zone.*—From the forests and jungle of the Bamboo Zone, it is a great relief to emerge on to the Alpine meadows above. These appear to be divisible into two zones, the Lower and Upper Alpine zones.

The Lower Alpine Zone occurs on Kenya between 10,000 and 11,500 feet. In its lower part it contains a few small bamboos and scattered forest trees, outliers from the zone below. Its most conspicuous features are shrubs and meadow plants. The former consist largely of heaths, belonging to the genera *Erica* and *Blæria*, of which the species are not yet described. In protected situations these may grow to the height of 15 feet, but on exposed slopes they occur mainly as low bushes, about the height of gorse.

The most striking flowering plants are sessile thistles, yellow composites much like the dandelion, gladiolus (*G. watsonioides*), and meadow orchids (*Disa Gregoriana*, Rdle.) A tree lobelia of a species very different from those of the zone above, and in growth more like the *L. Deckenii* of Kilima Njaro, occurs here.

8. *The Upper Alpine Zone.*—This zone extends from about 11,500 to 14,000 feet. The most typical plants in it are the tree groundsels (*Senecio keniensis*, Baker fil.) which grow in groves along the slopes of the valley. These, though belonging to the same genus as our common ragwort and groundsels, are extraordinarily different in form, consisting of fluted trunks of soft brittle wood, which branch once or twice, and sometimes a third time. The branches each support a great cabbage-like rosette of leaves, from which arise long spikes of groundsel flowers. Two tree lobelias (*L. Telekii*, Schw., and *L. Gregoriana*, Baker fil.) also occur in this zone. The most conspicuous plants associated with them are the "everlasting flowers" of the Cape and Abyssinia, belonging to the genus *Helichrysum.*

9. *The Snowfields.*—The eighth zone now ends at the height

of about 14,000 feet, although it once extended higher, for dead trunks of the lobelias lay on the moraine of the glacier, several hundred feet above any living ones, and in the zone now covered by the snowfields and glaciers. There a few small patches of lichens and dwarf *Helichrysum* (the highest of which were seen at 16,600 feet) are the only representatives of plant life.

The vegetation of Eastern British East Africa thus consists of eight or nine floras, which are so different from one another that it is difficult to discover any features which characterise the flora as a whole. The severity of the struggle for existence in this region, however, is shown by a few points that may be worthy of remark.

The first characteristic of the flora that impresses itself upon one—and it does so in a very pointed and unpleasant manner—is its prickliness. Some plants seem to consist of nothing but a collection of prickles; these are developed on every part of the plant, on the stem, stalks, flowers, and seeds, while the leaves are often reduced to a few needles. The thorns and prickles help the plant in nearly every stage of life; for they scatter the seeds by clinging to the fur of passing animals, they protect the plant against animals that would devour it, and in some cases obtain for the plant the food it requires from the air. Some of the plants have the prickles arranged in a very ingenious way: thus the "wait-a-bit" thorn (the Wacht-ein-beet of the Dutch settlers of the Cape) has hooks pointing in opposite directions, so that a jerk backward to disentangle clothes caught by them, only impales these all the more firmly on another set. In some cases, as if the thorns were not sufficiently formidable, ants burrow into their bases; the plant increases the growth of tissue to bury the invaders, and thus the base of the spine is enlarged into a woody bulb, capable of inflicting a nasty wound. Most of the succulent plants are protected by an abundant crop of thorns and spikes. Some of them, such as *Sanseviera guineensis* and most of the aloes, have broad leaves with edges spiked like a saw; another species of Sanseviera (*S. cylindrica*) has bayonet-shaped leaves, which end in a point so hard and sharp that it goes through leather as easily as through paper. The leaves in this "nkonge," as the Zanzibari call it, are placed opposite one another in a plane which, by a peculiarly malicious arrangement, is twisted into a spiral. Hence from whatever side these "chevaux de frise" be charged, the unlucky traveller only plunges on to the pricks; and this happens with aggravating frequency when chasing wounded game through the scrub in the dim light of early dawn.

Another type of defence extensively adopted consists of circles of stout hairs pointing downwards, which prevent the more simple-minded of the ants from reaching and robbing the flowers. Nearly all the grasses are provided with these spiked collars, for their farinaceous seeds are especially attractive to ants. In many of the flowering plants the same purpose is effected by having the leaves placed in opposite pairs with their bases "connate," or growing together to form a cup, as in the English teazel; in the rainy season, when alone the plant is in flower, this cup is filled with water, and this natural moat prevents crawling insects from reaching the coveted stores of honey and pollen.

The method of reducing the waste of seeds in a plant by their germination on the parent is not unknown among our English plants, and cases occur more frequently in the upland meadows (or Alps) of Switzerland. In Africa this "viviparity" is very extensively adopted. The case of the mangrove has already been quoted, but the method is used in the plants of the "Nyika" in an opposite manner; for in these the leaves act as a kind of parachute, and scatter the young plants broadcast before the wind, instead of restricting them to a narrow belt.

The action of the wind in moulding the forms of trees is well shown in the open plains of East Africa. The few timber trees that occur there are, like the baobab, fixed by trunks of enormous bulk, so that the wind acting on the leafless branches has but little power over them. In other cases the leaves are narrow and spiny, and the branches expand in flat horizontal sheets, which present only their narrow edges to the wind. In none of these cases does the upper portion of the tree give the wind much leverage by which to overthrow it. The smaller trees are also shaped by the same agency. The acacias which form most of the scrub are low and flat-topped, expanding above like an umbrella. This form is apparently acquired as a protection against the tornadoes that sweep over them. The shape is not much use to an isolated tree, for the wind can then get underneath the branches and tend to wedge it up; but these trees usually grow packed together, and their flat tops form a surface, over which the storms sweep with little effect.

It is no doubt also as a protection against the wind that, as if to atone for the leaflessness of the trees, the flowering plants are woody. Thus on the plateaux we have to look for the main masses of blossom, not to the turf on the meadows, but to the clumps of shrubs, which often glow with the yellow of the *Dombeya*, the *Calodendrum*, and the St. John's Wort (*Hypericum*), or appear gray with the groundsel-like flowers of the "lelesha" (*Tarchonanthus camphoratus*, L.)

In spite, however, of such points as these, the flora is mainly of interest in connection with the problems of geographical distribution. For the contrast between the tropical flora of the coastal plain or the Tana delta and the temperate floras of the Alpine zone, between the dense jungles and forests of the lower slopes of Kenya and the sparse scrub of the sandy Nyika, is the most striking feature in the natural history of the region. To a geologist this is of especial interest, for the distribution of the plants is determined to a considerable extent by the geological structure of the country. The nine zones are no doubt largely due to the different temperatures of different heights. But altitude is not the only factor; the boundary lines of the zones do not exactly follow the contours. The variations of temperature due to height are overpowered by the influence of the amount of moisture in the air, which is often determined by the character of the rocks in the country to windward. Thus the western sides of Kilima Njaro and Kenya have a greater rainfall and broader snowfields than the eastern, as the lava country to the west is more fertile than the sandy wastes to the east. The geological factors, moreover, not only determine the amount of the rainfall, but whether this is used to the best advantage. The higher mountains, and the lowlands beside the coast and the larger rivers, have a luxuriant vegetation, because the former intercept moisture from the elevated currents of air, while the latter are watered by a heavy dew. These sources are perennial; but, on the open plains, the air is dry and the rains are concentrated into two rainy seasons; the plant life in these districts is therefore dependent on the capacity of the soil to retain moisture. Geologically the plateau region consists of an undulating upland of gneiss, covered in places by sheets of volcanic deposits. According to which of the two classes of rocks crops out at the surface, there are one of two types of vegetation. The volcanic rocks form very rich soil; they absorb moisture readily, and part with it slowly. Hence the lava tracts are clad in a rich green turf or forest, and have numerous springs. The gneiss, on the other hand, is very porous; and, as the foliation is generally vertical, the water that falls upon it rapidly percolates to a depth at which it has no influence on the vegetation. The tracts formed by this rock are therefore barren sandy wastes, supporting only the dry grasses, the soft-stemmed leafless trees, the dreary scrub, and the succulent shrubs, that constitute the desert flora of the East African Nyika.

CHAPTER XVI

THE ZANZIBARI

"Whip me? No, no; let carman whip his jade,
The valiant heart's not whipt out of his trade."
Meas. for Meas. ii. 1.

AMONG the many blessings which the Arab has given mankind, not one of the least is the Zanzibari. Long before Bruce, Mungo Park, and Denham had roused English interest in African exploration, before even Prince Henry the Navigator had sent out his fleets, or Vasco da Gama had rounded the Cape and quarrelled with the people of Mombasa, the Arabs had quietly found their way into the heart of the Continent, settled on the shores of Tanganyika, built trading stations on the hills of Unyamwezi, and discovered all the great lakes as well as the snow-capped summits of Kilima Njaro and Kenya.

Probably for more than two thousand years caravans, organised by Arabs, have been despatched from the east coast to the principal trade centres of the interior. This work can only be done by human transport; for the tsetse and other flies poison one set of transport animals, the uncertainty of water and the roughness of the road are fatal to another, while long caravans in single file are at the mercy of hostile natives unless the load-bearer can fight as well as carry. Hence the Arabs had to raise a supply of native porters, and the Zanzibari is a product of the trade which could not exist without his services. Indeed the term Zanzibari is only a name applied to natives from any up-country tribes, who have come down to the coast and adopted the trade of caravan porters. Without this class of men, geographical exploration in the great belt of Tropical Africa between the Sahara and the Zambesi would have been impossible. The expeditions which discovered the great lakes and tracked the rivers of the Equatorial regions have all had to rely upon the transport organised by the slave-raiders and ivory-dealers of the Suahili coast. Hardly a fact about the geography of the interior of the Continent has been discovered, except at the price of the lives of these poor natives; and our knowledge of the roads that lead there is in direct proportion to the number of Zanzibari whose bones bleach beside them.

In spite of this, the services of the Zanzibari are not regarded in Europe with the gratitude they deserve. Travellers have given very different reports as to the behaviour of their men; but the best known accounts of them have been unfavourable. Burton, for example, describes his porters thus:—

"The self-willed wretches demean themselves with the coolest impudence; reply imperiously, lord it over their leaders, regulate the marches and the halts, and though they work they never work without loud complaints and open discontent. Rations are a perpetual source of heartburning; stinted at home to a daily mess of grain porridge, the porters on the line of march devote, in places where they can presume, all their ingenuity to extort as much food as possible from their employers." [1]

Cameron tells us [2] that in his caravan "the majority of the men were thieves, and pilfered unceasingly from their loads"; and he gives support to the view that the Zanzibari must be handled sternly, for he tells us elsewhere [3]—"I found that we had treated our men with too much consideration, and they in consequence tried to impose on us, and were constantly grumbling and growling."

Mr. Joseph Thomson tells us that he had to select the men for his "Masai Land Expedition" from "a flood of vagabondage . . . the blind and the lame, the very refuse of Zanzibar rascaldom, beach-combers, thieves, murderers, runaway slaves, most of them literally rotten with a life of debauchery"; and he accordingly had to start with "a villanous crew." [4]

[1] R. F. Burton, *The Lake Regions of Central Africa*, vol. i. (1860), pp. 342-343.
[2] V. L. Cameron, *Across Africa* (1877), vol. i. p. 11.
[3] *Op. cit.* vol. i. p. 107.
[4] J. Thomson, *Through Masai Land* (1887), p. 23.

It is natural that stories of difficulty, mutinies, and crimes, and of the frequent failure of expeditions through the misbehaviour of the men, should leave a deeper impression on the mind than the records of faithfulness and quiet obedience. The prevalent belief in the utter villainy of the Zanzibari is expressed by Drummond in the following passage :—

"These black villains the porters, the necessity and the despair of travellers, the scum of old slave-gangs, and the fugitives from justice from every tribe, congregate for hire. And if there is one thing on which African travellers are for once agreed, it is that for laziness, ugliness, stupidness, and wickedness, these men are not to be matched on any continent in the world. Their one strong point is that they will engage themselves for the Victoria Nyanza or for the Grand Tour of the Tanganyika with as little ado as a Chamounix guide volunteers for the Jardin; but this singular avidity is mainly due to the fact that each man cherishes the hope of running away at the earliest opportunity. Were it only to avoid requiring to employ these gentlemen, having them for one's sole company month after month, seeing them transgress every commandment in turn before your eyes—you yourself being powerless to check them except by a wholesale breach of the sixth—it would be worth while to seek another route into the heart of Africa." [1]

I shared this opinion when I first went to Africa, and agreed with the other members of the expedition as to the great importance of the attempt to run a camel caravan on a large scale so far to the south. We thought that if camel transport succeeded, it would revolutionise the conditions of travel in Equatorial Africa by superseding the lazy, lying, expensive and mutinous Zanzibari. It was therefore a surprise to me when, on asking Mr. Piggott on the "Malda" about systems of transport, he spoke enthusiastically of the merits and virtues, efficiency and economy of the Zanzibari, and said that he doubted whether anything else could compete with them. I was the more astonished to hear the Mombasa men thus praised, as I remembered Johnston's warning to travellers—"Never, if they can help it, to engage porters at Mombasa. Independently of all questions of religion—Mohammedan and

[1] H. Drummond, *Tropical Africa*, 4th ed. (1891), pp. 5-6.

Christian alike—the inhabitants of the Mombasa district are a thoroughly bad lot. It is hopeless to win them by kindness, or infuse a spirit of discipline by sternness. They are liars, cowards, thieves, and drunkards." [1]

My first experience of Zanzibari porters was unfavourable. At Mkonumbi several were engaged for us, and failed hopelessly. They could not carry the regulation loads; they showed no knowledge of camp ways; they proved sullen and lazy. However, these were only a scratch lot—mostly freed slaves, who had been tempted from ordinary agricultural labour by the prospect of high wages and little work. After they had been discharged, I saw no more of Zanzibari, until our eighty porters under Omari came to the camp at Ngatana. Their attitude throughout was amusing rather than pleasant. They made no attempt to conceal their contempt for the Abyssinians and Somali, and I think we also came in for a share of this feeling. Whenever the porters arrived in camp from Witu with a fresh cargo of stores, the Abyssinians would at once drop their loads and fall exhausted beside them; the Zanzibari, on the other hand, simply to show their contempt for such feebleness, would dance and sing, and race round the camp with their loads on their heads, until they were ordered to stack them. The feeling of the Zanzibari towards the Somali was more definitely hostile; jealousy of the higher pay and privileges, which they thought the Somali did not deserve, added bitterness to their contempt. In their good-natured way the Zanzibari were generally ready to help the Abyssinians to make themselves comfortable in camp; but they seemed to gloat over the miseries which the Somali endured, owing to their ignorance of the camp dodges, in which they themselves were experts. The result was that in time the hatred between the two classes of men became so intense that, to avoid quarrels, the Zanzibari camp had to be pitched a few hundred yards from that of the Somali, and the men were kept as much separate as possible.

In other ways, however, the Zanzibari gave us no trouble. When they coveted Pokomo pumpkins, they got them for the asking from their friends the villagers, or they stole them so cleverly that the Somali had the blame. They never dis-

[1] *The Kilima-Njaro Expedition* (London, 1886), p. 43.

tressed us by dying, and were not even seriously ill. In the crowd of Somali and Abyssinians which came twice daily to the dispensary hut, when Mackinnon was "at home" to invalids, not a Zanzibari was to be seen, unless one were there for the fun of jeering at the feeble folk. The Zanzibari, in fact, did their work in a quiet, business-like way without fuss or noise. Moreover, they entertained us greatly by coming over to our camp to sing and tell funny stories.

But in spite of all this we did not like them. They assumed an air of bumptious arrogance that was not pleasing. It is not exactly gratifying to feel that your native servants despise you, as ours manifestly did. We felt we could not depend upon our men if we quarrelled with their headman. We knew they used to go to Omari, saying, "This is not a *safari!* Why should we waste our time here any more. Let us beat our drum and go!" Had we had any unpleasantness with Omari, they would have beaten their drum and gone. Thus a certain distrust was added to our consciousness of their aggressive self-assertion.

But when once we had crossed the Tana and begun our march, the Zanzibari had to work, and then everything was changed. They showed themselves adepts at camp ways, and proved most willing workers. When a halt was called, camp was pitched speedily and without confusion. Tents were up, water and wood brought, fires lighted, and cooking done, as if by magic. There was none of the old trouble to make the men perform these tasks, and do the other odd jobs required to render camp comfortable and orderly. The Somali had to be told everything, and only obeyed slowly and sullenly. Each Zanzibari, on the other hand, knew his special duty, and hastened to do it with willing obedience. We often had a forcible illustration of the difference in their demeanour, when we were trying to sleep at night after a hard day's work. The noise in our Somali camp was maddening, especially when we were ill with fever; Benett-Stanford would go out and order the men to be quiet. "We Englishmen," he would say, "work all day and try to sleep at night; but you Somali sleep all day, and then chow, chow, chow all night." The Somali would laugh and be quiet for three minutes, and then the hubbub would begin again. If the Zanzibari were too noisy we had only to send a message to Omari; his stern voice would ring through the camp, "Kilele, kilele" (Silence, silence), and every sound would be hushed.

The change from the Somali to the Zanzibari was so pleasant, and the behaviour of these latter so satisfactory during our return march to the coast, that I resolved to employ them exclusively on my second expedition. The Somali wanted to go with me, for they were returning to Aden penniless; many of them begged to be employed, and some even offered to serve for nothing. But I would not take one at any price. I trusted entirely to Zanzibari, and well indeed did they justify my trust.

When comparing experiences with others who have used Zanzibari, I feel that I was fortunate in getting a good set of men; and the fact that I had not a single deserter shows that I did not get any of the riff-raff of Mombasa or Zanzibar, for they would have bolted at the start. Yet my men were not by any means a picked staff; being in a hurry to start I had at the last to take any porters I could get, to supplement those chosen from the first expedition. That the men did not desert later on was probably due to the fact that the caravan was worked on a different principle from that often adopted. Several men of great experience had given me advice as to how to treat Zanzibari. Prince Boris Czetwertynski, for example, had told me that the only way to make a Zanzibari respect you is to flog him. Yet this system of management is not always successful. A caravan had gone into the interior just before me, the leader of which, a very experienced man, was reported to have adopted the "flogging" system. He had lost so many men by desertion, that he had been obliged to camp for two months in the "barra" waiting for reinforcements to be hired and sent up to him. I could afford neither the time nor the money involved in such a delay, and realised that it would be better to lead the caravan than to drive it. The result was we had very little trouble. The men made a march of unprecedented swiftness, and had a really hard time of it. When asked by Hobley how they had got on, upon their return to the coast, Omari's reply, "Teli khazi, *teli* khazi!" (Much hard work, very much hard work!); "teli mvua!" (much rain!), expressed the general opinion. Nevertheless throughout the men did their work, and put up with exposure with the greatest cheerfulness. Flogging was only necessary twice, and most of the time we were rather like a family party.

My first axiom in the treatment of my men was based on Sir Evelyn Wood's maxim, when organising his expedition against the Boers—the first thing to be considered is the health of the baggage animals. My porters were my baggage animals, and I made every arrangement I could for their health. I tried to adapt my ways to their habits, for my nature being more plastic than theirs I could more easily fit into a new environment. But I never allowed the work of the expedition to be interfered with, and if exposure were necessary the men had to stand it. On Kenya, for example, they were kept at a high altitude far longer than was good for them.

My second axiom was to know as much as possible about my men. I was interested in them as samples of African natives, and was also anxious to learn all I could of their language, habits, and temperaments, so that I might use the men to the best advantage. I often spent a few minutes with them at night, chatting by their camp fires, drawing from them stories of their past journeys and telling them reminiscences of my own. They may have credited me with a good deal more disinterested benevolence than I deserved; but I believe this close intercourse with my men was the chief source of our harmonious co-operation. Among other things, it impressed on me that the Zanzibari are not a tribe but a class, and must be dealt with accordingly.

Only twenty-four of the men of my caravan knew to which tribe they belonged, and these twenty-four represented no less than fourteen different tribes—Wa-konde, Wa-zaramu, Wa-digo, Wa-doa, Wa-makwa, Wa-degereko, Wa-gindo, Wa-ganda, Wa-biza, Wa-kame, Wa-nyamwezi, Wa-nyasa, Wa-hadamu, and Wa-pemba.

The national characteristics of these tribes differ so widely, that it is useless to treat the men as if they were all turned out of one mould; the difference of type, moreover, gives a diversity of capacity which a leader may apply to the various needs of an expedition. For example, the Wa-degereko are a domesticated and agricultural race, men of enormous physique but of a timid and essentially effeminate character, whose only theory of fighting is to live to do it on another day; they have no notion how to find their way in the jungle or how to track game, but they have an uncommonly keen eye for business and bargaining. The natives of some tribes, on the contrary, are born hunters and trappers, who have been trained from their youth to deeds of daring in war and in the chase, or, like the Wa-nyamwezi, have a high sense of union and discipline. Others, like the Wa-digo, are intelligent, courteous, but cowardly, and such excellent linguists that this tribe supplies interpreters to most of the up-country caravans. The Man-yema are a cannibal tribe, whose stupidity and roughness are scarcely redeemed by the dash and daring of first-class fighting men. The Wa-konde are brusque and business-like, and you can lay responsibilities on them for which the sullen, grumbling Wa-kame are totally unfit. It is evident that the methods of dealing with such varieties of character should, if possible, be as varied. A joke may do with one man what a threat alone could achieve with another; one may be most moved by a promise of reward at the coast, when another can be touched by an appeal to his honour. It is from caravan leaders who do not study the characters of their men that we hear most of the stories of mutiny and desertion, and that the only argument a Zanzibari can understand is the whip.

I may mention the following incident as illustrating the inspiriting example of a good leader, and the Zanzibari's capacity for following it. On one occasion, after being forty hours without water, we came upon a scanty supply, and my headman Omari, who had been trained by Stanley, refused his share. He shook his head at first, for so dry and stiff were our tongues that we could hardly speak, and then stammered out that he was not thirsty. When I insisted on his taking his small share, he quietly handed it to the porter who appeared most overcome with thirst. As I looked at his face, grimy with the labours of the long day, I could not help thinking of Kipling's lines—

> "But for all 'is dirty 'ide
> He is white, pure white inside."

Later on I asked him how he had denied himself. He said, "It was nothing! I've seen Bula Matari (Stanley) do the same thing lots of times, and if he could do it, Inshallah! so can I."

Like the rest of the world, the Zanzibari have their faults, and many of their failings are intensely irritating. Their greatest vice is a tendency to desert; they are inveterate liars, and frequently present an amazing mixture of timidity and foolhardiness. The Zanzibari have a hereditary fear of the Masai, and an undrilled caravan is liable to terror-stricken panic in case of attack. Yet the men persist in straggling behind in the most hostile country, and if the fancy takes them will go off single-handed to pillage a plantation, apparently totally indifferent to the fact that by so doing they are endangering the very existence of the whole party. Perhaps the most troublesome fault of all is that they cannot be trusted to keep awake on sentry duty. According to general report, the greater the danger the more soundly the sentries sleep. To find your Askari sentinels either dozing or fast asleep during the most risky hours, night after night, is a severe strain on the temper, and you are then apt to curse Zanzibari in general, and kick the sentries in particular. I always went round camp two or three times a night, and when in hostile country did not take off my clothes, except to change them, for days together.

The recklessness of the men about their food is another trying characteristic. At the commencement of a new stage in the journey we had to serve out ten days' rations, and some of the men would eat so much in the first few days that by the end of the week they had none left. But they can go for great distances on what appears to be the most insufficient food; some of my men carried loads of 110 lbs. from dawn to dusk, with only an hour's rest in the middle of the day, on a pound and a half of beans or Indian corn, and sometimes less than that. Whence their "foot-pounds" of energy were derived puzzled me, until I noticed that they became thinner and thinner. They illustrate the laws of compensation; for the amount of food the men can eat, when they have it, is simply phenomenal. When we reached the Kikuyu country on the return journey, I owed all the men arrears of food, amounting with one group of men to seven days' rations. I offered to give them beads or wire instead of the excess of food, that they might buy for themselves any little delicacies they fancied, such as chickens or ripe bananas. But they refused my offer. "You owe us seven days' food," they replied; "seven days' food we will have or nothing." Of course it was given them; but, in the evening, one of them came as a delegate from the rest to ask for medicine. He complained of severe internal pains, and seemed very uncomfortable. His power of diagnosing his symptoms being limited, I asked what he had been doing or eating. He calmly replied he had done nothing, and had only eaten the food that had been given him. Each of the men having obtained his seven days' rations had borrowed a big cooking-pot, made a great fire, and had cooked and eaten the whole of the $10\frac{1}{2}$ lbs. of beans. I was somewhat annoyed and declined to give medicine, telling the emissary the only expedient I could think of to prevent fatal consequences was a band of hoop iron; this we had not got, so he must tie himself together with my climbing-rope. When the story of this Milo's meal was circulated round camp it was regarded as a good joke, and the appetite of the six huge feeders was added to the list of stock-jokes of the caravan.

A more serious failing in the Zanzibari is his liability to violent paroxysms of passion. When the fit is on him he is hardly accountable for his actions; he will take offence at the merest trifle, and until he recovers is ready to desert in spite of any consequences to himself. The well-known Zanzibari tragi-comedian "Tom Charles," for example, threatened to desert from Teleki's caravan only two days before reaching Mombasa, owing to some tiff with the native headman; it was only Teleki's good-natured intervention that dissuaded him from forfeiting his two years' hard-earned pay and his reputation as a reliable servant, besides rendering himself liable to a long term of imprisonment. This tendency to passion greatly increases the difficulty of handling the men; they have to be humoured and coaxed like children; and when in anger, appeals to their self-interest are in vain, for the last persons they then think of are themselves.

To secure peace in a caravan, a leader must respect the customary organisation. The sole duty of a porter is to carry his load from camp to camp, and all other work is supposed to be done by special men. Stacking the loads, sentry duty, putting up the tent of the leader of the expedition, and collecting his firewood and fetching his water, is the work of the Askari. There is always one of these to every ten porters. They do not carry loads unless a porter is taken ill on the march, and then the Askari is at liberty, before taking up the man's burden, to give him ten blows with a stick as a safeguard against malingering. This is a recognised right, though the Askari do not as a rule insist upon it, unless in an obvious case of imposition. I never saw it enforced. On the contrary, in cases of illness the men were always ready to help one another, and several times I saw an Askari insist upon taking a load from a sick man pluckily struggling to bear up under it. The Askari are responsible to the "munipara" or headman, who holds an extremely important position, and upon whom to a large extent depends the peaceful working of a caravan. He distributes the loads and food, and must see that all loads are brought into camp. Hence he generally marches last, and must not leave this post without the permission of the European leader, who then generally takes his place. It is the absolute rule of the road that no man is to be left behind alone. If a porter is too ill to proceed, the headman stays with him, sending on a message by one of the Askari for a fatigue party to be sent back to carry the invalid. The porter has his rules to obey; *e.g.* he must not upon any account leave the track without leaving his load there, or giving some other sign to the headman. The only man I lost was sacrificed through his own neglect of this rule; he was tired, stole from the path, and went to sleep under a tree; when later on he tried to follow us he lost the way; our search for him was vain, and he, no doubt, was either starved to death or killed by exposure to the cold, unless the lions mercifully put an end to his sufferings. As a rule the men obey their unwritten code implicitly; it is, indeed, a matter of habit with them, and an Ethiopian will sooner change his skin than his habits; indeed that is not so difficult as one might imagine, for in cutting our way through the thorn jungle of the Sabaki or Lake Losuguta, the men changed their skin for sticking plaster at an appalling rate.

The whole ethic of a porter's religion is concerned with his load. His creed is very simple, consisting of two articles—(1) Thou shalt not drop or abandon thy load; (2) Thou shalt not steal from it. Stanley tells a good story which well illustrates the first rule. When he started on his search for Livingstone he was new to Equatorial Africa, and anxious to try new methods of transport. So he took out a light, narrow waggon, which was laden and hauled inland. The waggon, however, was always in mischief, and generally hours behind the rest of the caravan. Stanley at last got tired of it, and one day, when it was unusually far in the rear, he sent back an order that the thing was to be thrown into the jungle, and the men with it were to come on as quickly as possible. Hours passed and they did not appear. So Stanley marched back and met the stragglers, led by the strongest man staggering along with the waggon, axle, wheels, and all complete, upon his head. The porter pleaded that he could not leave his load, and that it was easier to carry it thus. But this was such a *reductio ad absurdum* of wheeled traffic, that the waggon was thrown on one side and there abandoned.

On Kenya I had a somewhat similar experience, when a porter, not by any means one of my best men (in fact one of the worst), nearly sacrificed his life in the effort to save his load. He had been caught in the snowstorm that broke upon us, the day of our arrival at our camp above the forests. He could not drag his load up the steep slope that led to the camp, and he would not go on without it. It was a mere matter of etiquette. The load would have been all right if he had left it, and there were no natives to steal it; but it was against the porter's religion to leave his load, and he sat upon it. After an hour's search I found him, half covered in snow, lying on his load, nearly frozen to death. A little brandy revived him, but he was too weak to stand, and I had to carry him up to camp. Next morning when he was better, but while I was still suffering from the irritation of having to hunt for him in the snowstorm, I told him he was a fool to have stopped there, and that he ought to have left his load and come on when he could have done so. "How could I leave my load without my master's order?" was the man's reproachful reply. Such is the stuff of which a good Zanzibari is made.

The regulation load which the Zanzibari is supposed to carry is 60 lbs. Many of the missionaries think this excessive, and refuse to allow any of their men to be hired, unless a promise be given that they shall not be required to carry more than 50 lbs. But as a matter of fact the weight is more often raised than reduced. The British East Africa Company allow loads of 65 lbs., and, at the outset of an expedition, this is the utmost that a man should be expected to carry. Later on, when the men have become used to their work, the amount can be increased. We often had loads made up of 50 kibaba of food which would weigh in all 75 lbs. The packing adds another 5 lbs., and the porter's ten days' rations 15 lbs. more. A rifle and 20 rounds of ammunition would raise the total to about 105 lbs. In addition to this the man has to take his turn at carrying a heavy copper "sufuria" or cooking-pot, and always has a roll containing his shelter tent and the cloth that serves as a blanket, and a few other trifles of his own. Porters with such loads as these are therefore really carrying 110 lbs. at least, instead of the regulation 60 lbs.; and the men seemed to regard this as quite moderate treatment. Teleki[1] on one occasion gave his men loads ranging from 110 lbs. to 148 lbs.

It appears almost as if, while you keep to their load, the porters will allow you to impose upon them *ad infinitum*, or at any rate as long as their strength will hold out; but order them to fetch you some firewood from a bush a hundred yards distant, to carry a load across camp, or to do anything that is an infringement of the established *dasturi* or custom, and it is quite another matter. If you respect *dasturi* and *ask* a man to perform the service, he will be flattered and do it at once; but the smallest demand upon him outside his recognised duty will put him in a bad temper for the rest of the day. At the beginning of an expedition, when there are all sorts of odd jobs to be done, the men are slack, and the leader perhaps not used to caravan ways, *dasturi* is continually invoked. If the appeal is not respected, the chances are the men will desert. Their natural view is, that if a leader will not respect their privileges near the coast when protection is at hand, what will he not do up country when his men are entirely at his mercy? The porters, as a rule, implicitly obey the laws of safari custom, and expect their leader to do so too.

Trivial infringements of *dasturi* are therefore bitterly resented, though the men appear generally ready to tolerate them, if really necessary. On Laikipia I had to withdraw

[1] Höhnel, *Zum Rudolf-See*, p. 698.

certain privileges, which are usually held sacred. The men protested; but I told them that the only *dasturi* on Laikipia was to be speared by the Masai and eaten by the hyenas. If they wanted that fate they could easily get it; if not, they had better not talk about *dasturi* till we reached a safer country. The men laughed, and we heard no more about the matter.

Greatly though the Zanzibari loathe any interference with their customs and privileges, they will tolerate even that rather than an act of injustice. This rankles in the Zanzibari mind, and is certain to be paid off some day. Amongst my porters was a Mgindo, serving under an assumed name, which we will suppose to have been Wadi. He was my most faithful personal attendant. When we were in a waterless camp at night, Wadi would wait till no one was looking, and then sneak my water-bottle and fill it up from his own small calabash. Several times when our food supplies were approaching exhaustion, and we were on short rations, he would tie up half of his in a corner of his loin-cloth to save it for me. I only found this out by accident in Kikuyu. As soon as the natives brought food into camp and the prospect of famine was averted, Wadi produced his store and began to eat it. I chaffed him about his caution, and he did not seem to like it. Omari afterwards remonstrated with me for my banter, for he said, "Did you not know that the food was all being saved for you?" Wadi had remarked to him a day or two before that if the Kikuyu would give us no food, "Mpokwa" would want the beans sooner than he would. Wadi was always doing things of that sort, walking miles after the day's work was done to collect herbs for food, and then giving them all to others. Nevertheless he was a man with a bad record. He had been one of the two ring-leaders in perhaps the worst act of mutiny that has disgraced the annals of British East Africa. Wadi was with a caravan in the interior. According to his account, the leader knew no Ki-suahili and was dependent entirely on his interpreters—a lot of mission boys. One of these committed an act of theft. The leader of the caravan said the thief must be found. Wadi was accused. His defence was falsely translated by the interpreter, who was a confederate of the thief. Wadi was condemned and flogged. A fortnight later there was a mutiny one afternoon in camp. Ninety porters suddenly seized their guns, some loads of ammunition and ivory, and marched off. They sold their ivory for food and fled back towards the coast. They stormed every village they came to which they could attack with safety. They captured women from one tribe and sold them for food or ivory to the next. Thus looting, massacring, and murdering, they swept down to the coast, leaving a blood-stained trail behind them. The whole story is a terrible one. The actual leader was a man named Wadi Kombo, who has long since been called upon to answer for his crimes. But there is good reason to believe that my faithful Wadi not only joined, but helped to instigate the mutiny. He had been wrongfully punished, the injustice rankled in his mind, and he took his revenge.

This had happened several years before, but Wadi's hatred of mission boys was still keen and bitter. Once a drunken native was rolling round the camp, trying to steal, and generally misbehaving. I threatened to have him kicked out. He pulled himself up in a ludicrous effort to assume an air of dignity, and said he was no mere up-country savage (*mshenzi*), for he had been to the coast. "Yes," remarked Wadi, who was standing by, "I thought he had been brought up at ——," mentioning a well-known mission station. On another occasion he caught my tent-boy drinking some of my water when there was none in camp, and Wadi had emptied his calabash into my water-bottle. He was very angry and kicked the boy about. Suddenly he remembered what a terrible breach of camp discipline he was committing. A headman has the right to flog any porter or Askari during the absence of the leader of the expedition, reporting it to him when he returns; but he may not flog the leader's personal servants, such as tent-boy or cook. If necessary he can put them in irons, but he must not flog them. For a mere porter, then, like Wadi to do this was a grievous sin. He suddenly realised the enormity of his offence and came half-sobbing to me to apologise. He explained that he had a little water that he did not want, so, rather than throw it away, he had poured it into my bottle. Then he had seen my boy drinking it, became angry, and hit him. He begged humbly for forgiveness, which was readily granted. Wadi saw I was not angry, so, as he left the door of the tent, he looked back and remarked, "You will forgive me, won't you, Bwana?

I know no better. You see I was not brought up at a mission school."

I have said that the childishness of the Zanzibari introduces one great difficulty in handling them, but this may be more than counterbalanced by appeals to their keen sense of humour. Almost anything can be done with them by chaff. Time after time incipient rows were nipped in the bud, awkward questions and just, but inconvenient, protests were evaded by the feeblest of jokes, or a little banter at the expense of the spokesman. I had a series of slits cut in the wall of my tent, and through these could see any little incident that happened in the camp. One day the guide or kiringozi, Wadi Hamis, upset the pot of food he was cooking into the fire, greatly to the wrath of the others. A few days later he came up as leader of a few men, to complain that they could not do something expected of them. "Of course *you* can't; all *you* can do is to capsize your cooking-pot into the fire," was my irritable and inconsequent reply. The sullen Wadi Hamis was completely taken aback, and the others rolled away, shaking with laughter, and spluttering out to the rest of the camp that Bwana said all Wadi could do was to upset the cooking-pot. Thus, in a burst of merriment, what might have been an unpleasant row was averted. Parties of porters were continually coming up with questions and complaints. As a rule these were fair and easily satisfied; but when they were not, it was only necessary to turn the laugh against the ringleader, who would slink away abashed, while the others would lose their grievance in their merriment.

It was not only little troubles that could thus be soothed. For several days before we reached the Kikuyu frontier, on our march across Laikipia, the danger of famine hung heavily on our minds. We all knew the difficulties before us, but we did not care to refer to them. At length a few of the malcontents plucked up courage, and came in a body to ask me what was to be done. The deputation respectfully pointed out that all our food would be gone before we reached the frontier, that the Kikuyu would not let us cross it, or sell us food until they had discussed the matter for some days among themselves. "What shall we eat then?" asked the spokesman. "Kulu mimi" (You may eat me), I replied with my blandest smile. My omnivorous propensities had been one of the standing jokes of the caravan. I was always teasing the men about their fastidiousness in food, and they replied by jeering at my readiness to eat anything. As Mohammedans they were extremely particular, and would rather die than indulge in cannibalism. When members of a cannibal tribe such as the Manyema join a caravan, they are constantly chaffed about their food, and are nicknamed "Kulu-watu" (Those who eat people). My offer of myself as food for the hungry was perfectly safe therefore, and was received as a splendid joke. The men went off doubled up with laughter, stammering out to their comrades, "The master says when we get to Kikuyu and have nothing else to eat, we may eat him." When the wit of the camp objected that I was too thin, the laughter was redoubled; the difficulties ahead were all forgotten, and no more was said about the approaching starvation of the caravan. Many other instances might be given of a funny answer (or what was deemed such) turning away wrath.

The quick sense of humour of the Zanzibari can not only be used to soothe, but also to inspirit them. When the Masai couched their demand for our trade goods as toll, in a request for payment for their hongo dance, the porters were alarmed and urged that the tax should be paid. I went forward and did a dance of my own, mimicking that of the Masai to the intense amusement of my men. I then put in a claim for hongo for my dance, and said the two claims would neutralise one another. The novelty of this proceeding so revived the spirits of my men that they decided we must not pay, and were perfectly ready to fight.

Besides appreciating my feeble attempts at humour, the Zanzibari could supply wits of their own. One of the most useful of my party was a Mdoa named Mwini Mharo, a man of very varied antecedents and a record by no means spotless. He had once been a headman and had been degraded. He had had a bitter quarrel with Lugard and deserted. After this he had been sent up to Uganda in the chain-gang. On his return, he had enlisted in the expedition with which I started. There he had mainly distinguished himself by being involved in an act of theft, which could not be absolutely proved against him, though the evidence was very strong. After his journey with me, he went inland again with an expedition, the leader of which gives him a woefully bad character. But I found him so useful that I made him my head Askari, and he was mainly responsible for the trade goods. He was such an expert thief himself, that he was up to all the dodges; and as he knew that if anything were stolen he would be suspected, he durst not steal himself. Thus practically nothing was taken. He was a confirmed liar, as a rule very indolent, and had very little control over the men. He was, moreover, liable to fits of murderous passion, in which he would draw his knife and try to slay the object of his wrath. So Mwini's character was not a perfect one, and I have to admit he had his failings. But, in spite of all, he was indispensable to me. Though a thief, he was not a kleptomaniac. When roused, his capacity for work was colossal. His sight was marvellously keen, and his knowledge of animals and their ways was very intimate. He could track a wounded antelope or follow up an obscure trail with unerring instinct, and without apparent effort. His courage was superb. But his main virtue lay in his rollicking good-humour. When the men were almost too weary and exhausted to drag themselves along in some double march across difficult country, and we were all gloomy, depressed, and surly, Mwini's spirits would rise to their highest. I would call him to the front, and soon some lively marching song would break from him, cheering our hearts in spite of ourselves, until we were all joining in the chorus, or shouting back the refrain. Unconsciously our pace would quicken, and as if by magic "the crooked would be made straight and the rough places plain." If we had to camp out in the plains or spend the night in some dismal swamp, crouching fireless and foodless in the pelting rain, Mwini would vary his songs with stories of the strange things he had seen and heard. At first the men would be angry and tell him to stop. Huddled up under the open fly of the tent, which I would share with them on such occasions, I have watched them, taking in sullen despair a malicious pleasure in the contemplation of their own sufferings, looking as if sheer misery were the only thing that rendered life possible, and scowling at Mwini's ribald irreverence to their solemn dignity of martyrdom. But one by one they would thaw to his jokes, till the whole camp was infected, and the danger of an epidemic of fever or complete nervous collapse through the exposure would be over. Mwini was often far more efficacious as a prophylactic against fever than any medicine.

A mystery hangs over Mwini's fate. It is thought by some, who know him only too well, that the capture of the donkey caravan with which he disappeared came about through some treachery of his own. They believe he is still safe enough somewhere, enjoying a share of the booty of this last escapade. But it seems to me far more probable that the capture was made by the Masai, with whom under no circumstances would Mwini have conspired, for, from some reason, he hated them bitterly. This explanation at least is more agreeable, and I prefer to remember rather how he cheered us when the way was weary, how he shared his rations with less provident companions, how tenderly he nursed our sick, how untiringly he trudged by me as we scoured Laikipia together, to find a way that would avoid the kraals of the Masai. I always looked with something like a blind eye at many of his little ways and weaknesses, for I found in him a faithful friend and an invaluable servant, and there will ever be in my heart a soft corner for my merry old comrade Mwini Mharo.

The study of the psychology of the Zanzibari is another source of interest. His mental attitude to the European is an interesting one, somewhat resembling that of a seventeenth-century Puritan towards the Deity. Nothing a European does now surprises him. His mind is absolutely *blasé*. The most cunningly devised European machinery fails to evoke the slightest expression of astonishment. There is an explanation ever ready to his mind. I once called one of them to a telephone at Witu, and let him hear the voice of a man whom he knew was at Melindi, five days' march away. He listened, and said he recognised the voice, as if it were all the most natural thing in the world. He was not in the slightest degree surprised. It was simply another European dodge, and there was an end of it, as far as he was concerned. Probably he would not have been surprised if his great-grandfather had been called from the grave to speak to him through the wire. The instrument was European: he was African. He could not hope that his finite mind would understand the infinite subtlety of the Mzungu, and why should he try? The

Zanzibari knows he cannot understand everything, so he does not attempt to, and the ways of Europeans no more surprise him than does the sunset. But failure to understand the European does not lessen his sense of obedience. My own men worked for me right loyally, though they never understood my object. At first they nicknamed me "Dudu," which may be interpreted as "bugs and beetles," a name finally replaced by that of "Mpokwa" or "bulging pockets." The fact that the men sometimes wondered at my object was shown when, in the Kikuyu country, one of them was dangerously ill — we thought dying. One night he seemed anxious to ask me something, though he hesitated to speak about it. I went away, but a porter soon recalled me to the sick man. Then, with a great effort, he asked me to promise that, if he died, I would not put his head in a box and take it home with me, like the skulls I had dug up at Tzavo.

From the relief my promise gave, not only to him, but to others, I could see that many of the men had felt that I should not hesitate to add their skulls to my collection, if they died on the march. But this had not prevented them from risking their lives time after time in my service; and when the farewells came at Mombasa, I felt that the Zanzibari are the real heroes of African exploration. They do their work without the stimulus of the incentive of exploration; they have no share in the interest of the scientific problems; they enjoy none of the credit of success. They only receive their scanty allowance of a pound and a half of grain a day as food when on the march, and a miserable pittance of ten rupees a month as pay on their return. Yet for these they have to endure hardships and privations, compared with which those of their European master, with his comfortable tent and store of tinned provisions, are for the most part trivial inconveniences. The very highest success in life they can hope for is only promotion to the rank of headman. Only a small minority can ever obtain the greater dignity and higher responsibilities of this position. For the majority, if they escape the natives, who are ever ready to murder a lonely porter for the sake of his load, and never fall ill during a period of double marches and half rations on one of the cold, inland plateaux, there remains but the oblivion of an early and unhonoured grave.

CHAPTER XVII

THE NATIVES OF EASTERN BRITISH EAST AFRICA [1]

"Dislike me not for my complexion."—*Merch. of Ven.* i. 2.

THE connection between the character of a people and the geographical and geological structure of the country in which

[1] *Note on the Literature.*—The study of the natives of British East Africa naturally began on the coast-lands. Attention was first given to the Suahili, and especially to their language; for knowledge of this we are mainly indebted to Krapf (1850 and 1882) and Steere (1870); Suahili folklore has been reported by Steere (1873), W. E. Taylor (1891), and indirectly by D. J. Rankine (1891) and A. C. Madan (1887). The best history of the coast towns and their inhabitants is in Burton's *Zanzibar* (1872), from which most later accounts have been drawn. For knowledge of the tribes near Mombasa, such as the Wa-nyika, Wa-giriama, Wa-pokomo, and Galla, we are indebted to Krapf (1858), Wakefield (1866), New (1873), Fischer (1878), Gedge (1892), and Taylor (1891).

Of the inland tribes the most interesting is the Masai; the first accounts of its language are those of Krapf (1854) and Erhardt (1857); Johnston (1886) has given a clear account of the language and discussed its affinities. The tribe was first adequately described by Fischer and Thomson; their accounts, both published in 1885, admirably supplement one another. Since then most travellers who have visited the Masai country have described the tribe, but with the exception of Peters (1891) and von Höhnel (1892) most of the work has been done in German East Africa: as by Johnston (1886), Meyer (1888 and 1890), French Sheldon (1892), Baumann (1894), etc. The Wa-kamba have been described by Krapf (1858), Hildebrandt (1878), and cursorily by later travellers across their country.

The tribes of the Nyanza basin have now a voluminous literature dealing mainly with the Wa-ganda and Wanyoro: it began with Speke (1863), Baker (1866), and Stanley (1878); was continued by missionaries such as Felkin, Ashe, Mackay, and Wilson; and has been scientifically treated by Stuhlmann (1893) and Baumann (1894). Lugard's account of the Wa-ganda (1893) must also be mentioned.

The problems in connection with the dwarf tribes rest on the scientific descriptions of Schweinfurth (1874), Hamy (1879), Flower (1888), and Stuhlmann (1893). The most readable general account of these races is by Quatrefages (*Les Pygmies*, 1887: English translation 1895); a list of the principal literature is given by Stuhlmann, and a valuable summary of the whole by Schlichter (1892).

A bibliographic index to the languages is given by Cust (1883), and an account of the African races generally by Ratzel (*Völkerkunde*, Bd. i., 1887). With these exceptions, the literature referred to deals only with the tribes of British East Africa; but the anthropology of that area cannot be considered without reference to the two monographs of Paulitschke (1886 and 1893), and the works of Stuhlmann (1893), Baumann (1894), and a series of memoirs in the volumes of the *Mittheilungen von Forschungsreisenden* they live, has been frequently pointed out by historians, ethnographers, and geologists. Buckle, for example, in a well-known passage in his *History of Civilisation*,[1] has drawn a most instructive contrast between the geographical conditions of India and of Greece, and shown how their influence can be traced in the characters of the inhabitants of the two countries. Thus he points out that in India nature is all-powerful; for example, the mountains are so high that they cannot be scaled, and the rivers so broad that they cannot be bridged; while the climate is an alternation of torrid drought and torrential rains, of prolonged calms and irresistible tornadces. The powers of man are feeble in comparison, and he is therefore dwarfed and overawed. In Greece, on the other hand, nature is moderate in all things: the hills are low, the streams small, the rains evenly distributed, and the climate temperate. Man's individual development is therefore stimulated, and not repressed. In India nature is colossal, and therefore man is puny. In Greece nature is mild, and therefore man is strong.

A similar connection between the temperament of the Semites and the physical features of their surroundings has been remarked by other writers. As Prof. Keane expresses it, the Semitic intellect is "less varied but more intense [than the Aryan], a contrast due to the monotonous and almost changeless environment of yellow sands, blue skies, flora and fauna limited to a few species, and mainly confined to oases and plains reclaimed by irrigation from the desert, everywhere presenting the same uniform aspect. Hence to the Semites mankind is indebted for little philosophy and science, but for much sublime poetry associated with many profound conceptions of a moral order."[2] Draper, again, has shown that Europe illustrates the same principle; for the climate of that continent progressively modified the Asiatic tribes that entered it, until they resembled the races among which they settled, and until a condition of "ethnical equilibrium" had been thus attained.[3]

und Gelehrten aus dem Deutschen Schutzgebieten and in the new *Zeitschrift für afrikanische und oceanische Sprachen.*

[1] H. T. Buckle, *History of Civilisation in England* (ed. 1885), vol. i. pp. 137-146.

[2] Art. "Semitic Race," *Cassell's Storehouse of Information*, vol. viii. (1894), p. 72.

[3] J. W. Draper, *History of the Intellectual Development of Europe*, vol. i. (ed. 1875), pp. 34-35.

The East African races afford a further illustration of the dependence of character on environment; since, as the geographical conditions are different from those of India or Greece, they have produced a different type of national character. This case is especially instructive, as the direct influence of the geological factors is so clearly shown.

As we have seen in Chap. XII., the most striking feature in the physical conditions of Eastern Equatorial Africa is instability. Geological evidence was there quoted to show that the climate of the country has undergone a great change at no distant period. The facts of zoological distribution quoted in Chap. XIII. confirm this. Statements made by the Arab governors of Mombasa and Mambrui, and the meteorological statistics collected on the coast, support the same view; they show that the climate is still characterised by excessive variation and uncertainty. The tables of rainfall occasionally show us the total failure of one rainy season, followed by a disastrous excess in the next; while such incidents as the famine at Njemps repeat the same story in more tragic terms.

Owing to the peculiar conditions of the African climate, the natives are more than usually dependent on the rainfall. Merensky[1] has pointed out that the maps of South Africa which show the distribution of rain correspond closely with those which show the distribution of population.

But rain is not the only factor that has to be considered. In a region of such unstable equilibrium as Eastern Equatorial Africa, we have to allow for changes in the structure of the country as well as in its climate. Great earth-movements have occurred there recently, and, it may be, are in progress still. These may at any time have altered the subterranean drainage of a district, dried up the springs, diverted the courses of the rivers, turned a garden into a desert, and wiped out a tribe. The traditions of such disasters haunt the people's memory, and occupy a leading position in their folklore. The terror of their possible recurrence exercises a disturbing influence on the native character. It keeps alive a disposition towards nomad life, alien alike to the growth of either a fatalism like that of India, or a culture like that of Greece. All the tribes, however,

[1] A. Merensky, *Beiträge zur Kenntniss Sud-Afrikas* (1875), pp. 78-79.

cannot become nomadic. Some of them are physically and mentally incompetent for the strain of such a life, and must be content with servitude or else submit to the ever-recurring raids of the more powerful tribes. The physical conditions of the country, therefore, help to divide the people into two classes: one consists of warlike conquering nomads; the other of feebler races, who either eke out a precarious existence on mountain summits, in forest clearings, and on islands in the vast malarial swamp, or else live as serfs and helots in subjection to the dominant tribes.

If this were all that had happened owing to the migrations of African tribes,—if the conquerors had exterminated all the conquered, if the new possessors had kept apart from the dispossessed,—the race characters might still have been preserved fairly pure. But this has not occurred. The bands of invaders probably contained a majority of men, while owing to polygamy an excess of women was required. The men therefore had to take to themselves wives from the people of the land in which they settled; hence mongrel races occur wherever such national movements have taken place, and abound from one end of the continent to the other. The Hottentots, for example, are now regarded as hybrid between the Bushmen and Bantu, instead of belonging to a distinct race, differing from either. The Suahili in the same way have resulted from the intermarriage of Arab and Negro. The whole of the great group of the Fulah is probably a product of the blending of Hamitic and Negro tribes. It is this miscegenation which has introduced uncertainty into the classification of the African races. There is no trouble in determining the affinities of the Bantu Zulu, the Semitic Abyssinian, the Hamitic Somali, or the aboriginal Bushman; but to estimate the systematic value, and to discover the relationships of hybrid tribes, which have merged the physique of two races, and perhaps adopted the language of a third, is so difficult that many ethnographers have abandoned the effort in despair.

Moreover, not only are the physical features of the people confused by intermarriages resulting from nomad life, tribal war, concubinage, and polygamy, but the language test has also been rendered unreliable, owing to the extent to which some tribes have adopted words and inflexions, or even the whole language, of an altogether different race-group. A well-trained sense of hearing gives many of the people a remarkable facility in the acquisition of new languages; and nomadic habits, national migrations, and intertribal commerce have given abundant scope for the exercise of this useful faculty. Slaves have introduced forms and phrases into the language of their owners; helot tribes have been forced to learn the tongue of their masters; and conquerors have adopted the speech of the conquered. On the one hand, for example, the Bantu Wasania[1] now speak a dialect of Galla, just as the Aryans of Phrygia have adopted the language of their Turkish conquerors, or as the Celts of Western Europe have in the main accepted those of their Teutonic or Latin neighbours. On the other hand, the Hamitic race which invaded Uganda has now adopted the language of the Bantu people whom it conquered, just as the Teutonic Normans and Lombards exchanged their own for a Latin tongue.

Owing to the extent to which both the physical and philological data have been affected, it is obvious that the study of the African people would be an especially difficult branch of the science of anthropology, even if all the available information had been obtained and—had been collected by experts. But we know nothing of many tribes whose evidence would be most important; while others, who might have supplied useful links of evidence, have been crushed out by famine or war, and left either no trace of their existence, or, like the Zimbas, live only as the memory of a once-dreaded name. Most of what is known, moreover, is due to the reports of travellers, whose opportunities of intercourse with the natives have been very limited, and who have brought no special training to the task of observation.

It might therefore seem wiser to attempt nothing more than the record of new facts in the tabular form approved by anthropologists; but the problems connected with this subject are so interesting, that we cannot dismiss the African native with a simple tabulation of nose lengths, and a few pages of comparative vocabulary.

In spite of all the difficulties of the subject, it is now generally agreed that the African tribes may be divided

[1] A tribe living on the Sabaki, and elsewhere on the East Coast; see p. 328.

among five of the principal divisions of the human race—viz. Negrillo, Negro, Hamitic, Semitic, and Mongolian. The following account of the natives of British East Africa will deal with

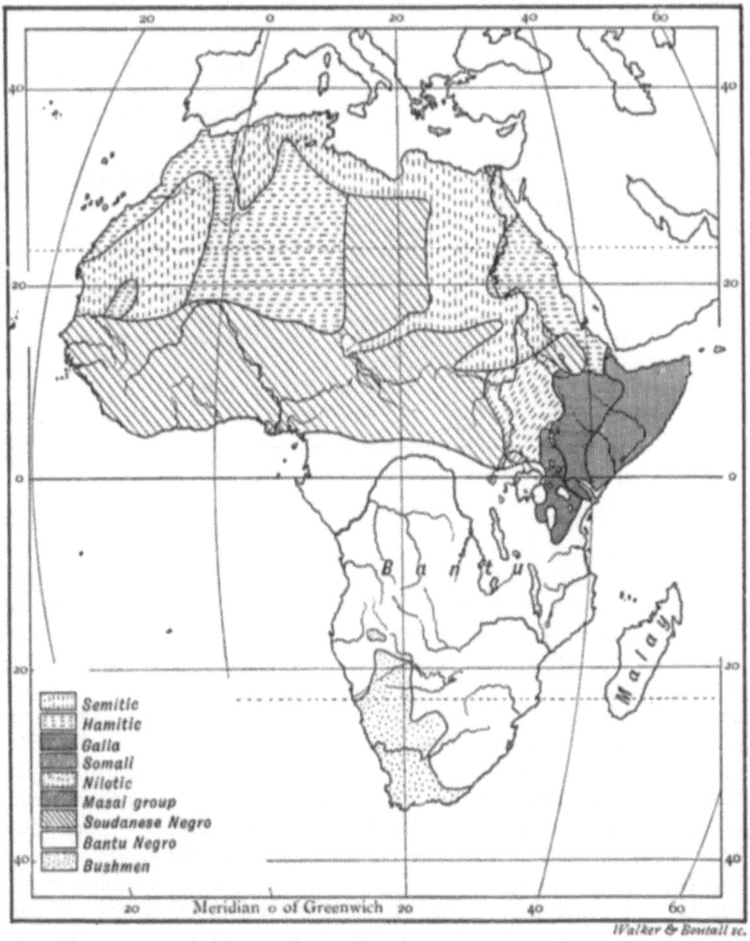

FIG. 18.—Ethnographical Map of Africa.

them in this order, so far as they are represented in this region; but before proceeding to this it will be convenient to summarise the classification adopted.

Main division.	Typical African tribes.	Representatives in British East Africa.
1. Negrillo.	Pygmies and Bushmen.	Doko of Laikipia.
2. Negro.		
Section A. Papuan.	...	Unrepresented in Africa.
,, B. African.		
Division 1. North-Western or Soudanese.	...	None in British East Africa.
,, 2. Southern or Bantu.	...	Pokomo, Wa-kamba, etc.
,, 3. Negroid.	Hausa and Fulah.	Masai, Wa-ganda, etc.
,, 4. Negrilloid.	Hottentots.	Some of Wanderobbo.
3. Caucasian.		
Section A. Hamitic.	Tribes of North-Eastern Africa.	Galla, Somali, etc.
,, B. Semitic.	Moors and Abyssinians.	None.
4. Mongolians : Section Malays.	Sakalava and Hova of Madagascar.	None.

Section A.—The Stone Age in East Africa

As we shall see in the following section, there is one tribe living in British East Africa which carries us far back in the history of the human occupation of this region. There are, however, traces of a still older race. It is well known that in Europe the earliest men were unacquainted with the art of preparing metals, and were only provided with weapons and implements made of stone, which occur in vast numbers in our river gravels. Similar tools have also been found in Northern Africa, and have been recorded from the Cape by Sir George Grey, Dale, Sanderson, Gooch, Penning, and other authors. Stone axes have been described from the Gold Coast by Winwoode Reade, Cameron, and Burton ; but in 1892 no record of old stone implements had been recorded from Equatorial or Eastern Tropical Africa. I was therefore interested, during an excursion up one of the side valleys of the Iveti Mountains, to find a chipped flake of obsidian. The flake was rough, but the chipping on its edge was unquestionably the work of man, while the material was absolutely different from any rock in the neighbourhood. I thought, however, that it might have been chipped by a Zanzibari for use as a gun-flint, and dropped on the caravan route which passed a few miles away. I therefore did not trouble any further. A few days later, on the Kapte plains, I found some more ;

but small fragments of obsidian occurred in the volcanic tuff on which these lay, so the evidence was again inconclusive. I found some more specimens lying beside the track in the Kikuyu country, but still the gun-flint theory seemed a possible explanation. It was not till I reached the highest point of the Kikuyu uplands, before beginning the descent into the Rift Valley, that I found final proofs of a stone age in Equatorial Africa. I was digging in some gravels, searching for fossil shells, when I found a broken obsidian implement of the type known as a "ridged flake." Its edge was chipped, so that there was no question of its artificial origin; and as it occurred in a patch of sand in a gravel, there was no doubt of its antiquity. Aft · this I found similar implements in many different parts of the country, including the platform on the Kikuyu scarp, the alluvial plain of Lake Suess, the plateau between Lakes Losuguta and Kibibi, the lake terraces around Lake Baringo, on the Sobat pass leading from this lake basin to that of Sukut, on the summit of the pass across the range of Subugu (the Marmanett Berge of von Höhnel), and in many places on Laikipia, as, for example, at Nairotia.

The best specimens came from a locality near the ford over the Gilgil river, where I found the site of an old settlement. As the implements were associated with hundreds of artificial chips, they were probably made on the spot. Fragments of rough pottery occurred at the same place, but I could find no bones, or tools made from bone.

The implements found are all made of obsidian, except a borer, which was of pitchstone. They consist of several different types, the principal of which are illustrated by the accompanying figures.

The commonest implements are flakes, which are either flat, ridged (Fig. 19, No. 1), or polygonal. A few with saw edges were met with, such as No. 4. The Gilgil site yielded several cores, from which obsidian flakes had been struck off (No. 5). Flakes with one end bluntly rounded to use as skin scrapers (Nos. 2 and 3) are also represented. Borers made from long splinter-like chips also occur, as in No. 7. Small knives, made by chipping away one or both edges of a flake, are fairly numerous, while the Gilgil camp yielded a triangular arrow head (No. 6).

I asked many of the natives whether any such implements as these were used by existing tribes; but they jeered at the idea that they could ever have been of service to any one. They said the only stones used were pumice, kaolin, and ochre. The first is obtained in many localities, and distributed owing to its value in sharpening spears. The kaolin, or china clay, and the ochre are both used as paints.

There is, moreover, geological proof of the antiquity of these implements. Both on the summit of the Kedong Pass, and in the old settlement at the Gilgil, the implements were found *in situ*, in deposits which must have been formed many centuries ago; in the latter they must be earlier than a gorge cut through the lava there. Those from Baringo must also have been used at a time when that lake was much larger than at present.

The forms of the implements do not give any definite evidence of great age. Sir John Evans has kindly examined some of them, and says they are of a neolithic type.

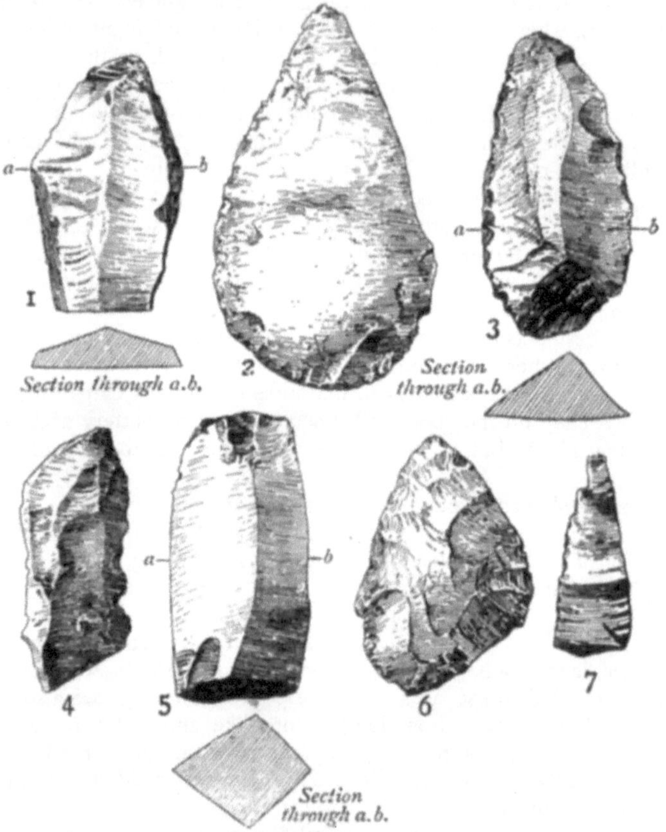

FIG. 19.—Neolithic Stone Implements from Masailand.

Recently three distinct descriptions of stone implements have come from Tropical Africa. M. Dupont has found some on the Congo; Dr. Jousseaume[1] has described a series from Somaliland in an article in *L'Anthropologie;* and Mr. Seton-Karr has brought back a large series from the same country, which have been exhibited (September 1895) at the Ipswich meeting of the British Association. The Somaliland

[1] Jousseaume, "Réflexions anthropologiques à propos des Tumulus et Silex Taillés des Çomalis et des Danakil," *L'Anthropologie*, t. vi. (1895), pp. 392-413.

implements are identified by Sir John Evans as neolithic, and as they are apparently all made of chert, their evidence is more satisfactory than that of tools formed of obsidian. A record of a stone spear from the Kilima Njaro region has also been published, but there are many points in the account of it which render the testimony doubtful.

Additional evidence of prehistoric people is afforded by the existence of stone cairns. These have long been reported from East Africa, but they have been explained as mere heaps of stones collected together in order to clear the ground for cultivation. I found two in the Sabaki valley, in positions where this explanation was quite impossible; but as my men were two hours ahead, they could not be recalled to open them. Our Somali used to tell us that in their country there are three sets of ruins, those of the Somali and Galla, and those of a nation earlier than either of the others. The Sabaki graves are not like those of the Galla, which have been well described by Paulitschke, and it is much to be desired that one of them may be opened, and that any bones therein may be sent to England.

SECTION B.—THE NEGRILLO OR PYGMY TRIBES

The Doko of Laikipia

One of the most interesting results from explorations during the last half-century in the interior of Africa has been the confirmation of many reports by classical geographers and early travellers, which had previously been disbelieved. The discovery of the sources of the Nile has verified the assertion by Ptolemy, that it rose in two great inland lakes; and Dr. Schlichter's detailed comparison[1] of Ptolemy's statements with present knowledge shows that the ancients were better informed upon the subject than the geographers of the last, and even of part of the present century. That the ancient Hindoos also had a certain knowledge of the continent is rendered probable by the facts reported by Rigby.[2]

Moreover, it has now been shown that the accounts of pygmies, given both by classical authors and by the pioneers of modern African exploration, are not mere travellers' tales. Schweinfurth has carefully examined pygmies in the very district in which Pliny reported their presence, and recent travellers have shown that Andrew Battel, the discoverer of the gorilla, was not drawing on his imagination when, in 1625, he described some dwarfs from the backwoods of Loanga. This ancient mariner, who, as the title of his story tells us, was, "sent by the Portugals prisoner to Angola, who lived there, and in the adioyning

[1] H. Schlichter, "Ptolemy's Topography of Eastern Equatorial Africa," *Proc. Roy. Geog. Soc.* new ser. vol. xiii. (1891), pp. 513-546.

[2] C. P. Rigby, "Remarks on the North-East Coast of Africa, and the various Tribes by which it is inhabited," *Trans. Bombay Geog. Soc.* vol. vi. (1844), pp. 89-90.

Regions, neere eighteene yeeres," has so clearly described the pygmies that his account is worth quoting. "To the north-east of Mani Kesock are a kind of little people called *Matimbas;* which are no bigger than Boyes of twelve yeares old, but are very thicke, and live onely vpon flesh, which they kill in the Woods with their Bowes and Darts. They pay tribute to Mani Kesock, and bring all the Elephants' teeth and tayles to him. . . . And of these [the women] will walke in the Woods alone and kill the Pongos [gorillas] with their poysoned Arrowes." [1]

The records of these dwarf races are now very numerous, and relate to all parts of Africa south of the Sahara. The works of Paul du Chaillu, Oscar Lenz, and Marche demonstrate the occurrence of numerous settlements of these people in the region of the French Congo; there they are scattered from the south of the Cameroons, through the Gabun and across the basin of the Ogowe, to the Congo near Stanley Pool. In the highlands to the west of Mwutan Nzige (the Albert Nyanza), which form the watershed between the Congo and the Nile, live other representatives of the Negrillo race, who have been carefully studied by many travellers.[2]

In the upper basin of the Congo and on the banks of its tributary, the Kasai, another group of tribes of dwarfs has been described,[3] which has received the name of Watwa, Batwa, or Batua. Still farther to the south are a few settlements of similar people, such as the Mossaro, in the upper basin of the Zambesi. These are of much interest, because they link the dwarfs of the Congo with the Bushmen of Bechuanaland and the Cape.[4]

Although the first reports of the existence of these dwarfs made during the present century came from British East Africa,[5] no positive proof of their occurrence there had been published when I landed on the coast in 1892. Dwarfs had been found in the Ruwenzori district by Stuhlmann, and a portrait of a boy published by Borelli[6] might have been regarded as proof of their existence to the south of Abyssinia; but Cecchi[7] described the Doko, the tribe to which the boy belonged, as "negroes of tall stature," and Leo Reinish,[8] the great authority on Hamitic languages, included the Doko in the Sidama, one

[1] Purchas, *His Pilgrimes in Five Books*, pt. 2 (vol. ii.), lib. vii. ch. iii. (1625), p. 983.

[2] Schweinfurth, Junker, Casati, Emin, and Stanley. They have also been recorded by several others.

[3] By Stanley, Wissmann, Grenfell, Bateman, and Wolf.

[4] A useful bibliography of the pygmy races is given by Stuhlmann, *Mit Emin Pasha*, 1893 (1894), pp. 473-475. Also in appendix to English translation of De Quatrefages' *The Pygmies* (1895), App. A and B, pp. 239-248. Schlichter's masterly summary of the literature is referred to on p. 332.

[5] T. Boteler, *Narrative of a Voyage of Discovery to Africa and Arabia, etc.*, 1821 *to* 1826 (1835), vol. ii. p. 212.

[6] Jules Borelli, *Éthiopie méridionale* (1890), p. 313.

[7] Ant. Cecchi, *Da Zeila alle frontiere del Caffa* (1885), vol. ii. p. 463.

[8] L. Reinish, "Das Zalwort vier und neun in dem chamitisch-semitischen Sprachen," *Sitz. Phil. Hist. Cl. K. Akad.*, Wiss. Wien. Bd. cxxi. Abh. xii. (1890), p. 4.

of the three members of the High Kushite group of Hamites. Hence the only evidence of the existence of dwarfs to the east of the great lakes rested on native rumours. These, however, were very precise. They were first reported in detail by Captain Harris[1] in 1844, who obtained the information in Shoa two years previously. He calls the dwarfs the "Doko," and says they are "a pygmy and perfectly wild race, not exceeding four feet in height, of a dark olive complexion, and in habits closely approximated to the beasts that perish." He describes their country as "clothed with a dense forest of bamboo, in the depths whereof the people construct their rude wigwams of bent canes and grass. They have no king, no laws, no arts, no arms; possess neither flocks nor herds; are not hunters, and do not cultivate the soil; but subsist entirely upon fruit, roots, mice, serpents, reptiles, ants, and honey. . . . Both sexes go perfectly naked, and have thick pouting lips, small eyes. The hair is not woolly, and in the females reaches to the shoulders."

This description was repeated by the Rev. J. L. Krapf in 1860,[2] who says he obtained the information from a native of Enarea when visiting Shoa on behalf of the Church Missionary Society.

Nothing, however, is said about this in Krapf's earlier record of his journey,[3] and it is impossible to compare the account of the Doko in his later book with that of Harris without concluding that Krapf's account is a plagiarism. This conclusion throws doubt on the value of his statement that he saw one of the pygmies in Barawa, and that it "accorded completely" with the description, which he had copied from Harris.

Evidence in support of the existence of the East African dwarfs, which, though based only on hearsay, was apparently more reliable than that of Krapf, was given about the same time by Père Léon des Avanchers. In 1859 this author recorded[4] the existence of the race of the Wa-berikimo, and marked them on a map as living to the west of a lake named El-Boo, or, as he called it in 1866,[5] Lake Baro. In this latter paper he reported also the existence of a race named "Cincallé," or "What a marvel," to the south of Abyssinia. Rigby, moreover, in 1844 mentioned that a tribe of dwarfs, the Berikimo, scarcely 3 feet in height, lived six weeks inland from Mombasa.[6] But the latest informa-

[1] W. C. Harris, "Particulars concerning the great River Gochol and the Countries adjacent thereto, from Native Information collected in the Kingdom of Shoa," *Trans. Bombay Geog. Soc.* vol. vi. (1844), pp. 63-64. This Gochol is the western of the two rivers said to enter the northern end of Basso Narok.

[2] J. L. Krapf, *Travels, Researches, and Missionary Labours, etc., in Eastern Africa* (1860), pp. 43-45.

[3] Jas. M'Queen, *Journals of the Rev. Messrs. Isenberg and Krapf, detailing their Proceedings in the Kingdom of Shoa*, London, 1843.

[4] "Esquisse géographique des pays Oromo ou Galla," *Bull. Soc. Géog.* sér. 4, t. xvii. (1859), p. 163.

[5] "Lettre à M. Antoine d'Abbadie," *Bull. Soc. Géog.*, Paris, sér. 5, t. xii. (1866), p. 171.

[6] C. P. Rigby, "Remarks on the North-East Coast of Africa, and the various Tribes by which it is inhabited," *Trans. Bombay Geog. Soc.* vol. vi. (1844), p. 80.

tion about this tribe, obtained by Paulitschke[1] from the Somali, placed it in the valley of the Sobat, a tributary of the Nile. Its position was, therefore, removed 400 miles to the north-west, and outside British East Africa.

Another statement that suggests the occurrence of dwarfs in this country is that, according to Tutschek, some of the slaves of the Galla on the eastern coast, the Wasania and Watwa, have "clicks" in their language. These clicks are very characteristic of the Bushmen and dwarf languages, and if the statement be true, then these two tribes must be Negrillo. It rests, however, solely on some remarks by Tutschek,[2] and is improbable. Fischer gave a short vocabulary of the language of the Wasania,[3] which shows that they now speak a dialect of Galla; but the tribe is certainly not related to the Galla, and is either Bantu or Negrillo. Herr Würtz, one of the missionaries at Ngao, had seen some Wasania, but had not heard them use any clicks; indeed, he told me that Wasania is only the Ki-pokomo name for the Waboni. This, however, is doubtful, for Fischer gives lists of words from two languages which he assigns to the Wasania and the Waboni, and they differ widely from each other. Gedge[4] has described the habits of the Waboni from information collected by the members of the Tana Expedition, and these resemble those of the Doko. Nevertheless it is still impossible to say whether the Watwa, Wasania, and Waboni are one tribe or more, and whether they are Negrillo or Bantu.

There was, therefore, no reliable evidence of the occurrence of pygmies in British East Africa, and in my hasty march I did not expect to see any.

On reaching the summit of the last pass over the chain of Subugu, to the south-east of Baringo, we saw in the distance the smoke of great prairie fires, which my men said were lighted by the Wanderobbo. I had tried to get information about these people in Mombasa, but could learn nothing satisfactory about them. Bird Thompson told me they were a tribe of "bastard Masai," but could not say which other race shared with the Masai the parentage of these people. The Rev. W. E. Taylor, the keenest ethnographer on the coast, told me that they were a Bantu race who now spoke Masai. Paulitschke,[5] on the other hand, says that they are a race with the physique of the Hamite, but speaking a true negro language. Mr. Taylor's view seemed such a natural one that I did not trouble much about the people, especially as my men said that the Wanderobbo are a wild race, who

[1] Phil. Paulitschke, *Ethnographie Nord-ost Afrikas* (1893), pp. 34-35.

[2] C. Tutschek, *A Grammar of the Galla Language* (1845), p. 6; but his statements do not bear out the claim rested upon them by later authors.

[3] *Zeit. für Ethnol.* Bd. x. (1878), pp. 141-144.

[4] E. Gedge, "A Recent Exploration up the River Tana," *Proc. Roy. Geog. Soc.* new ser. vol. xiv. (1892), p. 518.

[5] *Op. cit.* p. 33.

live in the fastnesses of the forests, and will not enter into communication with caravans. In one of the woods on Laikipia, a few miles to the north of the "Thelphusa swamp," we stumbled by accident on one of their deserted huts. It was a mere bivouac, formed of four upright logs, with the interspaces filled by brushwood, and covered by a pointed roof of twisted boughs. A few days later I had the good fortune to see some of the natives. I had left the path and was busily engaged collecting specimens of a tree St. John's Wort (*Hypericum Schimperi*, Hochst.), when I was recalled by a shout of "Watu" (People) from the men. I ran back to the point whence came the sound of my interpreter's voice, and forced a way through the bushes on to the path beside him. He was talking to a couple of naked natives, and I cannot say whether they or I were the most startled. They heard the rustle in the bushes and looked round, expecting to see another Zanzibari—and they saw me. Their hands dropped in horror, and one of them said to the other "Ngai" (God). I had expected and dreaded to find that they were Masai, instead of which they were Negrillos. They were both of them young men of about 4 feet 6 inches in height. They were of a brown colour, different from the copper brown of the Zanzibari or Masai. They had bent shins, rounded heads, longish hair, and protruding jaws, and the outline of their backs had the characteristic Bushman curve.

They spoke Masai, and readily consented to lead us to a ford over the Guaso Nairotia, and with much interest watched us pitch our camp. I sent one of them with a present to their father to ask him to visit us, and kept the other a hostage till he came. He arrived towards evening with three other men, of whom one was of normal size, and was clearly a half-breed Mkwafi. The bearing of the men was frank and unsuspicious; they shook hands at once, without any of the preliminary palaver practised by the Masai. I happened to strike a match, and the "elder" was so interested that I gave him one or two, with which he promptly burnt his fingers. But this did not in any way interfere with his friendly attitude, and he immediately told us that he wanted to sell us some elephant tusks. He insisted, however, that we should at once "eat muma," or go through the rite of blood-brotherhood with him. He declined to do this with me, but said any of my men would serve as a substitute. Ramathan was selected, and the two men sat down together upon the ground, the Askari's rifle and the native's bow being placed side by side upon their heads. Each of the men made a slight incision on his arm, and the blood thus obtained was mixed together, and each of them smeared half the mixture on to one of his fingers and licked it off. Meanwhile Alli, a porter who knew some Masai, drew a knife backwards and forwards along the weapons upon the men's heads, singing a wild Masai incantation, to the effect that the Mzungu and the Wanderobbo were thus made friends and brothers.

I gave the elder a present of some wire, and insisted on his giving me one of his arrows in exchange. When he knew I wanted it as a keepsake and not for use, he took it back, and rapidly and adroitly carved a rough design upon it with the head of another arrow (Fig. 20). I was delighted to see him do this, for skill in wood-carving and love of design are very characteristic of these Negrillo people.

FIG. 20. Ornamented Doko Arrow.

The elder left us at sunset, after arranging that some of his people might travel under our escort to the Kikuyu country to purchase food. I readily consented to this, hoping on the journey to find out more about them. I learnt, however, very little. They live in the recesses of the forests in small families or clans, scattered over an enormous extent of country. Their culture and habits are quite primitive. The pottery they have they buy from the Kikuyu, for they do not know how to make it. They do not cultivate anything, but live on wild fruits, roots, and the produce of the chase. They also collect honey and keep it in bags made from skins. They do not fish, and have no domesticated animals. Their only weapons are bows, arrows, and knives. Their dress merely consists of loose sheets of undressed skin, hung over the shoulders, but they are often prettily ornamented by designs of beads. Their personal ornaments are very simple, and they have none of the leg-rattles or finger-guards of the Masai. Earrings are worn in both the upper and lower lobes of the ear, and constitute their most elaborate adornments. The young men each had a "kipule"[1] formed of a string of ten white "pound-beads." One of the women who came with us, who was not a pure bred Doko, wore earrings in both lobes; in the upper was a string of red and white beads, from which hung a loop of iron chain, and from the lower a disc of coiled brass rod two inches in diameter was suspended by a leather belt. The elder of the clan had a pair of earrings, each made of two brass knobs on the ends of a bent piece of brass rod bound with iron wire.

I did not discover in this tribe any trace of independent religious ideas. Their term for God is that of the Masai, and the rite of blood-brotherhood they have apparently acquired from the coast traders, for they know it only by its Suahili name.

My interpreter, Ramathan Aperti, who had spent some time among the Wanderobbo of Rangatan Nyuki, on the western side of the Rift Valley, said the tribe has a language of its own, though they always use Masai when speaking to strangers. He said they would not allow any

[1] The word means a dangling earring in the upper lobe; a "kipini" is a stud-like ornament inserted in the same part.

foreigner to hear them talk in their own tongue, and added that the language was very simple, needing the help of signs to such an extent that the people cannot understand one another in the dark. This story is also told about the Aduyahs of Fernando Po by the late Colonel Ellis.[1] Ramathan could only tell me one word of their language, which was "lovoi" for water. The Masai term for this is "ngare." "Lovoi" has a certain resemblance to the words for water in other dwarf languages; thus Stanley[2] says that the Bakwa word for this is "libo"; while Stuhlmann[3] records the term in four dwarf languages as *o-u*, *u-u*, *ui*, and *owu*.

I listened attentively in the hope of hearing the men use clicks like the Bushmen, unpronounceable sounds that have been compared to those made by coachmen to quicken the pace of their horses. The Bushmen have six such clicks, which have to be represented in writing by conventional signs. Thus "/ kamap" is the word for fox, "/" indicating the "cerebral click"; "‡" represents the "palatal click," which may be imitated by pressing the tip of the tongue against the palate beside the gum and then suddenly withdrawing it; "*8*" is the "labial click" made when a word is spoken, while the tongue is being rapidly moved as in flute-playing. Clicks have been found among some of the equatorial dwarfs, but in spite of the closest attention I could not detect any trace of them in the speech of these "Doko." The clicks have, however, probably been lost from the languages of most of the equatorial dwarfs. It is a well-known law in philology that languages tend to throw off sounds that are difficult of pronunciation; thus in English the *p* in psalm is silent; *h* has become mute in French, and in Italian has been lost even in the written language.

When we reached the Kikuyu country, and the Doko guides had obtained from us all the protection they needed, they suddenly disappeared. I was disappointed to have found out so little about them. But it is quite evident that they are neither Bantu-speaking Masai, nor Hamitic people who have adopted a negro tongue. Though some of them were clearly mongrels of a Masai or Wa-kwafi parentage, the pure bred ones belonged to the dwarf Negrillo race.

Stuhlmann, however, has recently met with some "Wanderobbo," and says they belong to the Hamitic race, and in appearance are much like the Masai.[4] Mr. Scott Elliot also saw some men called by the same name, and says they were tall and slim, and therefore were probably Wa-kwafi. The name "Wanderobbo" is a Masai word meaning "poor fellows," and it is very likely applied by the Masai to any of the neighbouring tribes who live among the woods and have no cattle. It is only a term of contempt, analogous to that of "Washenzi"

[1] A. B. Ellis, *West African Islands* (1885), p. 76.
[2] H. M. Stanley, *Darkest Africa*, vol. ii. App. B, table facing p. 442.
[3] F. Stuhlmann, *Mit Emin Pasha*, p. 459.
[4] Stuhlmann, *op. cit.* p. 845.

or "savages," which the Zanzibari apply to any up-country natives. It is consequently of no scientific value, and must be discontinued as a tribal name.

I could not learn from the people the name by which they call themselves, and therefore suggest that they should be called the "Doko," for they agree in habits, appearance, and position with the tribe thus named by Harris and Avanchers. The Doko were said to occur on a high, cold, misty plateau, in the neighbourhood of great bamboo forests; their home is about six weeks' march from Mombasa, and between a snow-covered mountain called Obada and Lake El-Boo or Baro. The mountain must be Kenya,[1] and the lake Baringo. Hence it seems safe to conclude that the Doko or Wa-berikimo of Harris, Avanchers, Krapf, and Rigby are the elephant-hunting Negrillo on the plateau of Laikipia and the district to the north.

The use of the term Negrillo for these Doko raises the question of its scientific value. The first striking point in connection with the African pygmies is their occurrence in numerous small, scattered, isolated colonies. This may be explained in two ways. "Discontinuous distribution" such as this is usually regarded in biology as suggestive of great age, and this has led to the view that the dwarfs are the survivors of a race that once occupied the whole continent south of the Sahara. On the other hand, it is possible that they may have arisen independently by degeneration from different negro tribes. The latter view is supported by Sir William Flower[2] from a detailed description of some Akka skeletons sent home by Emin, and he accepts for the dwarf tribes Hamy's[3] name of Negrillo, regarding those of Africa and Polynesia as "parallel ethnical elements." Dr. Schlichter,[4] however, in a masterly review of the literature of the whole subject, expresses his belief that the skulls in the College of Surgeons Museum, with which those of the Akka were compared, are possibly not those of pure Bushmen. Stuhlmann's[5] admirable photographs of the dwarfs from the west of Ruwenzori show that the "steatopygy" characteristic of the Bushmen is well marked in some of the equatorial dwarfs, and this removes another strong objection to the theory which regards them as the African aborigines. This theory has also been recently supported by

[1] Kenya is not a native name for the mountain. Obada may possibly be based on Ebor or Ebar, which is the Masai name of Kenya. The fact that Léon des Avanchers inserts both El-Boo and Baharingo on his map suggests that the former should be regarded as Basso Narok; but his Baharingo is the Victoria Nyanza, and his El-Boo cannot be Basso Narok, for he says it is surrounded by Wa-kwafi, and Baringo is the only lake of which this is true.

[2] W. H. Flower, "Description of two Skeletons of Akkas, a Pygmy Race from Central Africa," *Journ. Anthrop. Instit.* vol. xviii. (1888), pp. 3-19, Pl. i.-iii.

[3] Hamy, "Essai de coordination des matériaux récemment recueillis sur l'ethnologie des négrilles ou pygmées de l'Afrique equatoriale," *Bull. Soc. Anthrop.*, Paris, sér. 3, t. ii. (1879), p. 100.

[4] H. Schlichter, "The Pygmy Tribes of Africa," *Scott. Geog. Mag.* vol. viii. (1892), pp. 289-301, 345-356.

[5] F. Stuhlmann, *Mit Emin Pasha*, Pl. xvi. xvii.

the discovery of clicks in the language of some of the tribes.[1] No doubt the colonies have long been isolated, and have been forced to live in situations where the conditions of life are unfavourable; thus considerable differences between the various tribes have been developed. Nevertheless the several descriptions agree in three points. First—the people are small in stature, the measurements that can be relied on giving them a height of from 3½ to 5 feet. Second—they are lighter in colour, being described either as olive brown, yellowish brown, or chocolate-coloured. Third—their habits and mode of life are of extreme simplicity. Reports that go into greater detail, moreover, show that the dwarf races have large rounded heads, are prognathous (*i.e.* have protruding jaws), and have short legs.

It is accordingly claimed by some ethnologists that the equatorial dwarfs agree with the Bushmen of Cape Colony in physical features, colour, sounds, and mental characteristics, and therefore belong to the same race group. The earliest writers on the subject attributed the origin of these light-coloured, primitive Bushmen to a settlement of a colony from over the sea. This is improbable, for whatever the Bushmen are, they are not sailors. It is far more likely that they entered Cape Colony by a migration from the north, for there is historical evidence to prove that they (and their half-breed allies, the Hottentots) once lived much farther toward the north than they do at present. Now they are limited to the south of a line leaving the western coast in latitude 22° S., which runs northward to the Zambesi valley in lat. 17° S. long. 20° E., and then turns southward, and passing Lake Ngami and through Bechuanaland, reaches the eastern coast near Port Elizabeth. In the eighteenth century, however, both they and the Hottentots were met with much farther to the north, as in 1767 a Dutch ship under "Corporal Thomas Hobma" found them all along the west coast up to 12° 47′.[2] Philology affords further evidence of this fact. Thus the Bapedi call the country to the west of them "Boroa," which means the "country of the Bushmen," though at present the district is inhabited by Bantu people.[3] Fritsch,[4] in fact, in his great monograph on the South African races, concludes that the Bushmen once extended throughout South Africa up to the Zambesi, and perhaps beyond it. Beke would carry them farther, for he compared the drawings with which they ornamented their rock shelters to the hieroglyphs of the Egyptians, and placed them among the Hamites as relatives of the Copts. Others, on no better grounds, made them close allies of the Semitic Phoenicians, or even of the Jews. The drawings of the Bush-

[1] For example, the tribe reported by Serpo Tinto from the region between the Kuando and the Kubango, two of the head streams of the Zambesi.

[2] A. Merensky, *Beiträge zur Kenntniss Sud-Afrikas*, p. 77.

[3] Ratzel, *Volkerkunde*, Bd. i. (1887), p. 118, and Merensky, *op. cit.* p. 78.

[4] Gustav Fritsch, *Die Eingeborenen Sud-Afrikas, Ethnographisch und Anatomisch* (Breslau, 1872), p. 386.

men, however, far more nearly resemble those made by the aborigines of Southern Australia than they do those of Egypt,[1] and similarities in the folklore of the two races have been frequently recorded.[2] For example, the Australians say that light originated by an emu's egg being thrown into space; the Bushmen have the same legend, except that a man is substituted for the egg. In the Milky Way, the Australians see the smoke of fires of the old race that preceded them, and the Bushmen smoke made by a girl throwing wood ashes into the sky. The Magellan clouds, according to both races, are a pair of animals; but the Australians call them birds, and the Bushmen steinbock. The star Arcturus supplies both people with food, giving the larva of the wood-ant to the Australians, and rice to the Bushmen.

Moreover, the physical features, weapons, and domestic implements of the Bushmen and the equatorial dwarfs are similar to those of the aborigines of Malaysia. Some anthropologists, such as Professor Keane, regard this as so well established, as to give valuable support to the theory of the existence of the hypothetical continent of Lemuria over the site of the Indian Ocean, and even to show that this land area must have been contemporary with man. According to this view a pygmy race once extended throughout Africa, south of the Sahara, and through India, Malaysia, and Australia, from which the natives of the Andaman Islands, the aborigines of Australasia, the Akka of the Congo, the Doko of Laikipia, and the Bushmen of the Cape are all descendants.

This is no doubt a very attractive theory, for it offers a simple explanation of the resemblances between the aborigines of Africa and of Polynesia. Should it be proved that the main physical characters of all these dwarf races are the same, then the theory will probably meet with ultimate acceptance. But if, as Sir William Flower maintains, the skulls of the Akka are those of degenerate negroes, and differ fundamentally from those of the Bushmen and of the Polynesian Negrito, then the theory cannot be upheld. In that case, the group proposed by Hamy for the tropical African dwarfs under the name of Negrillo must be retained. It will rank as a section of the African negroes, instead of being kept as a distinct group, or being included with the Bushmen and Polynesian pygmies, in one great aboriginal race —the Negrito.

SECTION C.—THE NEGRO RACES

The term Negro is defined in Walker's Dictionary as "a native or

[1] Compare those shown in illustration of the recent paper by R. H. Mathews, "Aboriginal Rock Paintings and Carvings in New South Wales," *Proc. Roy. Soc. Vict.* new ser. vol. vii. (1895), pp. 142-156, Pl. viii. ix.

[2] See, for example, W. H. I. Bleek, "On Resemblances in Bushman and Australian Mythology," *Cape Monthly Mag.* new ser. vol. viii. (Feb. 1874), pp. 98-102.

descendant of the black, woolly-headed races of man in Africa," a definition which probably correctly expresses the popular significance of the term. This is not, however, the sense in which it is used by most ethnographers. As popularly understood the term includes the Bushmen, who are Negrillo, and the inhabitants of Madagascar, who are Malays; while it excludes all people not inhabitants of Africa, except the descendants of the slaves in America, and the "Sidi" colonies of India. But in ethnology the term Negro is used in three different ways. (1) One school restricts it to a group of tribes which inhabit a comparatively narrow belt of Africa, that stretches with certain breaks from the mouth of the Senegal river (18° N. lat.) on the west coast of Guinea, to near the Equator on the Upper Nile. Their most striking features are a long narrow head, black woolly hair, dark brown or sometimes black skin, and thick lips and broad nose. But the tribes are not united, either by their features or languages, into a homogeneous group. (2) Ratzel distinguishes by the name all the dark, woolly-haired Africans from the light-coloured or long-haired races, such as the Bushmen of the Cape, and the inhabitants of Northern and North-Eastern Africa.[1] (3) A third use of the name, and the one which I have followed, includes under it all people with very dark-coloured skins, frizzly hair, and protruding jaws, and most of whom have also broad noses and thick lips.

According to the use of the term here adopted, the people included as Negro are divided into two groups—the frizzly-haired Papuans of Polynesia, and the Negroes of Africa. The former may be at once dismissed, as they are unrepresented in Africa; the latter can be divided into four sub-groups—

1. The North-Western or Soudanese.
2. The Southern or Bantu.
3. The Negroid.
4. The Negrilloid.

The Soudanese group includes the typical tribes of Negroes, to which Keane and some other ethnographers restrict the term. The members of this group have certain physical characters in common, although these are not easily defined. The languages, however, are extraordinarily numerous, and belong to many distinct groups. This division of the Negro race need not concern us, for although representatives of it are met with on the western bank of the Upper Nile, none are known to occur in Eastern British East Africa.

The second group, or the Bantu, occupies nearly the whole of Africa south of a line drawn from the Cameroons to the mouth of the Juba. Here and there in this vast area occur isolated tribes who are not Negro at all. Some of these, like the Bushmen and the dwarfs, are remnants of the aboriginal inhabitants; others are Hamitic invaders

[1] Ratzel, *Volkerkunde*, Bd. i. p. 129.

from the north. Unlike the Soudanese group, the Bantu is characterised by the intimate relations of all its languages, while the physical features of the people are diverse. The Bantu group is therefore based on linguistic, and not on physical grounds; whereas the Soudanese group is physically simple and philologically complex. The Bantu is the most important race group in British East Africa, including such tribes as the Wa-kamba, Wa-taita, Wa-nyika, Wa-pokomo, and Suahili. To it also belong the Kaffirs of the Cape, the Zulu, Matabili, Mashona, and various tribes around Nyasa, and most of the natives of the Congo basin.

The two remaining groups are purely artificial. They contain races which are mongrel between either the Soudanese or Bantu, and various non-Negro elements. The first group, or the Negroid, results from intermixtures of Negro and Hamitic races. Its most important representatives are combinations of Hamites and Soudanese, such as the Nuba of the Nile Valley, and the Fulah of the western and the Haussa of the central Soudan. Corresponding mongrels of Hamites and Bantu are common in British East Africa, and include the Waganda and Masai.

The Suahili of the eastern coast ought perhaps also to be included among the Negroids; altogether the foreign element is Semitic instead of Hamitic. But as the amount of Semitic blood in the tribe is small, it is most convenient to leave it with the Bantu.

The fourth or Negrilloid group includes the tribes which are intermixtures of Negro and Negrillo. They are far less important than the Negroid, owing to the comparative rarity of the Negrillo element in the continent. The Hottentots of the Cape are the best known representatives, but the Wambuba and some other less known tribes, and some of the "Wanderobbo," have a similar origin.

(*a*) THE BANTU OF BRITISH EAST AFRICA

1. *The Suahili*

The Bantu people constitute the basis of the population of British East Africa, and, with the exception of a sprinkling of Negrillo or dwarf races, were probably at no very distant period the only inhabitants of that region. The number of tribes in the country is still considerable, though their power has been broken and their range reduced by the invasion of the Negroid Masai, and of the Hamitic Galla and Somali.

The first of the Bantu people with whom the traveller comes in contact are the Suahili, a hybrid race, resulting from the intermixture of Arab colonists with the Bantu natives of the eastern coast of Africa. For centuries, probably for tens of centuries, a steady stream of immigration has flowed from Arabia to East Africa. The Arabs intermarried with most of the tribes on the coast, and with many of the people from the interior who were carried there as slaves. The offspring of these unions are the Suahili. In some cases they are normal half-breed Arabs; from this type a gradual transition can be traced to that of the children of parents of different Bantu tribes, in one of whom an obscure trace of Semitic blood occurs.

As to the date of the first entrance of the Arab settlers there is no sufficient evidence. I had, however, owing to the kindness of Mr. Bird Thompson, the pleasure of meeting at Witu an old Suahili named Sherifu ben Abdullah, who had been secretary to the Sultan of Witu, and had long been interested in the past history of his race. In a series of conversations, in which Thompson acted as interpreter, he told me some of the current traditions about the various settlements on the coast. According to his information there were three main introductions of outside influence. Originally the coast was occupied by natives of whom Sherifu knew nothing. They were no doubt Bantu, and probably spoke Ki-ngozi, an archaic form of the Suahili language now used only in poetry. Then, most likely long before the Christian era, the Arabs opened a trade with Africa, and their vessels sailed down the coast. According to Bent it was an Arabian race that erected the ruined buildings of Zimbabwe, and first worked the gold mines of Mashonaland. That traces of this influence still survive in that district is shown by the fact that the natives still ornament their utensils with a crescentic and herring-bone design of a characteristically Arabian type. Bent's conclusion has not passed unchallenged, though the evidence in its favour is very strong. If it be accepted, then it is possible that the Arab settlements on the eastern coast were made in connection with this trade. There is, however, no doubt that Phœnician and Egyptian merchants sailed down the Red Sea and along the African coast several centuries before the time of Christ. They were either preceded or soon followed by Arabian traders, who established settlements along the coast. The first of these were probably in what is now Somaliland, whence they gradually extended southward.

The second stage in the history of the East African coast was entered about 700 years ago, when a settlement took place of which traditions are still current, confirmed it is said by manuscripts in the possession of some of the old Arab families at Lamu. There was a great fight in Muscat between two factions, one of which, the Nabahani, was defeated and its members expelled. Under their leader Saif the exiles settled on the island of Patta, opposite Lamu.

A couple of centuries later the Portuguese began to colonise the district. They worked their way along the coast from the south, and, finding Lamu a convenient point whence to sail eastward to India, they made their first permanent settlement on that island. They built a station, which is now the town of Shella, by the entrance to the harbour. An old church, now used as a mosque, and some ruined buildings still remain as relics of this occupation.

The Portuguese were finally expelled from the coast by the united efforts of the Mazrui (the leading family of Arab settlers), the natives of the coast, and fresh bands of Asiatic emigrants. The Portuguese had, however, left their mark on the language, and, if my informant be correct, in some places also on the physical character of the people. Sherifu knew that the Suahili language was indebted to the Portuguese; thus before they came there was no word for "table," and the Portuguese *meza* was therefore adopted. He also said that the people of Siyu, an island of the Lamu group, were really Somali, changed by intermarriage with the Portuguese settlers.

After the expulsion of the Portuguese, the whole coast gradually became subject to the Sultan of Muscat, or (as he is more correctly named), the Iman of Oman, and it continued to receive fresh contingents of Arabian settlers.

There were thus three periods in the growth of Semitic influence on the coast—the long prehistoric immigration; the settlement of the Nabahani; the conflict with the Portuguese, and the final rule of the Iman of Oman.

One of the proofs that the rise of the Arabian power dates back to pre-Mohammedan days is supplied by the religion of the Suahili, which is Islamism of a modified and very tolerant type. My friend Sherifu told me that for a long time the east coast Arabs declined to accept Islam, although at the cost of a desperate struggle with their fanatical compatriots. When at length they bowed the knee to Mohammed, they introduced many modifications into the creed, and have ever acted up to their motto, "Kafiri akufaæ, si Isilamu asiyekufaa" (Better a useful infidel than a useless believer).

The following are the chief points of difference between Arab and Suahili Mohammedanism:—

1. If a slave woman bears a child to an Arab owner, the child only is free, whereas both the woman and child are free according to Suahili law.

2. If an Arab touches a European when on his way to mosque, he has to wash his face and hands before entering, whereas a Suahili regards this as unnecessary.

3. The Suahili will eat with Europeans, and take food killed by them, though not by a native of another tribe. A Muscat Arab, however, will not take any food unless killed by a Mohammedan.

4. The Suahili, on the other hand, are more particular over contact with dogs. If an Arab touches one, he is purified by washing once; whereas a Suahili has to wash six times in water and then once with sand.

5. When travelling, the Suahili merge the middle day and afternoon prayers into one, and also those at sunset and at eight o'clock. But after staying in a place for eighteen days, the Suahili have to make the four prayers separately; whereas the Muscat Arabs do not revert to the ordinary ritual until their return to Muscat

6. If a Suahili free a slave who subsequently becomes rich, upon the death of the slave the property passes to his old master; whereas that of a slave freed by an Arab is at the absolute disposal of its owner.[1]

Sherifu said there were other differences, but that these were the chief. He laughingly added that even these were only matters of ritual, and it was absurd to make so much fuss about them. He said that the two sects are excessively bitter against one another; they both claim to be the true interpreters of the Koran, and say that the others are liars. They carry this sectarian bitterness to such a length that they will not worship in the same mosque, and would even prefer to allow a European to enter one than a member of the rival sect.

A point of some interest in connection with the Suahili race is the belief on the east coast that marriages between them are either barren, or productive only of one or two children, who are themselves invariably sterile. This was told me by several residents, whose evidence is especially reliable, as they were not aware of its bearing on any biological theories. The statement was first reported to me by Bird Thompson, and subsequently by Herr Würtz of Ngao. The races, however, in the towns of the coast zone are so mixed, that the true light-coloured Suahili are scarce, and such marriages are unusual. The people who call themselves Arabs are generally half-breeds, and it is very rarely that either parent of a Suahili is of pure Arab origin. It is, however, asserted on the coast, that if the child of an Arab by any of the coast women (*i.e.* a true Suahili) marry one of the offspring of a similar union, then the marriage is either barren or the offspring are sterile.

If this could be proved, it would have an interesting bearing on the question of the unity of the human species, for the main argument in support of this is, that all the races of mankind are capable, when crossed, of yielding fertile offspring.

The most interesting point, however, about the Suahili is their language, the *lingua franca* of Equatorial Africa, a knowledge of which enables a traveller to make himself understood by most of the Bantu tribes of the interior, and even as far west as the Congo Basin and the Atlantic coast. The language, indeed, is more important than the race, being one of the six great languages spoken at the present day. It may therefore be advisable to refer to the three main characters by which it differs in structure from the languages of the Indo-European group.

The first peculiarity of the Bantu language is its use of prefixes, instead of suffixes, in declension. As an example take the word for "good." In Latin the various forms of the word are *bonus*, *bona*, *bonum*, *boni*, *bono*, etc. In Ki-suahili they are *ngema*, *njema*, *jema*,

[1] It should be remembered that these rules are not now strictly obeyed, owing to European interference and native religious indifference.

mwema, *wema*, *mema*, *pema*, *chema*, *vyema*, and *kwema*. The root is here the final, instead of the first syllable. Any one used only to European languages would at first think *ngema*, *pema*, and *vyema* to be altogether different words; and, on the other hand, from analogy with *bona*, *boni*, *bono*, would expect *mta*, *mti*, *mto*, *mtu* to be different forms of the same root. But *mta* means "town," *mti* "tree," *mto* "river," and *mtu* "man." Thus it is the ending *-ema*, *-ta*, *-ti*, *-to*, *-tu* that is the fixed and essential part of the word, while the first syllable is variable and less important.

The changes in the first syllable serve several purposes. They express number in nouns, bring adjectives and pronouns into agreement with the words they qualify, and indicate the moods and tenses of verbs. Thus *mtu* means "a man," and *watu* "men"; *mti* "a tree," *miti* "trees"; *kilemba* "a turban," *vilemba* "turbans"; *chombo* "a dhow," *vyombo* "dhows." Again, *kupata* is the infinitive mood of "to get" (*ku* being the preposition "to"); *mapata* is the present tense of this verb, *napata* the past, and *tapata* the future.

Another set of prefixes are abbreviations which enable ideas to be expressed in one word, which in English require several. Thus *M-* prefixed to the root of the name for a district is short for *mtu*, and means a native of the tribe which occupies it; *Wa-* is an abbreviation for *watu*, and means two or more of the natives; similarly the prefix *U-* attached to the root indicates the name of the country, and *Ki-* either the language of the tribe or the adjectival form of the name. Thus *U-kamba* is the name of a district in East Africa; *Wa-kamba* is the name of the tribe which inhabits it; *M-kamba* that of a single member. *Ki-kamba* is either the proper adjective relating to the tribe, or, as a noun, is the name of its language. Similarly *U-nyika* is the Nyika country, the *Wa-nyika* the people that dwell therein; an *M-nyika* is one member of the tribe, and *Ki-nyika* its language.

In other divisions of the Bantu the same variations are expressed by similar, though sometimes slightly different prefixes. Thus *Isizulu* is the Zulu language. *Bu-ganda* is the name of the country generally known in England as Uganda; a native of the country is an *M-ganda*, of which *Wa-ganda* is the plural, and *Lu-ganda* is the language.

The adjectives, numerals, and pronouns, and in some cases the verb also, have to agree with the noun. They are all consequently declined in accordance with a set of rules known as the "concord," which is a second interesting feature in Ki-suahili. The nouns are divided into eight classes, according to the formation of the plural. All the variable parts of speech are declined in the same way as the nouns, though this is sometimes masked by contractions. This is shown by the following examples:—

Mtu mema = "a good man."
Watu wabili wema = "two good men."

Mti mwema = "a good tree."
Miti mibili mema = "two good trees."

To take a longer case, "My good child must have food" is

Mtoto mema mangu mana jocula.

Conversion into the plural changes it to

Watoto wema wangu wana jocula.

This concord, though it makes the grammar rather difficult, renders the language musical by its adoption of "apt alliteration's artful aid." This is also increased owing to every word ending in a vowel, a rule which is even applied to words of foreign origin; so that a "man-of-war" becomes *manowari*, and a "gin" *jini*.

A third feature in the language strange to a European is the method of agglutination or combination of several words into one, so that a sentence is sometimes converted into a single word, as in the Suahili riddle, *Hausimiki, hausimami.* A more instructive case is such a compound as *Nitakipataje*, meaning "how shall I get it," which is constituted as follows:—

Ni-	ta-	ki-	pata-	je.
I	sign of future tense	it	get	how.

The pronominal inflections, concord, and agglutination shows that Ki-suahili is a language that has undergone many changes, but it is of interest to note that it still shows many traces of a primitive condition. Thus it is essentially a spoken language, for although it has long been written in Arabic characters, this has not affected its structure. Hence there are no degrees of comparison, for these can be expressed by modulations of the voice. Thus *mbali* means "far"; but *mbāli*, pronounced slowly, with a strong emphasis on the middle syllable, implies "very far"; if, again, it be spoken in a high-pitched voice, it means "a very great distance away." Similarly *msuri*, when applied to food, means "nice"; *msūri* is "very nice"; and if the word be said with a smack of the lips and a rolling of the tongue it means "delicious."

The language in this respect is poor, as it is also in terms; but in some cases it is richer than English. Thus it has three words for "handful"—*konzi*, the amount that can be grasped in the hand; *ukufi*, that which will rest on one; and *chopa*, that which will lie on both hands together. The precision in the use of these terms is, however, only a result of the crudeness of the native system of weights and measures.

The abundance of imitative words is also a primitive character. Thus "fly" is *ntzi*, the sound of which is a suggestion of its buzz; *kuku* is "fowl," and *tufu-tufu* is "boiling." *Kiboko*, the hippopotamus, reminds one of the short, deep grunt of this animal, just as *paa* does of the bleat of the small antelope, of which it is the native name. *Uma* (to ache), *-anana* (gentle), and *-ororo* (soft), are less obvious illustrations of the same type.

The frequent reduplication of words in order to emphasise their meaning or to express the plural is another primitive feature, which in English is preserved only in such survivals as *pp.* for "pages," and the double *l* for "laws" in the symbol LL.D. *Poli* is "slow," but so sluggish are the habits of the ordinary Suahili that the word is nearly always used in its more emphatic form of *poli-poli*. *Kupasuka* is "to be torn" (*ku* being the infinitive preposition "to"); *kupasukapasuka* is "to be torn into shreds." *Pale* is "that"; *palepale* means "that one there." This reduplication is like that of the Zulu, in which *kulu* being "great," *nkulunkulu*, "the great-great," is the name for God.

Cases of repetition of either one or several syllables in a word are almost innumerable, such as *mkoko*, "the mangrove," or *mguruguru*, the native name of a species of lizard.

The homely baby words which occur in every language are, of course, met with in Ki-suahili. *Ababi*, "my lord," is the same as the "abba" of many Oriental languages. *Mama* for "mother," and *titi* for "a nipple," remind us of the use of these words in English; *ku lala*, "to go to sleep," and *ku kwenda tata*, "to toddle" (*ku* being the preposition "to" and *kwenda* the verb "go"), also agree with English usage; but in Ki-suahili *lala* and *tata* are the formal words and not merely baby language. In other cases, however, these primitive sounds have different meanings in English, and these are suggestive of differences in the social customs of the people. Thus *dada* means "a sister," which is an indication of the importance of the elder girls in the nursery. *Bibi* for "grandmother" reminds us that, owing to early marriages and to women sharing in the hard labour of the field, grandparents play a more active part in the care of children than they do in England. *Baba* stands for "father," while *papa* is a "shark," the meaning of which is not obvious.

The language of this people is therefore worthy of at least passing consideration, for it introduces us to a type of grammatical structure which is strikingly different from that of the Indo-European group, and it retains so many primitive features as to afford suggestive illustrations of the growth of language and of grammar.

2. *The Wa-pokomo*

In striking contrast to the Suahili of the eastern coast are the Wa-pokomo of the valley of the Tana. While the former have been affected by many foreign influences, the latter, living in their swamp-surrounded homes, have preserved their primitive simplicity. In spite of the proximity of the tribe to the coast, very little has been written about it. Krapf[1] published a vocabulary of the Ki-pokomo in 1850, and New[2] in 1873 gave a short list of words in the language, and a brief description of some of the villages nearest the coast. Fischer[3]

[1] J. L. Krapf, *Vocabulary of Six East African Languages* (1850).

[2] C. New, *Life, Wanderings, and Labours in Eastern Africa* (1873), pp. 526-527.

[3] G. A. Fischer, "Die Sprachen im südlichen Gala-Lande," *Zeit. fur. Ethnol.* Bd. x. (1878), pp. 141-144.

No. XVIII. TWO WA-POKOMO OF THE TANA. *Page* 343.

in 1878 gave a more reliable comparative vocabulary, which showed that the language is closely allied to Ki-suahili.

The Wa-pokomo may be at once distinguished from their neighbours by their powerful physique. They are taller than the Suahili, and their massive herculean frames are strikingly different from those of the thin-limbed Somali and Galla (compare Pl. XVIII. and XIX.). The second feature in the people that impresses itself upon a visitor is their affectionate, kind-hearted, happy-go-lucky nature. They seem far more attached to one another than African natives usually are, and the domestic virtues are developed in them to an unusual extent.

The position held by women in the tribe is exceptionally favourable. According to Bird Thompson[1] the Wa-pokomo are actually monogamous. In some of the smaller villages this really appears to be the case; but it is probably owing to the paucity of women, from the extent to which they have been carried away by raiding parties of Somali and Suahili. Herr Würtz of Ngao assured me, however, that a man may have as many wives as he can buy, while many men are compelled to be polygamists against their will, owing to their inheritance of the wives of a deceased brother. Most of the elders in the villages on the Lower Tana appear to have the right to treat the wives of their sons as concubines. The morality of the tribe is, therefore, not so high as has been stated. The position of the women in other ways is, however, undeniably superior to that of other tribes. The girls are often betrothed when six years old, and their lovers then begin to make presents to them; but marriage does not take place until the women are adult. In some districts early marriage is prevented by the rule that no man may marry until he has killed a crocodile, and given a part of the flesh to the woman to eat; but this does not seem to be enforced in the villages of the Lower Tana. The Wa-pokomo say that it is owing to the late date of marriage among them that they are so much stronger than the Suahili.

The women spend most of their time in the villages and do not work much, either in the fields or on the river. During the pressure of work at sowing time and harvest they help the men; but the latter do most of the hard work, such as hoeing the ground, while the women sow the seeds or transplant the young plants. When travelling in canoes, the paddling is left to the men. A point still more to the credit of this tribe is that the women share the pleasures of the men. Thus they join in the dances on equal terms, and take their meals with the men, in spite of Drummond's remarkable assertion[2] that "no African would ever demean himself by eating with a woman."

[1] Mr. Thompson's opinion on this point has been previously quoted by E. Gedge, "A Recent Exploration up the River Tana," *Proc. Roy. Geog. Soc.* new ser. vol. xiv. (1892), p. 516.
[2] *Tropical Africa*, 4th ed. p. 59.

The valour of the Wa-pokomo is, however, not equal to their domestic virtue. A more cowardly race of men it would be difficult to find. Their timidity was impressed on me owing to the trouble I had to get into communication with them, and then to allay their suspicions and fears. They have for so long submitted to being raided and oppressed by the Suahili, Somali, and Galla, that they appear to have completely lost any idea of how to fight. They have spears, but these are only used as paddles and against crocodiles, reed-rats, lizards, and fish. In spite of their great personal strength, they apparently never attempt to use their weapons in self-defence. When a Somali raiding party appears on the Tana, it orders the Wa-pokomo to provide "maus" or canoes; and though the Wa-pokomo know that these will be used against themselves, they weakly obey. But it is hardly fair to expect men who do all the manual labour of the tribe to show the same military prowess as those who specialise in war and the chase, and leave all the work of the village and of the fields to the women. An animal cannot be both a war-horse and a dray-horse, and the constant danger of slavery, death, or famine is the price the Wa-pokomo women pay for the lack of specialisation in labour, or, as it might otherwise be expressed, the equality of the sexes.

The Pokomo religion is a fetichism of which very little is known, for the rites are secret. Each man carries about with him a charm, and most of the villages have a little shed built to protect some small article, such as an empty bottle, a used cartridge case, or an old meat tin, which is buried in the ground, as a protection against the Somali. Herr Würtz of Ngao has found out most about the religion of the people, and the following information was mainly given me by him. The Pokomo god is named "Auri mwantya dsongo ngombe auri kinemu"; but this name is secret, and Herr Würtz has never been able to find out its meaning. The god is always spoken of as the "Old Man of the Woods"; he is held in great horror, and formerly no young M-pokomo would go into a forest for fear of him. The elders of the village are the priests, and form an order, Ngadsi, the members of which are all sworn to secrecy. In each village there is a small oblong hut, very different from the beehive-shaped houses of the people. I asked at Vuju what it was, and being told that it was the prison I went inside. It was quite empty, and there seemed nothing in it to explain why the crowd of natives should have fled screaming to their canoes and to the swamps. It appears, however, that this hut was "Ngadsi," and no one but the most sanctified priests ought to have been allowed to enter it.

The Lutheran missionaries at Ngao say that this order is the only real governmental power in the Pokomo, so that they do not care to destroy faith in it. The elders, however, certainly greatly abuse their trust. The people are compelled to place contributions of food in

"Ngadsi" for the "Old Man of the Woods," who is supposed to come at night and fetch them. In this way the priests live in plenty, even during periods of famine. They are certainly conscious impostors, for by means of a peculiarly-shaped drum, which is beaten and blown at the same time, they make a noise described as louder than the roar of a lion. This they say is the voice of the Old Man of the Woods.

Bird Thompson told me that the Wa-pokomo have a certain idea of immortality. One evening, as he was chatting with his canoemen round the camp fire, he asked them respecting one of their number who had been killed during the day by a crocodile. The men said that their dead comrade knew that they were sitting round the fire and were thinking of him. Their ideas of the future life, however, were very hazy; they said their friend was very uncomfortable, and that the present life is better than the next.

The Wa-pokomo are fond of singing, and have many songs which are a mixture of Arabic, Galla, Ki-suahili, Ki-pokomo, and Ki-boni.[1] They have also a number of sacred songs, called the songs of Gadsi, which are, however, kept strictly secret.

As regards social government, the Pokomo system is patriarchal. Mkururu, my old friend at Ngatana, claimed to be the chief of the Lower Tana Wa-pokomo, and said there were three other men of equal power farther up the river. He seems, however, to exercise no authority except in the villages of his immediate neighbourhood; and even in these it is very restricted. A village consists usually of about five families, each of which has a headman. These with the *Wa-gangana*, or "medicine men," form the members of Ngadsi and rule the village. At Ngao, which is a large settlement, there are nine families, including 100 men, and a larger number of women; but in this case there are only four elders.

The knowledge of arts among the Wa-pokomo is rudimentary. The dress consists only of a narrow cotton loin-cloth in the case of men, and a short petticoat, made of several flounces, for the women. The principal ornaments used are articles made of brass wire and neck laces of large white beads. A man who has killed a crocodile wears for some days afterwards a long strip of the skin. To supplement their limited clothing, and to protect themselves against the cold and wet, the natives rub their bodies with castor-oil; while they often stain themselves bright red by mixing ochre with the oil. The hair is thick, and dressed with mutton fat and castor-oil into curls about three inches long, which hang round the head like the ends of a mop.

The principal food of this tribe consists of the banana, plantain, beans, cassava, pumpkin, sugar-cane, and maize. The castor-oil plant and tobacco are also extensively cultivated, but the latter is used only as snuff. The people have few manufactures. Their iron

[1] A collection of these songs has been issued by Würtz ("Lieder der Pokomo," *Zeit. afrik. ocean. Sprach.* Bd. I. Ht. 4) as these pages pass through the press.

spear-heads, hoes, and knives they buy from the coast. The principal industrial achievement is the carving of their maus or canoes. These are made of the wood of the sycamore,[1] a tree known to the natives as the *mkuyu*. The canoes are keelless, and are simply made by burning and digging out straight portions of tree trunks.

From the figs of the sycamore the natives make an intoxicating beer, of which they are very fond. At the season when the fruit is ripe, all the inhabitants of a village are sometimes found in a state of drunken stupor.

The Wa-pokomo suffer severely from skin diseases. One, which the Suahili call leprosy[2] (*ukoma*), is very prevalent. I once went for some hours down the Tana in a canoe with eight natives, five of whom were disfigured by this malady. The Somali who ought to have gone with me refused to enter the canoe, as they said the natives were unclean, and we should catch the disease.

The Wa-pokomo are fond of dancing and singing, but do not play at the more elaborate games popular on the coast. The children amuse themselves with "seki," which consists of knocking over two opposite rows of grass stalks, by spinning a teetotum made of a piece of gourd rind fastened on a wooden peg. In a game I watched at Ngao, supposed to be a fight between Galla and Pokomo, the four grass stalks on each side represented the chief, a headman, an ox, and much cattle. The children were surrounded by a circle of adults, who cheered every success of the Pokomo champion, and failure of that of the Galla.

As to the affinities of the Wa-pokomo, there can be no doubt that the tribe is a branch of the Wa-nyika, which occurs along the coast opposite Mombasa. The Wa-pokomo now occur in a narrow belt on each side of the Tana from Kidori (1° S.) to within ten miles of the sea at Charra (2° 30′ S.) No doubt at one time they occurred to the mouth of the Tana, and extended for some distance along the coast in either direction. One colony of them still lives in two or three villages at Wasagu, on the island of Pemba; while a creek named Pokomoni at the back of the island of Manda indicates their former extension to the north.

3. *The Wa-kamba*

The study of the Suahili and the Wa-pokomo is subject to the disadvantage that neither tribe has been allowed to develop naturally along its own lines. The Suahili have been affected, both physically and intellectually, by external influences; the Pokomo spirit has been

[1] This is the true sycamore (*Ficus Sycomorus*, Linn.), not the maple, to which the name is given in England.

[2] Most of the people in the coast towns who are said to be lepers are only suffering from leucodermia. True leprosy, however, also occurs.

crushed by centuries of oppression. The tribe of the Wa-kamba is therefore of interest, as it illustrates a third type of Bantu character, for it has kept comparatively free from foreign blood, and has been sufficiently powerful to maintain its independence.

The tribe is reported to have entered British East Africa from the south. Stanley has referred to the existence of a district named U-kamba to the east of Tanganyika, and this is possibly a former home of the tribe.[1] At present it occupies a somewhat pear-shaped tract of country in British East Africa, including the basins of the Athi and the Tiva, and a small strip of country in that of the Tana. Much of this area is uninhabited, and the Wa-kamba are mainly grouped in three districts — Kikumbuliyu on the south, the Iveti Mountains on the north-west, and Mumoni and Kitui on the north-east. The population of Kikumbuliyu is very limited in number; the late Dr. Charters of Kibwezi estimated that the inhabitants of the district around that station only numbered about 600; and the two other divisions of Kikumbuliyu (namely Masongaleni or Kikumbuliyu Chakati, and Kikumbuliyu Mwesho) probably do not together contain more than twice as many. The district of Mumoni and Kitui is larger in area, but the population is reported to be sparse, and at present the Iveti Mountains form the main home of the tribe. This section has been carefully studied by Mr. Ainsworth, to whom I am indebted for most of the following information.

The Wa-kamba of the Iveti Mountains are the most powerful and warlike of the British East African Bantu. But in spite of their courage and strength, they are not able to keep their country to themselves. A generation or more ago a colony of the Kikuyu succeeded in establishing itself in the very heart of the Iveti Mountains at Kilungu, and the site of the station of Machakos was occupied by Masai. There are men still living who took part in the struggle which expelled the latter from the district. The Wa-kamba surrounded the Masai kraals by night, and stormed them at daybreak. The Masai rallied on the banks of a stream, and there a battle raged all day. Two thousand of the Wa-kamba were slain, including the brother of the present chief Nzibu, who fell after performing prodigies of valour, which the natives still remember with pride. Enraged by his death, the Wa-kamba warriors made a final charge; the ranks of the defenders were broken, and a massacre of men, women, and children cleared Machakos of Masai.

The Wa-kamba, however, still dare not cultivate the open meadows that are exposed to attack. The shambas occur only on the hills that form an amphitheatre around Machakos, and the fertile meadows on its floor lay fallow until the British East Africa Company built a station and planted a garden there; for though the Masai have made no attempt to re-establish themselves in the district, they continually

[1] *How I found Livingstone* (ed. 1890), pp. 275-276.

send raiding parties to attack it. Two months after I had passed through Nzaoi on the journey into the interior, a large force of El-Moran dashed into the valley and obtained a rich booty in cattle, and in the following year (1894) a Masai party attacked the British fort; so that it is still hopeless for the Wa-kamba to attempt to use the rich plains upon their frontier.

If they could do this, the population would no doubt materially increase. Mr. Ainsworth estimates that of the Iveti Mountain district (*i.e.* the rectangular area between Nzaoi, Kavaluki, Machakos, and Maka) at about one million, giving an average of only 150 to the square mile. This is more than the average for Africa, which is estimated by Ravenstein[1] at 12 to the square mile, but it is far less than the country could support.

The political system of the Wa-kamba is based on the family. The people live in kraals or villages, each of which contains, as a rule, about fifty inhabitants, and is ruled over by an elder or *mwanto mere.* Each kraal contains practically a single family, and several are grouped together under a chief. Some of the chiefs have only two or three kraals, but others have more. Nzibu, for example, who is the *Mwanto Mineni*, or big chief, of the Machakos district, is lord over fifty.

Each kraal has its own plantations, the boundaries of which are marked by hedges, banks, ditches, or occasional heaps of stone. The ditches also act as irrigation channels, which are especially necessary for the fields of sugar-cane. The whole of the produce is the property of the kraal, and each member has a share of food served out from the common stock. The surplus supplies are sold, and the goods received in exchange belong to the kraal. There is apparently no definite ownership of private property, except clothes and weapons. The industrial system is therefore socialistic rather than patriarchal. The family is certainly the unit, but the head of the family has less absolute power than in typical patriarchal communities.

Order is kept in the tribe by the enforcement of a kind of common law, the unwritten rules of which are applied by a jury of elders. In trials for crime, precedent is rigidly followed. Capital punishment is apparently only inflicted for a second act of rape. The accused is tried by a jury of his fellows, before a judicial committee of elders. The latter sit on their stools in a group; the warriors or *mwaniki* stand round in a circle. The accused is confronted with his accuser, and is allowed to make any defence he can. If the crime is proved, the elders ask the *mwaniki* if the sin is bad, and they reply "Very." The elders then ask if the man deserves punishment. "He does," say the *mwaniki*, and the elders tell the warriors to flog him, an order which is carried out until some of the criminal's bones are broken. For a second offence, the man's throat is cut.

[1] In J. S. Keltie's *Partition of Africa*, 2nd ed. 1895, p. 521.

At present the Wa-kamba are an agricultural race, though they keep as many cattle as they dare. They are, moreover, fairly expert huntsmen, and kill the hippopotamus in the Thika-thika and the Athi, and the rhinoceros on the Kapte plains. Their staple food is grain, such as dhurra, wimbé, maize, and millet (*Panicum italicum*), and also beans, bananas, and pumpkins. They grow tobacco, which they use as snuff, though a few of the people have learnt to smoke. They possess a few iron hoes, but most of the agricultural work is done with wooden implements; the ground is first dug up by young men with hard pointed poles; the women then break the clods with curved sticks, while the children gather the grass and weeds into heaps.

Other arts are more advanced, especially metal-working. This is done by special smiths or *fundi*. The iron ore is collected from the beds of streams, and smelted by charcoal in small forges of the "Catalan type." The iron is then wrought into swords, spears, arrow-heads, and knives. There is, therefore, little demand for iron wire, which is the main source of iron to such tribes as the Kikuyu and the helot races subject to the Masai. Working in brass is more advanced than in iron, though the raw material is obtained as wire from the coast. This is heated and hammered out by an iron mallet on a stone or iron anvil. Many of the brass ornaments are pretty and elaborate. A small circle is frequently worn in the middle of the forehead, or let into the centre of a stool. Rings are made by twisting brass wire round a core of either brass or iron. The aprons of the married women are generally covered with large brass beads, which are made by the native smiths. The husband gives the smith the necessary amount of brass wire and a goat as payment for the work; the wire is cut up into short lengths, hammered into thin oblong strips, which are bent into cylindrical tube-like beads. Earrings of many different designs are worn, but brass rings are the commonest. Lip and hair ornaments are not in fashion, but the hair is greased and dressed.

Clothes consist either of cotton cloth obtained by purchase, or of leather aprons made from roughly tanned skins. The men, however, often wear only a flap of skin over the shoulders, in order to protect the lungs. The clothes are frequently stained with red and white earths. The body is generally kept rubbed with oil and decorated with streaks of paint; the commonest design is a white band, stretching across the face from ear to ear, and surrounding the eyes. The upper incisor teeth of boys are filed into pointed fangs at the time of circumcision, which is performed at the age of about seven or eight.

Pottery for cooking purposes is made from coarse clay, and baskets of plaited glass are used for many domestic purposes. String is usually made from the fibres of the leaves of the aloe and the *Sanseviera;* a superior quality, of exquisite finish, which is used for binding arrows, is made of plaited zebra hair.

The Wa-kamba are very keen business people, and not only enjoy bargaining with passing caravans, but themselves conduct expeditions to the coast. They carry down grain, tobacco, ivory, and gum, and drive with them herds of cattle, goats, and sheep. These native caravans go to Melindi or Mombasa, where they exchange their goods for beads, brass, and cloth. During this trade they have acquired a good many customs and ideas from the Suahili. Thus they use the "mkono" or hand, *i.e.* the length from the elbow to the finger-tip, as their unit of measurement. They count up to ten, but always use the graphic symbols at the same time; thus the numbers from one to five are counted on the fingers of one hand, the figures from six to ten are shown by clasping the extra number of fingers in the other hand.

The language of the tribe, or Ki-kamba, is very imperfectly known. A dictionary by Krapf[1] was issued by the Church Missionary Society, from information supplied by a native at Mombasa. This man first gave Krapf a Ki-taita vocabulary, and was then urged to continue with Ki-kamba. He appears to have supplied one out of his imagination. Mr. Ainsworth can get no recognition of the words from the natives of his district, and he points out that many of them have a suspicious resemblance to Ki-taita. The language is Bantu, and the following comparative list of geographical terms (generally among the most primitive words in a language) show that Ki-kamba is very different from Ki-suahili, Masai, and Kikuyu :—

	Ki-suahili.	Ki-kamba.	Ki-taita.	Kikuyu.	Masai.
Mountain .	mlima	kiima	mugondibwa	ngongo	doenyo
Hill . .	kilima	kitundwa	shungol	ngongo	narol
River . .	mto	utzi	mweta	luyi	guaso
Spring . .	chemchem	kithima	nizio	luyi lunini	njoro
Plateau . .	...	ulu	...	mwer	rangatan

In some respects the language is rich, as shown by the following list of names of the different stages of man, supplied me by Mr. Ainsworth :—

Kana.	Baby.
Evezi.	Lad.
Mwant' o' theli.	Middle-aged man.
Mwaniki.	Warrior.
Mwanto mere.	Elder.
Mvinto.	Fashionable young lady who has had no children.
Gungu.	Woman who has had one child.
Mwintu maka or kibeti.	Elderly woman who has had more than one child.

The religion of the Wa-kamba is primitive and indefinite. They have a certain hazy notion of a god, whom, like the Masai and Kikuyu, they call *Ngai;* but they know little about him, and very seldom invoke his aid. If they want rain they place offerings of

[1] *Comparative Vocabulary of East African Languages*, 1850.

banana, grain, and beer under trees; but that is apparently their only religious rite. Circumcision is practised, but possibly only for sanitary reasons.[1] The natives possess charms, but no idols. They do not protect their fields by the elaborate fetiches used by the Wa-nyika and Wa-giriama of the coast. Nor do they, in dealing with strangers, insist on the complicated religious observances practised by the Kikuyu.

They appear to have some idea of immortality, though I could learn nothing at all definite upon this subject. The dead bodies of chiefs are not thrown to the hyenas as with the Masai, but they are carefully buried. A householder is always buried under his hut, and his wife in front of the door, after which the hut is abandoned. The bodies of less important members of the tribe are simply thrown to the hyenas.

The Wa-kamba make very plucky soldiers. Armed only with inferior spears, and trusting mainly to bows and poisoned arrows, they hold their mountain fastnesses against the Masai and Somali, of whom the latter are beginning to press them from the north. The Wa-kamba themselves at times undertake raids upon their inveterate foes, the Masai and Kikuyu, and also upon the inoffensive and defenceless Wa-pokomo of the Tana. But as a rule they are peaceful law-abiding people, who prefer cultivation and trade to war and pillage. If only granted adequate protection they will no doubt steadily increase in numbers and wealth, furnish an inexhaustible supply of cheap food and efficient labour, and become one of the most important agents in the development of the eastern portion of our East African dominions.[2]

(*b*) The Negroid Races

1. *The Kikuyu*

The Kikuyu occupy a narrow belt of hilly forest country extending from the southern slopes of Kenya, south-westward to the edge of the Rift Valley at the hill of Ngongo Bagas. They thus serve as a buffer state between the Wa-kamba and the Masai of the district of Naivasha. They are usually regarded as normal Bantu; Stuhlmann,[3] for example, places them with the Wa-kamba and the Wa-taita in the "Younger Bantu" group. They appear to me, however, to be more nearly related to the Masai, and to be a race containing both Bantu and

[1] It should be noticed that this is entire and not performed as in the Masai and Kikuyu. As the natives have continually to ford streams and wade through swamps abounding in the larva of *Bilharzia hæmaturia*, the rite no doubt lessens the danger of incurring hæmaturia.

[2] The Wa-taita, Wa-daicho, Wa-nyika, Wa-giriama, and their allies, are the principal remaining groups of Bantu in Eastern British East Africa. Respecting these I have nothing material to add to previous descriptions.

[3] F. Stuhlmann, *Mit Emin Pasha*, p. 848.

Hamitic elements. They are therefore here included among the Negroid races.

Von Höhnel[1] has given such an admirable account of this tribe that I do not propose to describe them in detail, but only refer to the points that serve to indicate the infusion of Nilotic blood.

Their physical features are in many ways intermediate between the Masai and Bantu; the chin, for example, is less prognathic than in the latter. In their mode of life they more resemble the Bantu, as they are mainly agriculturists. They have some food plants, which are not possessed by the ordinary Bantu tribes, and which they cannot apparently have acquired from these. Thus they grow a millet which Schweinfurth identifies as *Panicum italicum*, L., and which he thinks must have been introduced from India before the time of Mohammed. It is true that the Wa-kamba now also cultivate it, but it is possible that they obtained it from their Kikuyu neighbours.

In their arms and equipments they resemble the Masai more than the Wa-kamba. In fighting they trust more to the spear and shield than to the bow and poisoned arrow, though these are used by the natives of the southern part of their country, who have much intercourse with the Wa-kamba. Kikuyu spears differ in shape from those of the Masai, for they are ovate rather than linear-lanceolate; they agree in this, however, with the Njempsians, who are certainly Nilotic Hamites and not Bantu. Their shields, swords, and trinkets are almost the same as those of the Masai. They dress in the same manner, the women being clad in roughly-tanned skins, and the men wearing flaps of leather over the shoulder, or hanging like a tail behind. The method of circumcision—a point of great systematic importance among these tribes—is the same as that of the Masai.[2]

Another point by which they differ from most of the inland Bantu tribes in East Africa, and resemble the Negroid and Soudanese races of the west coast, is that they have an elaborate series of religious rites, to which they attach considerable importance, and the obligations of which they scrupulously fulfil. Some of these are described in Chap. XI. (pp. 189-194).

The Kikuyu language is very little known. It is certainly not Bantu, and is allied to Masai. The list of geographical terms on page 350 shows how different the language is from either Masai or Ki-kamba.[3]

The temperament of the people, with their excitable, irritable ways—their thievish propensities, their hostility to strangers, and their

[1] *Zum Rudolf-See*, pp. 386-395.

[2] It has been described in a Latin footnote by H. H. Johnston, *The Kilima-Njaro Expedition* (1886), pp. 412-413.

[3] Von Höhnel quotes one of the elders as addressing his comrades as Wa-kekoyo. The Wa- is generally a Bantu prefix, but is also Hamitic. In this case it may have been attributed to the language by a misunderstanding, or been adopted from Suahili caravans.

loyal devotion to pledges when ratified by blood-brotherhood—all remind one of the Masai rather than of the Bantu tribes, such as the Wa-kamba and Wa-pokomo. The more complex social organisation, the higher authority of the chiefs and elders, and the very inferior position of women in the tribe, are further points in which the Kikuyu differ from the Bantu of this part of Africa.

2. *The Masai*

One of the minor difficulties that beset the traveller in Tropical Africa is the frequency of a sudden passage from one tribe to another, whose language belongs to an altogether different race group. In a limited tract of country there may be several tribes, whose languages are all more different from one another than are the extreme forms of those included in the Aryan group. This difficulty is greatest among the tribes of the Fulah group of the Western Soudan; but it also occurs in British East Africa, for this region is crossed by the line of division between the Bantu and Hamitic races. A single day's march is in places sufficient to carry a traveller from Bantu people to the kraals of the Nilotic-speaking Masai. Pronominal inflections, concord and agglutination, are all left behind, and the traveller has to carry on his intercourse with the natives in a language more different from the last he used than English is from Sanskrit.

The fact that after traversing the Masai country one can as suddenly jump back into Bantu tribes, suggests that the Masai are intruders. There is no doubt that this is the case. The Masai have entered the country by force, and they are enabled to stay there only by continual theft. The whole of their elaborate social and military system, which has been graphically described by Joseph Thomson in *Through Masai Land*, is based on the necessity for constant war raids. The Masai do not till the soil, and they live solely on meat and milk. As they do not breed sufficient cattle to supply the demand, they are bound to replenish their herds by seizing those of their weaker neighbours.

The men in this tribe do no regular work, but simply specialise in the art of cattle lifting. The women do all the hard work of the kraals, and carry the household goods from place to place during the migrations rendered necessary by the exhaustion of pasturage; boys herd the cattle, and members of helot tribes make the spears and chain ornaments. The young men, or El-Moran, simply learn how to dance, brag, and bluster when at home, and to conduct raids upon more peaceful tribes.

The Masai have been described so often that there is no need here to give any account of them. Reference need only be given to the careful observations of G. A. Fischer,[1] the graphic sketch of

[1] *Mitth. Geog. Ges.*, Hamburg, 1882-83 (1884), pp. 60-74, 224-236, Pl. iv.-vi.

Mr. Joseph Thomson,[1] and the discussion of the affinities of the language by Mr. H. H. Johnston.[2]

There is now no doubt that they are a mongrel race between the Nilotic group of Hamites and the Negroes. Their social organisation and language is Nilotic, but their physical characters indicate a considerable intermixture of Negro blood. It has been suggested that the Masai ought to be included among the Hamites; but, as in that case the Njempsians and the Kikuyu would have also to be transferred, it seems best to leave them all in the intermediate Negroid group.

3. *The Njempsians (Wa-kwafi)*

Hopes are held out by some travellers who have had experience of the Masai, that they may be induced to cease from raiding, to take service as a native police, and thus help to civilise the regions to which they are now nothing but a curse. It is not impossible that by firm and judicious administration this may be done. Mr. Hall, the present superintendent of Fort Smith, has engaged a force of Masai, which has settled beside the fort and helps him to keep the Kikuyu in order. That the Masai may change their habits is also thought possible from the present condition of the people known as Wa-kwafi, who are generally regarded as a section of this tribe.

Joseph Thomson[3] tells us that they are "unquestionably Masai in race, and only separated from that tribe through the loss of their cattle, and the consequent necessity of breaking their cherished convictions by cultivating the soil." According to this, which is the popularly accepted theory, the Wa-kwafi, only some five and twenty years ago, were a powerful clan of Masai, which lived in Laikipia. Then there is said to have been a war between them and the other Masai, in which the Wa-kwafi were defeated, deprived of their cattle and pasturage, and expelled from Laikipia. They are said to have then settled at Njemps, and adopted an agricultural mode of life.

There seems to be no doubt that in some places, as in the Kilima-Njaro district, Masai have accepted such a change of life; and where they have done so they are called Wa-kwafi. It is probable, however, that this name is applied by the Suahili to several distinct races in the Masai district, which till the soil and speak languages allied to Masai. In that case the term, like Wa-shenzi and Wa-nderobbo, has no scientific ethnographical value. Krapf, moreover, used the term as a synonym of Masai. The only so-called "Wa-kwafi" I met were those at Njemps, and to them I propose to refer as Njempsians, which leaves their relation to the other "Wa-kwafi" an open question.

It is quite possible that the story of the origin of the Njempsians

[1] *Through Masai Land*, chap. x.

[2] *The Kilima-Njaro Expedition*, pp. 446-477, 501-520.

[3] *Through Masai Land*, 4th ed., p. 450.

contains an element of truth, for their ancestors may have been pastoral nomads. But it seems quite clear that this change has not happened within so recent a period as is supposed. The people now differ from the Masai in physique, character, habits, and language.

Every traveller who has visited Njemps and recorded his experiences of the people, bears testimony to their friendliness, and frank, open-hearted ways. Thomson truly tells us that they are "singularly honest and reliable," and again, that they are "most pleasing natives," characterised by "their honesty, their unassuming ways, and their charming, unsophisticated manners." He further describes the people as physically unlike the Masai; he says they are degenerate, and that "this was especially noticeable among the women, who had lost their slender, genteel shape, and acquired the ill-proportioned, unwieldy contour of the negress."[1] The Njempsians are certainly taller and slimmer than the robust, well-built Masai. Their jaws are more prognathic, and the people are far less intelligent. They have a few religious prejudices, such as refusing to eat zebra, or to allow any part of this animal inside their villages, while the seeds of their crops are in the ground; but they are less particular than the Masai, for they eat fish and even rats.

Their dialect is also so different from Masai that some of my men, who knew the latter fairly well, were quite unable to understand the people of Njemps. If, therefore, the popular theory be true, then the Njempsians have changed alike their physique, character, habits, social organisation, and religion, in the course of about twenty or twenty-five years. A similar view was once held as to the origin of the Bushmen, who were said to be only Hottentots who had become degenerate owing to the theft of all their cattle by the Boers.[2] But the idea that such fundamental changes can take place so rapidly has long since been discredited.

Fortunately, however, in the case of the Njempsians we are not left to rely only on general probabilities, as historical evidence is available. In a memoir by Léon des Avanchers in 1859,[3] the Wa-kwafi are said to live around Lake El-Boo or Baro (Baringo), on the very site of their present home. This map records only the leading facts in the geography of the interior, as reported on the east coast by the Suahili traders; it is, therefore, not likely that the settlement of the Njempsians in the Baringo basin was then a recent event. It is more probable that they preceded the Masai instead of being descended from them, for, as Thomson's descriptions show, they have been more influenced by intermixture with Bantu. They have probably only been allowed to retain their present home, because its swampy

[1] *Op. cit.* p. 450.

[2] W. H. I. Bleek, "Scientific Reasons for the Study of the Bushman Language," *Cape Monthly Magaz.* new ser. vol. vii. (1873), p. 149.

[3] Léon des Avanchers, "Esquisse géographique des pays Oromo ou Galla," *Bull. Soc. Géog. Paris*, sér. 4, t. xvii. (1859), map, and p. 164.

jungles, rank grass, and scrub-covered plains are of no use to a pastoral people. When, therefore, the Masai invaded the country, they swept southward past Njemps, to the rich grazing lands around Naivasha and on Laikipia.

Section D.—The Hamitic Races

(*a*) *The Galla*

By some authorities the Masai are included in the Hamitic group, but we have only to compare the features of a member of this tribe with those of a Galla (Fig. 21) to realise the predominance of the negro element in the former. The aspect of the pure Hamite differs altogether from those of the Bantu and Negroid races. The accompanying portrait of a Galla presents no correspondence with the conception usually formed of an African native. The forehead is high and square instead of low and receding; the nose is narrow, with the nostrils straight and not transverse; the chin is small and slightly pointed instead of massive and protruding; the hair is long and not woolly; the lips are thinner than those of the negro and not everted; the expression is intellectual, and indicates a type of mind higher than that of the simple negro. Indeed, except for the colour, it could hardly be distinguished from the face of a European. These characteristics prepare us for the fact that the Galla are not African, but immigrants from Asia. This was impressed on me at the outset by their folklore, some of which had been collected by Bird Thompson. He told me the Galla story of the creation of the first man, whose name was Zadami—obviously a variant of Adam, so that apparently they still remember some of the primitive traditions of Western Asia. The history of the tribe is referred to in the following chapter, and it is unnecessary to describe the people, for this has recently been done in an elaborate monograph by Paulitschke.[1]

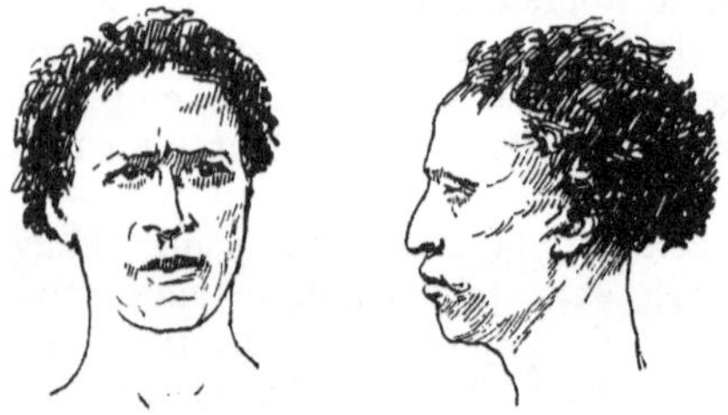

Fig. 21.—The Head of a Galla. (After Paulitschke.)

[1] Ph. Paulitschke, *Die Ethnographie des nord-ost Afrikas*, Wien, 1893. Some remarks in New's *Life, etc., in East Africa*, p. 274, suggest that the Galla have a certain belief in Totemism. It does not seem necessary to refer to the earlier theories in regard to the Galla. The first were stated by C. T. Beke, "On the Origin of the Gallas," *Rep. Brit. Assoc.* 1847 (1848), pp. 113-118. Speke regarded the Wa-huma settlers in Uganda as Galla, whereas Emin (quoted by Jephson, *With Emin Pasha in Central Africa*, 1890, p. 63) thought the Galla the descendants of the Wa-huma.

No. XIX. A SOMALI IN NATIONAL DRESS. *Page* 357.

(TOBE AND SANDALS, AND ARMED WITH SPEAR AND SWORD.)

(*b*) *The Somali*

I had little personal intercourse with the Galla, but I often saw more than was pleasant of their close allies, the Somali. This tribe also is of Asiatic origin. It claims descent from a certain Sherif Ishak (Isaac) bin Ahmed who crossed from the south of Arabia about 500 years ago, and died at Mait, near Burnt Island, where his tomb is still shown.[1] Some writers accept this tradition literally; but others, such as Miles,[2] regard Sherif Ishak as the leader of the last of a long series of migrations. Paulitschke quotes the Arabian historian, Ibn Said, to the effect that in the thirteenth century the Hawija were settled in fifty villages on the east coast around Merka. This was earlier than the time of Sherif Ishak, and the settlement was apparently even then of considerable age.

It has also been stated by no less an authority than Burton that the Somali race has absorbed a good deal of African blood, and are half-caste. The head, he says, "belongs equally to Africa and Arabia."[3] There is, however, no trace of Negro influence in the language,[4] and very little in the physical structure of the race. Burton does not, however, appear to have based his sketch on typical Somali; and parts of it read like a caricature, though showing, as usual, his remarkably keen insight into character. He describes them[5] as having "all the levity and instability of the Negro character; light-minded as the Abyssinians,—described by Gobat as constant in nothing but inconstancy,—soft, merry, and affectionate souls, they pass without any apparent transition into a state of fury, when they are capable of terrible atrocities." In this, perhaps, Burton was less unjust than he sometimes was. The behaviour of some of the Somali during the expedition to the Tana would have justified his severest judgments. Most of the men we had were passionate, lazy, and impudent, and some of them, despite the high reputation of the Somali for courage, proved to be contemptible cowards. Our experience was not exceptional, for all the officers of the British East Africa Company who had had men of this tribe in their caravans, disliked them exceedingly. Major Williams told me that the timidity of one of his Somali gun-bearers excited the ridicule of the Zanzibari, who are usually themselves accused of cowardice. Our trouble with the men was possibly due to the fact that they were all Aden Somali, who may have been corrupted by contact with civilisation, for travellers in Somaliland who have employed the natives are often enthusiastic in their praise. During the Tana Expedition we certainly saw Somali under unfavourable circumstances; they were

[1] F. M. Hunter, *Grammar of the Somali Language* (1880), p. xiv.

[2] S. B. Miles, "On the Neighbourhood of Bunder Marayah," *Journ. Roy. Geog. Soc.* vol. xlii. (1872), p. 68.

[3] R. F. Burton, *First Footsteps in East Africa* (1856), p. 107.

[4] See, *e.g.*, A. W. Schleicher, *Die Somali-Sprache*, th. i. (Berlin 1892), pp. vii.-xiii.

[5] Burton, *op. cit.* p. 109.

taken to a district very different in climate from their own, where malaria sapped their strength, and mosquitoes soured their tempers.

The men's fanatical devotion to their religion—a corrupt form of Islam—increased the difficulty of handling them. A favourite observance was a weird dervish dance around a fire, which would last until several of the men had thrown themselves into the flames in a fit of frenzy, or fallen into them in a trance. These dances had to be suppressed as they were injurious to the health of the men, but for this a subterfuge was necessary; they could not be directly prohibited, as they were said to be an essential part of the men's religion, and it had been guaranteed that this should not be interfered with. The fidelity of the men to their creed was extraordinary. At first we thought one half of this was affectation and the other half fraud. The men would not carry a bottle of whisky from one part of the camp to another, though they had to lift cases which were full of it; they would continually leave their work to pray, and seemed to accumulate arrears of prayer and pay them off whenever an unpleasant task had to be done. But when at Ngatana some of them died rather than take food which would have saved them, because it had not been killed according to Mohammedan rites, we were bound to respect, as well as to regret, their sturdy adhesion to their faith.

The headman Wasama was also the priest of the Somali. His chant was the music which marked time at the dervish dances, and his the exhortations that roused the dancers to their wildest fury. At sunrise and sunset he stood on his praying-mat in front of the line of Somali, and led their devotions. But Wasama would never have preferred death to defilement; when I mixed brandy with medicine for the sick men, he would give them the dose and swear that the "dowo" contained nothing unholy. He had been for some years an interpreter on a man-of-war, and there he had learnt not only "fo'castle English," but also the differences between an esoteric and an exoteric faith. He preached the one, but was always ready to practise the other, and his sermons were delivered in language of appalling profanity. I remember once appealing to him when a Somali tried to shirk some work on the excuse that he was bound to go and pray. Wasama expounded the orthodox Mohammedan rules for prayer in language that would have scandalised Billingsgate. Often, when at his devotions in one corner of the camp, his keen eye would detect a man doing something that he ought not to do; Wasama would at once leap from his mat and hurl at the culprit a volley of blood-curdling oaths, and then drop on to his mat again to conclude the interrupted prayer. But in spite of his language, Wasama was a man who did high credit to the Somali race. He was kind-hearted, devoted, energetic, courageous, intelligent, and skilful. He and a few others like him did their best to redeem the character of the rest. Wasama's brother-in-law, the famous Dualla Idris (Pl. XX.), is a Somali of the same type; by his services

Aja Achmet Wasama. Dualla Idris.

No. XX. TWO SOMALI HEADMEN. *Page* 358.

in Stanley's first descent of the Congo, and subsequently as headman during the James Expedition into Somaliland, Teleki's march to Basso Narok, Stanley's Emin Pasha Relief Expedition, and Lugard's administration in Uganda, he has done inestimable service in the cause of African exploration. Men such as Wasama and Dualla show that the Somali character is capable of great development; and it was impossible to compare even the roughest of our unruly crew with the Zanzibari or Pokomo, without feeling that the race has great capabilities. At first sight the people seem to be truly African in their levity, passion, and fitfulness; but scratch the Somali and you find the Arab, with Arabian capacity for devotion and discipline, and latent powers of organisation, which promise for the race a great future in the development of East Africa.

SECTION E.—THE SEMITIC RACES

The Abyssinians

While the Somali and the Galla represent one type of Asiatic invaders, the Abyssinians represent another. The two first are unquestionably Hamitic, while the Abyssinians in the main are Semitic. There are in the country several tribes representing the Hamitic races, which occupied Abyssinia before the invasion of the Semites: such are the Beja of the north, the Saho around Massowah, the Agau of the Central Plateau,[1] and the Sidama of Kaffa in the south.[2] But the dominant race throughout historic times has been Semitic. Among the men engaged for the Lake Rudolf Expedition, however, there were a large proportion of Hamites, since, being mainly drawn from the shop-boys of Massowah, the Saho were well represented.

True Semitic Abyssinians are probably not yet residents in British East Africa. The southern Abyssinians raid across the frontier into the British sphere of influence; but they do not seem to have founded permanent settlements there, and it is probably only the Hamitic tribes that conduct these raids. It may therefore appear superfluous to refer to these Semites. But as the introduction of Abyssinian porters into Equatorial Africa was an interesting experiment, which cost the Lake Rudolf Expedition a good deal of money, it may be as well to record the result.

Over eighty Abyssinians were engaged at Massowah, but unfortunately there was some misunderstanding as to the terms of their engagement. They maintained that they had been enlisted to go to Lamu and thence to India, where they were to act as escort to a native prince. When, therefore, they were landed on the beach at Melindi,

[1] L. Reinisch, *Sitz. ber. phil. hist. Cl. k. Akad. Wiss.*, Wien, Bd. cxxi. (1890); Abh. xii. p. 4.

[2] L. Reinisch, "Die Kafa-sprache in Nordost-Afrika," *ibid.* Bd. cxvi. (1888), p. 68.

given loads and told to carry them, they professed to be disappointed. At Witu they struck, and one or two were flogged until they consented to carry loads. No amount of flogging, however, would give the men the necessary personal strength. They broke down under the work; malaria at Ngatana found a suitable soil in their weakened bodies; they died, or were so enfeebled by disease, that the whole contingent had to be sent back to the coast, as it was useless to insist on their going on.

The costly experiment, therefore, failed absolutely; this was probably due to the fact that the men were badly selected, but it is very doubtful whether Abyssinians would ever compete with the Zanzibari in efficiency, and they certainly could not in economy.

The reason for the employment of an Abyssinian contingent seems also to have been fallacious. It was thought that, as the expedition would pass near some of the so-called "sacred cities" of Somaliland, the priests might preach a holy war against us, and our Somali escort refuse to fight. Recent experience, however, seems to show that such a danger was imaginary. The Somali themselves are tolerant of Christians, though fanatical in the practice of their own creed. Moreover, the various tribes of Somali are divided against themselves; hence an escort from the coast can be relied on, both from friendship and interest, to make common cause with their European leader, instead of deserting him at the bidding of their compatriots. In fact, coast Somali would probably have no more objection to fighting inland tribes of their own race, than of any other, provided the risks were not too great and the prospects of booty good. Less, however, was known about the interior of Somaliland in 1892 than is known to-day; and the organisers of the expedition did not think it advisable to entrust the whole defence of the caravan to Somali. Soudanese could not be obtained; Abyssinians were selected, mainly because they are Christians, and therefore were expected to be all the more faithful to us. Later on, however, we found that, as they themselves expressed it, they were very different Christians from ourselves; and it is doubtful whether the religious bond would have counted for anything. They refused to eat meat we had shot, or to drink beef-tea made from material prepared in England. Several of them died rather than take such food, at a time when they were so ill that they could take no other. In many ways they afforded us an illustration of Buckle's contention that "in every page of history we meet with fresh evidence of the little effect religious doctrines can produce upon a people, unless preceded by intellectual culture. . . . The characteristics of the creed are overpowered by the characteristics of the people; and the national faith is, in the most important points, altogether inoperative, [if] it does not harmonise with the civilisation of the country in which it is established."[1]

[1] H. T. Buckle, *Hist. Civilis. England* (ed. 1885), vol. i. pp. 261, 266.

In fact, just as the Somali have introduced into their Mohammedanism a mass of pagan practices, and the Suahili have softened their originally fanatical creed into harmony with their indolent and tolerant natures, so the Abyssinians have degraded the religion they profess, until it allows the systems of polygamy and slavery to flourish unchecked, and permits the retention of barbarous habits and cruel rites.

CHAPTER XVIII

THE NATIONAL MOVEMENTS AND FUTURE PROSPECTS OF BRITISH EAST AFRICA

"Usunye káutsuka hiraka"
(Slavery is not quickly eradicated).
Ki-nyika Proverb (W. E. TAYLOR).

"Slavery is not so easy to be abolished; it will long continue, in spite of acts of parliament. My friends, I have come to the sad conclusion that slavery, whether established by law, or by law abrogated, exists very extensively in this world; and, in fact, that you cannot abolish slavery by act of parliament, but can only abolish the *name* of it, which is very little."—CARLYLE.

AS we have seen in the introduction to the seventeenth chapter, the people of East Africa are essentially a race of nomads. Individuals may be stationary; settlements may be made and held for generations; but slowly, irregularly, and unconsciously, the tribes themselves are on the move. The predominant migration is a flow from north to south. The ancestors of the Masai worked their way along the Nile valley,[1] and entered what is now British East Africa from the north and north-west. The Somali have spread over the "Horn of Africa," and are encroaching from the north-east, parallel to the Indian Ocean shore. This statement of direction, however, is only true in a very general sense. Taken in detail the movement has been spasmodic and erratic, and has travelled in waves rather than in a direct and constant stream. Here, the back-wash of a broken wave has enabled a tribe to advance in the direction opposite to that of the main current; there, an eddy has thrown another at right angles to it. Occasionally

[1] C. W. Wilson, "The Tribes of the Nile Valley," *Journ. Anthrop. Instit.* vol. xvii. (1888), pp. 3-25.

a strong position has enabled a tribe to hold its own, as on a rock, while the flood has swept by it and isolated it from its fellows. Thus the Wa-kamba have been carried northward from their old home in the Ugara country, and drawn into their present position in the British protectorate; while the Doko of Laikipia and the Wa-ruguru of Kenya have been left in their forest homes, surrounded by alien races.

These national movements have in one way lessened the scientific value and interest of the study of the African natives; for they have confused the racial characters and introduced great uncertainty into the relationships of the tribes. They compensate for this, however, by serving as a great object in history, as we can now watch in progress before us a series of national migrations, similar to those inroads of the barbarians which destroyed the Roman Empire.

The movements in the present case are of two types; the Masai furnish an example of the one and the Somali of the other. The Masai make continual raids from their main kraals upon their neighbours, whom they distract and impoverish. At last, the position of these victims becomes unendurable and they move away, leaving their oppressors free to roam over the deserted district.

The Masai raids in British East Africa are made from two main centres—the district around Lake Naivasha, and the region of Nyiri and Matumbato to the north of Kilima Njaro. The raiding parties travel great distances, especially toward the east.

It is said that they used to go right up to Mombasa and even send spies into the town to ascertain if caravans were preparing to start for the interior. In 1889 they certainly raided within sight of the harbour, and in 1888 they killed a headman of Teleki's at the Mwachi river, only a day's march from Mombasa. Whole tracts of the Giriama country, a few miles north of the British East African capital, and almost close to the sea, are still left uncultivated on account of their liability to Masai attack. The extent and frequency of the raids are shown in Fig. 22, in which are marked the principal war-paths, and the localities where the doings of this tribe came under my personal observation, between February and August 1893. South of Merifano on the Tana, Harris and I found the Galla driving their flocks and herds across the river to escape the marauders, and saw the smoke of the burning villages whence the natives had fled. At the Kiboko river I found the dead bodies of some Wa-kamba who must have been surprised and murdered in their sleep, as their arrows were still in their sheaths, and their simés in their scabbards. Two days' journey north of this place the road was littered with the debris of broken boxes captured from a caravan taking stores to Sir Gerald Portal's party in Uganda. Again, on the Kapte plains near Bondoni, during our second march south from Machakos, we encountered a small party of El-Moran, who were on their way to attack some Ki-kamba villages. On the plains of the Thika-thika we met some Kikuyu refugees from Igeti; their country had been ravaged by the Masai army which we had seen enkraaled on the shores of Lake Naivasha, and the district, for two days' march in length by one in breadth, had been cleared as if by a hurricane. The fugitives described the sudden attack, the massacre, the devastation of the plantations, the capture of the cattle, and the burning of the villages. And yet, as we listened to this sickening story, we realised that this was merely one incident in a continuous series of such horrors.

At Nzaoi, in April, on our way inland, we had seen large herds of cattle, while on our return at the end of July the valley had been devastated and the herds captured. But on this occasion the brave old warrior Kiketi, chief of the district of Maka, told us how, having heard of the raid, he had summoned his fighting men, had attacked and routed the Masai on the return march, and recaptured the cattle. But victories such as this are rare, and even old Kiketi dare not cultivate the fertile lava plain that stretches westward from the foot of the hills in which he lives. It is only from the less fertile soil of the mountain fastnesses that his people can hope to reap the crops that they have sown. Throughout British East Africa, to the east of the Masai wedge, one can talk to none of the peaceful agricultural tribes without hearing sad stories of Masai raids. The people tell you of the impossibility of cultivating exposed districts—generally the most fertile ones—and they complain bitterly of the uselessness of keeping cattle, which serve only as an incentive to Masai attack.

It is frequently said that the disease which swept off most of their cattle has broken the Masai power, while during the past twelve months internal dissensions have destroyed their unity. The Masai, however, have suffered similar misfortunes and been engaged in civil war before; but they have recovered, and can no doubt do so again. While actually suffering from starvation, the tribe was doubtless crippled. Starving men cannot march hundreds of miles, and then patiently wait for the right moment at which to attack. The Masai were greatly reduced in numbers by famine and by epidemics, the result of eating diseased meat; but as soon as the crisis was over, the scarcity of food probably drove the warriors to greater activity, and compelled them to make more frequent raids. The Naivasha Masai had certainly acquired enormous herds of cattle by May 1893; it was too soon after the plague for these herds to have been bred, and the majority must have been captured. Lastly, the threatened attack on the powerful caravan of the Railway Survey in 1892, and the attack on the fort at Machakos in 1894, shows that their aggressive spirit has not been changed; and the recent murder of Dr. Charters of Kibwezi and Mr. Colquhoun adds one more to the long list of misdeeds, which make it necessary to curb the power of the Masai.

It will be noticed in the map that the proposed route for the Uganda Railway crosses the chief war-roads of the Masai, and nearly coincides with the main caravan road into the interior; it thus naturally runs from one Bantu settlement to another, for these are the trade centres and food-producing regions. One of the many incidental benefits that the Uganda Railway will confer upon the country is that it will enable a small force of men to watch the roads and protect the peaceful agricultural tribes from the attacks of their warlike neighbours.

In places the Masai have temporarily ceased from raiding, and some have even taken service with other tribes as herdsmen. According to the commonly accepted theory, the Wa-kwafi were originally a tribe of Masai who have permanently changed their mode of life, and have become an agricultural people. Some men who have lived much among the Masai, as for example Mr. George Wilson, have a high opinion of their intelligence and capacity for learning. It is

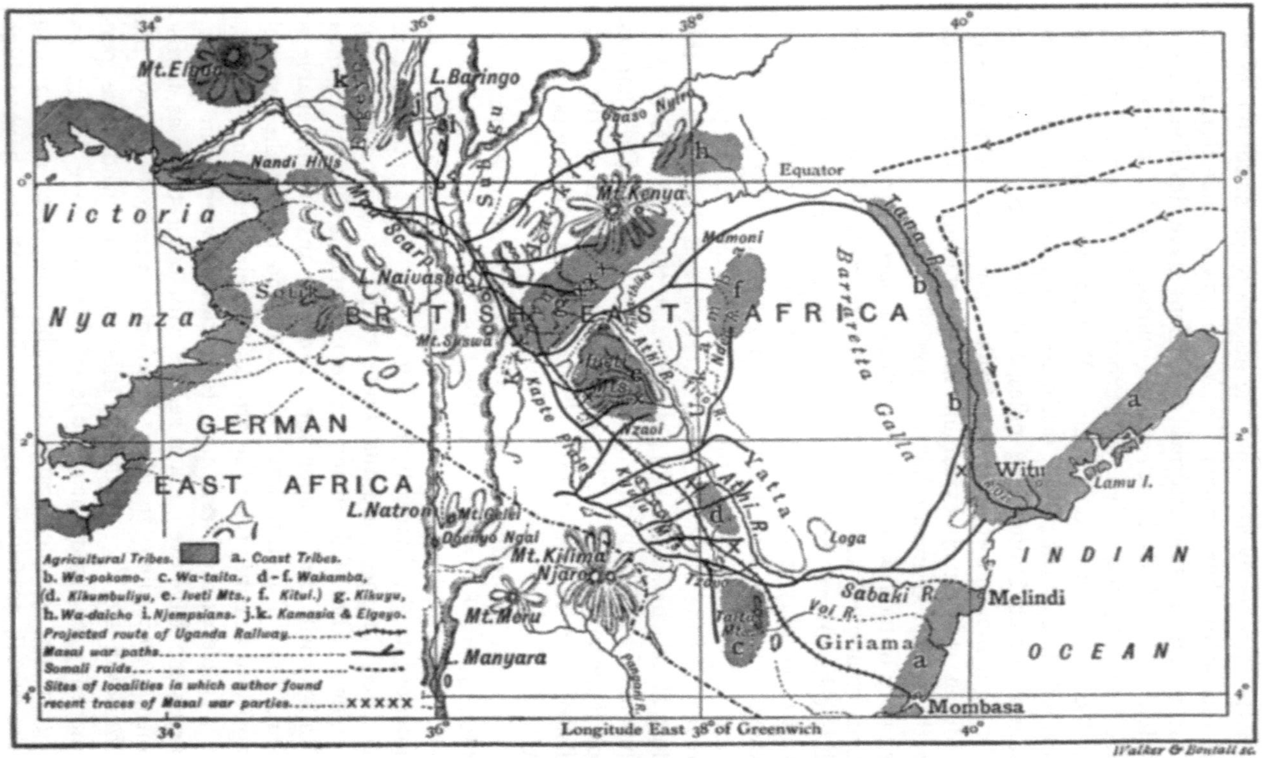

FIG. 22.—Map of British East Africa, showing Masai War-Paths.

therefore not impossible that this great tribe may acquire those habits of peaceful industry, which alone can save it from extermination. But this is not the view that is taken by most of the travellers who have seen much of the Masai. Proud and indolent, with an inherited contempt for honest work, and an ingrained hatred of steady exertion, it is doubtful whether the race will be able to accommodate itself to a settled life, or ever acquire the habits of discipline necessary in European service. Whether the Masai will be able to adapt themselves to altered conditions or not, it is certain that they will not do so until their military power has been broken. In British East Africa they play the part which was formerly taken by the Zulu and Matabili in the south, and by the Manyema on the Congo, a part which is still played by the Somali and Abyssinians farther north. Each of these held as their sacred vested interest the right to help themselves to the property of their weaker neighbours. While this claim is maintained, there is no hope of security or peace in the regions which they raid.

The southward advance of the Somali proceeds in quite a different manner from that of the Masai, for though they make raids, these have less effect than their slow, persistent encroachment upon their neighbour's territory. The power of the Somali is of quite recent growth. In the days when Vasco da Gama sailed up the coast of East Africa, they had not reached the Juba, and the Galla lorded it over the whole country from Abyssinia to Mombasa.

The Galla themselves were no doubt originally intruders into Africa. They appear to have crossed from Asia in the neighbourhood of Aden and the Straits of Babel-Mandeb. Thanks to their superior intelligence, and the skill with which they use their long broad-headed spears and round leather shields, they spread southward, driving the former Bantu inhabitants before them, and further reducing the few remnants of the aboriginal Negrillo race. Here and there in the Galla country, tribes of the older people have been allowed to survive, owing to their skill in managing canoes, as the Wa-pokomo, or in hunting, as the Walangulo or Waboni. They were, however, all reduced to the position of helots. But to-day the Galla power is broken. In the north the raids of the Abyssinians have destroyed their herds and crushed their spirit; in the south they have been practically exterminated since 1872, when their foe, the chief of Kau, called in the Somali to his assistance. The united forces engaged the Galla in a great battle at Kau, on the Ozi, and after a fight which lasted for three days

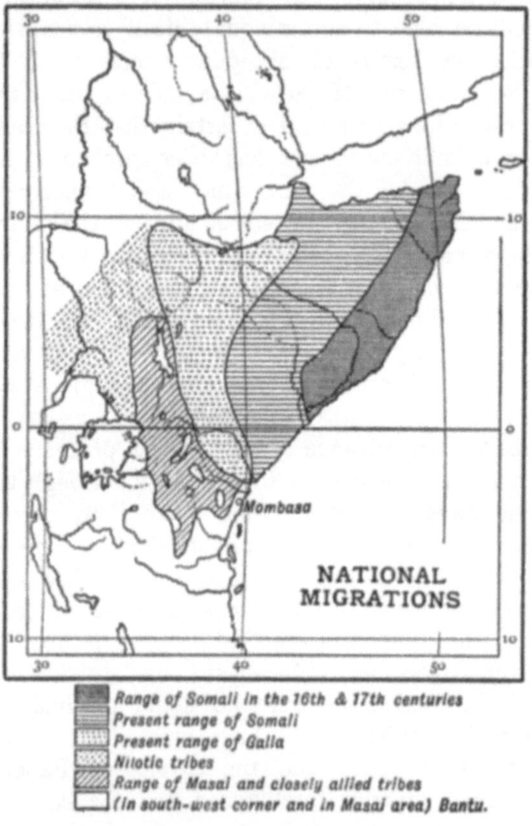

FIG. 23.

the Galla forces were annihilated. Krapf, sent in 1858 as "a missionary to the Galla," found them still the dominant race in the Tana valley, and estimated their numbers at between six and eight millions. Brenner estimated the number of the southern Galla at 20,000; but in 1892, Mr. Edmonds, the Wesleyan missionary to the Galla, calculated the number in the same district as only 250 individuals. The first and last estimates may have been extreme; but the Galla of the Tana valley now live only in weak isolated communities, though they still retain their old pride, and cherish in sullen impotence the memory of their former greatness.

The Somali, who have succeeded to their position, crossed from Asia at a later date, and were at first confined to the strip of land along the eastern coast of Africa, to the south of Cape Guardafui. The Somali are certainly closely related to the Galla; they have similar high foreheads, straight nostrils, thin lips, and long clinky hair. According to Somali traditions, the two tribes were cousins; but the Somali accepted light from Islam when they were offered it, while the Galla clung to their ancestral paganism. This is not unlikely to be the source of the present difference between their positions. Their religion has given to the Somali a power and motive force, which has enabled them, in spite of lack of cohesion among the clans, to advance southward and westward. Galla, Wadone, Wasaniya, Walangulo, Pokomo, have all gone down before their resistless march, and now only the agricultural tribes of the Wa-kamba and Kikuyu remain as buffer states between them and the Masai. Unless England intervenes these peaceful tribes seem doomed, and then the Masai and the Somali, left face to face, will commence a fight for the mastery, which will turn East Africa into an even greater pandemonium than it is to-day.

It is impossible to travel among the more peaceful tribes of East Africa and note the pathos of their lot, beneath the horror of the ever-threatening scourge of barbaric war, without feeling a keen personal interest in the people, and hoping that European intervention will prevent the present drama being played out to its bitter end. The time has passed, however, when one country could be expected to enter upon a purely disinterested crusade for the sake of another. No state would be justified in increasing the burdens of its people by undertaking the administration of a country without some hope of securing a return, either as an outlet for its surplus population or a new market for its goods. Unless British East Africa offers some prospect of doing one of these, the natives will be left to cut each other's throats without the intervention of the British policeman. They will be allowed to remain in poverty without any attempt to improve the resources of the country by British administration, and to starve in periods of famine unfed by British charity. It is this feeling which leads me to devote my last pages to the consideration of how far it would be right and wise for England to undertake this task.

The first essential for English administration is a climate in which Englishmen can live. It must be admitted at the outset that the sanitary record of the country is very bad, and recent meteorological observations render this only too easy of explanation. The death-roll of East Africa is simply appalling. When I landed on the coast in November 1892 the Wesleyan Missionary Society had five missionaries in British East Africa; when I returned to the coast in the August of the following year, there was but one: two had died and two had been invalided home. In November 1892 there were seven Englishmen in the districts administered from Lamu and from Melindi; there are only two left: two have been killed by natives (Mr. J. Bell Smith and Mr. Hamilton), and three have died of disease (Dr. Rae, Mr. W. Bird Thompson, and Mr. Edmonds). When I started from Mombasa in March there were twelve Europeans on or near the 300 miles of road between the coast and Fort Smith; before I returned to the coast five months later, fever had killed three and sent one back as an invalid to Europe; of the remaining eight the Masai have murdered one, and disease has carried away two more, so that only five of the twelve are left. Moreover, it must be remembered that even this fearful mortality—a rate of 23 per cent per annum—is the death-rate of picked men in the prime of life.

Nevertheless, in spite of this terrible record, which has been impressed upon me by the fact that so many of those who have died were friends, I believe that the climate is better than the facts suggest. It is not fair to test a climate by its action upon men enduring the hardships of camp life. If people in England slept in swamps and lived on bad food; if, when ill, they had no shelter but a draughty tent and had to make a daily march of ten miles or more across sandy plains, the death-rate would be more than one in twenty. To show the influence of such conditions on the deadliness of a disease, I may quote the experience of a friend who had to take a caravan of 300 men from Uganda to the coast. A few days after leaving the last food centre in Kavirondo, smallpox broke out among the porters. Every attempt was made to isolate the invalids and check the spread of the disease, but in vain; it swept through the caravan. The men had to make long marches every day across a high plateau, which is generally cold and wet. The only available food was beans and banana flour, and at night the men had to sleep in the open air. Under these conditions an attack was almost necessarily fatal. The men died off like flies; the baggage, including a valuable collection of natural history specimens and sporting trophies, had to be abandoned, and only a mere handful of men straggled into the next station on the road.

A better state of things might be expected in the permanent stations; but unfortunately for the reputation of East Africa, these are, as a rule, situated in unhealthy places. Most of the coast towns, such as Lamu and Zanzibar, are built beside tidal swamps, from which twice a day arise poisonous malarial exhalations. The inland stations are mostly in the valleys beside rivers and lakes, and the mission settlements seem especially placed in unhealthy positions. For example, there are three mission stations in the Tana valley, and they are each situated on the river bank and beside swamps. The Wesleyan station at Golbanti lies in the broadest tract of swamp on the Tana, and in the rainy season the little island on which it stands can only be approached by wading waist deep through slush and water. The temptations of a beautiful prospect, shady walks, and cool clear springs led to the establishment of the Scottish Industrial Mission at Kibwezi in the middle of a forest. It has in consequence proved disastrously unhealthy at certain seasons of the year. The only station on the Uganda road which is situated on open moorland, away from swamp and forest, is Machakos, and this has justified Lugard's choice by proving the most healthy station in British East Africa.

There is plenty of ground in the country like Machakos, where Europeans, with proper precautions, ought to live and prosper. Open moorland at the height of from 4000 to 5000 feet above the sea offers the best chances of successful European settlement. At higher elevations, though the air is cooler, the daily variation in temperature is excessive; and these heights are not even free from fever, for Brehme's recent work on

Kilima Njaro shows that it is endemic there at the height of over 7000 feet.

That the whole of Equatorial Africa is to be lightly dismissed as uninhabitable appears to me to be absurd. Discussions as to whether European colonists can settle and live there are now purely academic. The fact has been finally proved by the success of Blantyre in Nyasaland, where a colony of Scotch planters has founded a number of coffee plantations, which are in a most flourishing condition. We need only refer to that bright little journal, the *British Central Africa Gazette*, look through its columns of advertisements, and read the reports of its cricket matches and tennis parties, to realise how well the district must be adapted to the needs of a British settlement. As the colony, moreover, rests on an agricultural and not on a mining basis, its permanence seems assured. Granting, then, that with proper precautions and the exercise of greater care in the selection of sites for stations, Europeans can live in Equatorial Africa, the question remains whether there will be sufficient temptation to them to do so in large numbers.

The answer to this depends in the first place on the possible resources of the country, and in the second on the existence of an adequate supply of native labour. In discussing this subject East Africa is frequently compared with India, but the countries are so absolutely different that the comparison is useless. India has always teemed with inhabitants, most of whom are peaceful, industrious, and skilful, while in point of civilisation they stand nearer to the European than to the African. India has always been a country rich enough to tempt foreigners to enter it for loot, and has maintained many native arts and industries. When Englishmen went to India they found there an established civilisation, abundant agricultural and mineral resources, and an unlimited supply of cheap labour. The country only needed pacification and organisation.

Eastern Equatorial Africa, on the other hand, is poor, the population is thin, the natives are generally on the verge of starvation, and for the most part ignorant and indolent. The only district which can at all compare with India is Uganda, and the higher civilisation and greater intelligence of its people are due to the infusion of Hamitic blood. Uganda, however, is only a small area, and its power was on the wane and its population decreasing when Stanley first visited it. Since then it has lost ground materially. The East African Nyika, which occupies the largest proportion of British and German East Africa, resembles Beluchistan rather than India; and the Beluchistan protectorate is not usually regarded as a very important addition to the wealth and strength of the empire.

The natural products of British East Africa are neither numerous nor valuable. Corn and barley have been proved to grow well on the inland plateau, and maize is one of the staple foods of the country. Average wheat and barley are worth respectively 100s. and 107s. the ton; while freight from Mombasa to London is 25s. a ton, whereas from Chicago to Liverpool it is only 22s. 9d. So even when the railway is made to the coast, there will be little chance of competition with the grain-fields of America.[1]

The rice-fields of the delta of the Tana are extensive and prolific, and could easily be made to yield sufficient for local consumption; but they would hardly compete with Egypt and India in European markets. Fruits grow well, and something might be earned by preparing chutnee from the mangoes and tinning the pine-apples which are now wasted. The extensive grazing plains of the interior might be utilised for cattle ranches; but it is very doubtful whether these would pay at a time when, in spite of their superiority in climate and position, so many ranches are bankrupt in La Plata and the United States. In East Africa, moreover, cattle disease of a very bad type sweeps periodically through the country; and until its nature is understood, its ravages will continue and add greatly to the uncertainties of African cattle farming.

The only products which at present appear to pay for export to Europe from East Africa are ivory, rubber, orchilla weed (the lichen used for dye), copra (the dried kernel of the cocoa-nut), kichi-kichi (the seeds of the oil palm), chillies, and cloves. Cloves and copra are the most important, and could be increased indefinitely, but their growth is limited to the coast lands. The only important article of commerce from the interior is ivory, and the supply of this is uncertain. Fibre-

[1] The prices of the grain are taken from the *Mark Lane Express*, 2nd Sept. 1895 (vol. lxxiii. p. 307). For the rates of freight I am indebted to Messrs. Gray, Dawes, and Co., and to Mr. J. W. Young of Messrs. Simpson, Spence, and Young.

yielding plants abound all over the country, but the amount used is insignificant. Cotton, coffee, and tea are grown on the German plantations in the Witu protectorate, in German East Africa, and in Nyasaland. The British East Africa Company attempted to grow cotton, but the effort did not succeed, though the causes for this failure are preventible. The cotton from the German plantations at Ras Tchagga, in the British protectorate, is a very good staple, but the quantity raised is very small, and the amount exported from Zanzibar to Europe in 1893 was worth less than £3. There are, however, in the interior, wide tracts of country well adapted for cotton plantations, but it is difficult to see how this cotton could compete with that from the Southern States of America.

Still, the success of the settlement at Blantyre shows that some areas in Tropical Africa can be made to yield a considerable quantity of produce of sufficient value to pay for carriage to England. The profits are not likely to be great, and, so far, it is only Scotchmen who have succeeded in this enterprise; but the Blantyre colony has proved that men who have sufficient farsightedness and capital to be able to work and to wait, and to be contented with "small profits, slow returns," may reasonably hope for ultimate success.

Patience and perseverance, however, are indispensable to the development of the agricultural resources of the country, and the only chance of sudden prosperity in East Africa is the discovery of one or more of the precious metals or minerals in considerable quantity. I have no intention of making any definite prediction on this subject, for I did not go to Africa as a mineral prospector; and having no time to search properly, I thought it better to leave this work alone than to risk prejudicing the country by misleading negative results. The difficulties of discovering metalliferous veins and of estimating their value without extensive borings are so great, that a five months' traverse of so wide an area can furnish no ground for an opinion. Mining predictions are proverbially erroneous, and the value of negative evidence and superficial scraping in a hasty traverse is shown by the two following instances.

Climbing one afternoon, in 1891, over the moraines of the upper Arkansas River, I tried to fix the position of one group of boulders by a bearing from the spire of the town-hall of Leadville. I was using a geological map, on the comparatively large scale of four miles to the inch, which had been prepared by the United States Geological Survey.[1] It had only been issued in 1878, nevertheless Leadville was not marked in it. It was not till I recognised that a group of ruined shanties at the entrance to California Gulch marked the site of Oro City, that I could determine the position of the wealthiest town in the Rocky Mountains. The country had been surveyed by the official geologists of the United States, and a number of practical miners had lived for years in the district, without discovering that beneath the very slope on which their camp was pitched, lay what is now the most productive silver mine in the world.

The history of the Comstock Lode[2] is even more instructive. Some alluvial gold was found beside it in 1850; in 1851 colonies of miners settled in the district and worked there; but it was not till 1858 that it was accidentally discovered that they were living beside a silver lode, which subsequently proved to be the richest metalliferous deposit known in mining history.

It would therefore be absurd at the present time to express any opinion on the mineral resources of British East Africa. As gold is one of the most universally distributed of metals, it will no doubt be found. Many of the quartz reefs look promising, but most of the rocks which have been described as such are really quartz pegmatite veins. Silver is more likely to occur in the volcanic region, for in the United States it is generally associated with the "propylites," which are very extensively developed in East Africa. Silver and lead exist in the hills near Mombasa, where some trial borings and shafts have been made, and I found some thin seams in the Sabaki valley. Graphite is widely distributed, and near Tzavo it is common; in one place I found a few crystals of cassiterite or tinstone, but it is not probable that either mineral is sufficiently abundant to pay for working. Iron ores occur in considerable quantity all over the country, but there is no proper fuel with which to

1 "Geol. and Geog. Surv. of Territories," *Geol. and Geog. Atlas of Colorado*, United States.

2 A good summary of the history of this lode is given in the fourth chapter of Ed. Suess's *Die Zukunft des Silbers*, Wien, 1892, pp. 67-82. The standard authority is Lord's *Comstock Miners and Mining*.

work it, and except for limited local purposes it is valueless. Some of the tribes collect grains of hematite from streams, and smelt it by charcoal in furnaces of the "Catalan" type. But except in a few places, the native smiths cannot compete even in their own villages with iron from Birmingham, brought up country by trading caravans in the form of wire. I am indebted to my friend Professor Tate, late of the Royal College of Science, for assays of three iron ores collected by Mr. Scott Elliot in Usoga, which yielded respectively 61.69, 41.08 and 43.21 per cent of iron. But the richest ore contains so much phosphorus and sulphur that it is useless, and Mr. Tate states that the poorest ore is really the most valuable, as it might be used as an ochre.

The second factor on which the successful administration of East Africa depends is the supply of native labour, and the outlook here is not generally considered hopeful; the contempt in which men like Burton and Cameron held the African, and their failure to make their men work properly, still influences English opinion. We are repeatedly reminded that the African will not work, and cannot be made to work, except in slavery. For my own part, I take a more favourable view of the prospect, and consider it definitely good.

It has been repeatedly shown that the possibility of inducing the African to work is simply a question of tact and management. The average Bantu is lazy, but he likes to be comfortable. His vanity is easily roused, and his capacity for imitation is marvellous. Few people can be taught to want more easily, and his good-nature and great physical strength render him a most devoted and useful servant. His habits of indolence, it is true, are deeply ingrained, and cannot be overcome at once. Stanley's first efforts to make the natives of the Congo undertake regular work were failures. But he succeeded at last, and the Congo Free State has now over 100,000 native porters in its service, with an army of several thousand men enlisted from some tribes which were formerly the most degraded and ferocious savages in Africa. On the other side of the Continent a similar success has been gained, though on a smaller scale. Mr. Ainsworth at Machakos has so won the confidence of the Wa-kamba, that he has entrusted them with the defence of the fort, and the postal service for half the journey to the coast. On passing the Taita hills on my way inland, I found that Mr. George Wilson had just settled there, and was determined to persuade the natives to help him to make his road through their country. The mission station there had just been withdrawn on the ground that the Wa-taita are "an awful lot of sweeps," and the commander of a caravan had recently hung a few of the people as a warning to the rest. I therefore went up country pitying Mr. Wilson in his hopeless task. When I met him again on my return, I found, to my amazement, that his tact and patience had been successful; a couple of hundred Wa-taita were merrily working on the road, and as many more as were wanted could be had for the asking.

Patience, however, is indispensable. It required years of persistent labour by a band of self-sacrificing enthusiasts on the Congo to enlist the natives of that district, and in British East Africa the population is now extremely sparse. The sketch-map on p. 366 shows how large a proportion of the country is uninhabited, or occupied only by a few nomads. Yet granted security and peace, the suppression of war raids, and a guarantee that the crops will be reaped by the sower, then the Bantu will multiply like the rabbit, and spread throughout the land. I believe the problem of converting the next generation into steady, capable workmen is only a question of patient administration by men whose hearts are in their work, and who know how to treat the natives with tact and sympathy.

The present condition of the West Indian islands shows that the negro can be made to work as a freeman, and that the permanent retention of slavery is unnecessary. But the history of these islands during the first half of the present century also shows how much harm can be done by the reckless suppression of even so great an evil as that system. The sudden emancipation of the West Indian slaves furnishes a useful warning; it resulted not only in the temporary ruin of the colonies, but in the enormous increase of the slave-trade.

T. Fowell Buxton,[1] one of the foremost advocates of the Emancipation Act of 1833, confessed, six years after it had passed, that "the number [of slaves] annually landed in Cuba

[1] T. F. Buxton, *History of the African Slave Trade* (1839), p. 173.

and Brazil, etc., is 150,000, being more than double the whole draught upon Africa, including the countries where it had ceased, when the slave-trade controversy began! Twice as many human beings are now its victims as when Wilberforce and Clarkson entered upon their noble task; and each individual of this increased number, in addition to the horrors which were endured in former times, has to suffer from being cribbed up in a narrower space, and on board a vessel where accommodation is sacrificed to speed." To quote Alison's summary of Buxton's admissions: "Thus the effect of the emancipation of the negroes has been to ruin our own planters, stop the civilisation of our own negroes, and double the slave-trade in extent and quadruple it in horrors throughout the globe." [1]

We have already twice experienced in East Africa the disastrous effects of precipitate interference with slavery. The revolt in the Soudan, the rise of the power of the Mahdi, the massacre of Hicks Pasha's army, the loss by Egypt of the whole of her equatorial provinces, and the death of Gordon, are now generally recognised as the result of our attempts to suppress slavery before supplying anything to take its place. We tried to destroy that which was an integral part of the social system of the ruling race, without realising how vast a cavity would be produced and how fatal it would seem to them. Similarly the guerilla warfare, which has continued in British East Africa ever since the British occupation of Witu, is mainly due to the same cause. It is true that the fighting began by an expedition which was sent to avenge the massacre of a party of German traders. But the power of the Witu rebels was largely due to the secret assistance of the coast Arabs. They foresaw the ruin of their plantations through a continuance of British efforts to suppress slavery, emphasised as it was by the action of the missionaries, who afforded shelter to runaway slaves in spite of their promises to the contrary. The British officials at Zanzibar and Mombasa, and especially that statesmanlike officer, Captain Rogers of Lamu, are fully alive to the possibilities of an Arab rebellion.[2] At Melindi, at Mambrui and Magarini, in fact all along the coast, extensive plantations

[1] A. Alison, *History of Europe*, Continuation, vol. v. (1864), p. 54.

[2] Since this chapter was written, the simmering discontent has broken out into a serious rebellion.

are being abandoned owing to the impossibility of obtaining sufficient labour. The Arabs see their property being ruined, and are naturally hostile to British rule. Here and there one of them has raised the standard of revolt, and expeditions, such as that now in progress against Mbaruk bin Rashid, have continually to be undertaken. Our power of punishing the rebels is very limited. If we bombard the coast towns, we only destroy the property of our peaceful subjects, the Hindu banyans, who live there. If we storm the rebel stronghold, the chief escapes and builds another on the next forest-covered hill. If we ravage his plantations, we are really punishing the Indian banker at Zanzibar to whom they are mortgaged.

It is no doubt utterly distasteful to British sentiment to tolerate slavery in one of our protectorates, even for a time, with the knowledge that it is lessening. Our feelings are so strong on this point that we may easily be hurried into a course of action which will cause far more injustice and more suffering to the innocent than will be saved. British opinion on this question is a survival from the time when slavery was in force in the Southern States of America, in Cuba, and Brazil. There negroes were compelled to work under European masters, with European regularity and rigour. Much cruelty was no doubt used in these countries to force the negroes into compliance with methods of work for which they were constitutionally unfit. But African slavery as it exists in the British protectorate is a very different matter. The slaves cannot be bought or sold, and can only be transferred by direct inheritance from father to son. Their children are free. The slaves can hold private property, and it is not at all rare for a man who is a slave to own other slaves himself. They have two days a week on which to work for themselves. The owners cannot force their way unbidden into the slaves' houses, cannot flog them, and is bound to provide them with food, clothes, and lodging. The masters themselves are very indolent; and as there is no attempt to run plantations at high pressure, the slaves are not overworked. They have a much easier time than the freemen who work for a monthly wage in a European caravan, and they generally appear contented and happy. The system is certainly not slavery if we accept Cicero's definition as true—"Slavery is the obedience of a broken, abject spirit, deprived of free choice"

(*Servitus est obedientia fracti animi et abjecti et arbitrio suo carentis*). It is vain to search in an ordinary East African plantation for a broken, abject spirit in the slaves. The system is simply a serfdom, and it has been as necessary in the development of East Africa as feudalism was in Europe. Under the laws now in force the slaves will gradually die out and be replaced by freemen, and the change had far better come slowly than by a sudden revolution. It is objected that these laws have been in existence for years, and if they had been enforced slavery would now be extinct, and therefore all existing slaves should be freed at once. But the fact that the laws have not been enforced in the past, under native rule, is no proof that they will not be so, now that the country is under direct British administration.

Reference need only be made to the recent report of Mr. D. Mackenzie, the Special Commissioner sent to East Africa by the British and Foreign Anti-Slavery Society, to show that the practical grievances to be redressed have become both comparatively and actually small. The Commissioner has no sensational stories of diabolical wickedness to relate. Instead of finding that the slave-owners are Legrees, he describes them as benevolent-looking old gentlemen, who are good friends with the naval officers engaged in the suppression of the slave-trade; they received him with open hospitality, and their only act of unkindness was occasionally asking awkward questions. The only case quoted, except on rumour, of cruel treatment of slaves, is one of imprisonment for trivial offences; and the Commissioner tells us that his evidence as to the nature of the offences was the statement of the prisoners themselves. If a slave-owner of Zanzibar asked the inmates of Newgate why they were there, he would probably conclude from the answers that justice in England is blind indeed. The Commissioner appears to have been aided most ungrudgingly by the local authorities, both by the British officials, "whose kindness and attention during my stay in Zanzibar surpassed anything I could have expected," and the officers of the Sultan of Zanzibar, who "furthered my plans in every possible way." It was apparently only the missionaries who did not help, and the Commissioner's account of their attitude is one of the most significant things in his report. A few years ago some of the missionaries did serious injury to the British cause by their injudicious and illegal interference with the slave-trade. Mr. Mackenzie now complains that "it does not appear to me that missionary societies engaged in missionary work in Africa give that help which they might do to forward the cause of freedom." [1] He also expresses great disappointment at his interview with Mr. Piggott, who as the Administrator of the Company which has purchased the freedom of thousands of slaves, and as the Treasurer of the Financial Committee of the Church Missionary Society's headquarters at Freretown, is not likely to be accused of lack of sympathy with the natives. But the Commissioner reports Mr. Piggott as saying [2] that "the British East Africa Company had liberated a good many slaves, but the result was unsatisfactory, as they would not work. He was opposed to the abolition of slavery, as the slaves seemed to be perfectly happy, and in his opinion they seemed only fit for bondage. This tale was poured into my ears on several occasions. Mr. Piggott assured me that many missionaries were of his way of thinking, and from what I heard some of them say, his assertions were correct as to their opinion."

The missionaries, in fact, seem now to recognise that domestic slavery is not the unqualified evil they once thought it, and that it is better by the enforcement of existing laws to eradicate the system gradually, rather than by a sudden social revolution to risk the horrors of an East African mutiny. They seem also now to realise the truth that slavery may be regarded as a larval appendage, necessary in a certain stage of national development, and, later on, to be absorbed like the lobes of a larval star-fish, or thrown aside like the shell of a bird's egg. Rudely to tear off the bands of the young star-fish is fatal to its development, and to help an immature chicken by breaking open its shell is not likely to add to the happiness of the creature, whom the action was intended to aid.

I regret, therefore, that I cannot represent British East Africa as a land flowing with milk and honey, rich in precious metals, and crowded with a capable and industrious population. On the

[1] *A Report on Slavery and the Slave Trade in Zanzibar, Pemba, and the Mainland of the British Protectorates of East Africa* (1895), p. 22.

[2] *Op. cit.* p. 13.

contrary, I must admit that no great commercial prosperity can be immediately expected. In this, however, East Africa is not exceptional. I am not aware that any purely agricultural colony has been a financial success in the first or second generation. The United States were founded as a refuge from oppression; the Cape of Good Hope was occupied for strategical purposes as a half-way house to Batavia; Australia made no important progress until it began to yield gold. There is no reason, therefore, why we should expect from Equatorial Africa what these richer countries of the temperate zone have failed to do. The development of the country must proceed slowly and tediously, and will require to be carried on with caution, perseverance, and self-sacrifice. But it is not a task to be shirked by the country which has succeeded in the more difficult feat of ruling India and of spreading her colonies over the globe. England ought at least to beat any rival in this field. She has the best opportunities, the truest colonising instincts, and by remarkable good fortune has secured the most promising districts in Africa. No one who has travelled through the German and British protectorates in East Africa has failed to be struck by the fact that in the partition of this region England has obtained the better share. British East Africa includes more fertile, healthier, and more thickly populated districts than those of the German sphere; and in Uganda England holds the most important strategical position in the Continent. German East Africa contains nothing so rich as the Kapte plains or so fertile as the Tana delta, and has no such intelligent and comparatively civilised people as the Waganda. The fact that more progress has been made in the German protectorate than in the British, speaks volumes for the energy of our rivals and the wisdom of her administrators.

England, however, can well afford Germany her start, for our position is a strong one, and the construction of the Uganda Railway will secure our unrivalled supremacy in Central Africa. The French attempts in the Niger and on the Congo have so far led to nothing but heavy burdens and light glory. The successes achieved in the German sphere are hardly commensurate with the enormous sacrifices of men and money. The Congo Free State has certainly lost ground since the dismissal of the English staff. Of all the countries which have joined in the recent "scramble for Africa," England alone has so far gained materially from her African possessions.

With Canada crippled by the competition of the United States, and Australia passing through a serious commercial depression, South Africa is to-day the most flourishing of all our colonies. As it has only attained this prosperity after seventy years of steady work in a healthy invigorating climate, it would be absurd to be disheartened by slow progress on the Equator. But if we cannot reasonably anticipate an immediate return for our expenditure, we may hope to brighten the lives and secure the greatest happiness of the greatest number of the downtrodden inhabitants of our East African dominions. England may at once put a stop to the horrors of barbaric tribal warfare, as well as to the cruelties of slave raids, and by the gradual substitution of free labour for the present wasteful system of domestic slavery, arouse more enterprise, independence, and intelligence among the people. If England only continues her work in this region also, she will change chaos to order, transform swamps into fields, and irrigate deserts into gardens. Having found the inhabitants ignorant, oppressed, and poverty-stricken, she will give them industrial training, fixity of tenure, and freedom from disturbance. And when Britain's services to human progress are finally added up, the civilisation of her East African possessions will be counted as not one of the least important of her contributions to the happiness and order of the world.

APPENDICES

APPENDIX A

LIST OF LITERATURE ON EXPEDITION AND ITS COLLECTIONS

Baker, E. G.—"A new Tree Senecio from Tropical Africa," *Journ. Bot.* vol. xxxii., 1894, pp. 140-142.

"African Species of Lobelia § Rhynchopetalum," *ibid.* pp. 65-70, pls. 340-341.

Beddard, F. E.—"A Monograph of the Order Oligochæta," Oxford, 1895 (p. 612, *Polytoreutus gregorianus*, etc.)

Bell, F. J.—"Note on three Species of River Crabs of the genus Thelphusa, from Specimens collected in Eastern Africa by Dr. J. W. Gregory, Mr. H. H. Johnston, C.B., and Mr. F. J. Jackson," *Proc. Zool. Soc.* 1894, p. 166.

Boulenger, G. A.—"Third Report on Additions to the Batrachian Collection in the Natural History Museum," *Proc. Zool. Soc.* 1894, p. 646, pl. xl.

Butler, A. G.—"On a Collection of Lepidoptera from British East Africa, made by Dr. J. W. Gregory between the months of March and August 1893," *Proc. Zool. Soc.* 1894, pp. 557-593, pls. xxxvi. xxxvii.

Godwin-Austen, H. H.—"Notes on Trochonanina and other Genera of Land Mollusca, with Reference to the Generic Position of Martensia Mozambicensis and other Species," *Proc. Malacol. Soc.* vol. i. pt. 6, 1895, pp. 281-287, pl. xix.

Gregory, J. W.—"The Natural History of East Equatorial Africa," *Nature*, vol. xlix., Nov. 1893, p. 12.

"An Expedition to Mount Kenya," *Fortnightly Review*, Mar. 1894, pp. 327-337.

"Mountaineering in Central Africa, with an Attempt on Mount Kenya," *Alpine Journ.* vol. xvii., 1894, pp. 89-104.

"On Zoological Distribution in Africa," *Proc. Zool. Soc.* 1894, pp. 165-166.

Gregory, J. W.—"Contributions to the Physical Geography of British East Africa," *Geog. Journ.*, Oct.-Dec. 1894, vol. iv. pp. 289-315, 408-424, 505-524.

"Contributions to the Geology of British East Africa." Part I.

"The Glacial Geology of Mount Kenya," *Quart. Journ. Geol. Soc.* vol. l., 1894, pp. 515-530.

"Expedition in East Africa," *Geog. Journ.* vol. i., 1893, pp. 456-457; vol. ii. pp. 326-327.

"The Natives of East Africa," *Phonogr. Quart. Rev.* vol. i., 1894, pp. 8-13, 72-82.

Günther, A.—"Report on the Collection of Reptiles and Fishes made by Dr. J. W. Gregory during his Expedition to Mount Kenia," *Proc. Zool. Soc.* 1894, pp. 84-91, pls. viii.-xi.

Hahn, F.—"Gregory's Reise zum Baringo-See und Kenya," Petermann's *Mitth.* Bd. xli., 1895, *Beil.* p. 168.

Moore, Spencer Le M.—"New Acanthaceæ from Tropical Africa," *Journ. Bot.* vol. xxxii., 1894, pp. 129-139, pl. 343.

Rendle, A. B.—"A Contribution to the Flora of Eastern Tropical Africa," *Journ. Linn. Soc. Bot.* vol. xxx. pp. 373-435, pls. xxxi.-xxxiv.

"New African Convolvulaceæ," *Journ. Bot.* vol. xxxiv. pp. 36-39.

Schlechter, R., and Rendle, A. B.—"New Tropical African Asclepiads," *Journ. Bot.* vol. xxxiv. pp. 97-99, Pl. 356.

Sharpe, R. B.—"On Birds from Mount Kenia," *Bull. Brit. Orn. Club*, vol. iii. p. ix.

Smith, A. L.—"East African Fungi," *Journ. Bot.* vol. xxxiii., 1895, pp. 340-344.

Smith, E. A.—"A List of the Land and Freshwater Mollusca collected by Dr. J. W. Gregory in East Africa during his Expedition to Mount Kenia, with Descriptions of a few new Species," *Proc. Malacol. Soc.* vol. i., 1894, pp. 163-168.

Thomas, O.—"Descriptions of two new Species of Macroscelides," *Ann. Mag. Nat. Hist.* ser. 6, vol. xiii., 1894, pp. 67-70.

"Description of a new Species of Reed-Rat (Aulacodus) from East Africa, with Remarks on the Milk-Dentition of the Genus," *Ann. Mag. Nat. Hist.* ser. 6, vol. xiii., 1894, pp. 202-204.

APPENDIX B

CATALOGUE OF PLANTS COLLECTED

PART I.—POLYPETALÆ

By E. G. Baker, F.L.S.

Species.	Author.	Locality.
RANUNCULACEÆ.		
Clematis simensis	Fres.	
var. dentata	O. K.	Ndi, Taita Mts.
Ranunculus pinnatus *var.*	Poiret	Maji Chumvi, nr. Mombasa
Thalictrum rhynchocarpum	Dill. and Rich.	Upper Alpine Zone, Mt. Kenya
CRUCIFERÆ.		
Cardamine Johnstoni	Oliv.	Kenya at lower edge of Bamboo Zone, and on terminal moraine in Lower Alpine Zone
,, africana	L.	Kenya, terminal moraine in Lower Alpine Zone
Arabis albida	Steven	Upper Alpine Zone, Kenya
Crambe hispanica	L.	Kikuyu Scarp of Rift Valley
CAPPARIDEÆ.		
Cleome *aff.* ciliata	Schum. & Thonn.	S. bank of Kiroruma
Gynandropsis pentaphylla	DC.	Ndi, Taita Mts.
VIOLACEÆ.		
Viola abyssinica	Steud.	Kenya
POLYGALACEÆ.		
Polygala Petitiana	A. Rich.	Ndi, Taita Mts.
,, abyssinica	R. Br.	Ndoli, Iveti Mts.

Species.	Author.	Locality.
CARYOPHYLLACEÆ.		
Drymaria cordata	Willd.	Kithungulu, Kikuyu
Stellaria *aff.* media	L.	Lower Alpine Zone, Kenya
Cerastium *sp.*	...	,, ,,
PORTULACACEÆ.		
Talinum cuneifolium	Willd.	Sabaki Valley, by Second Stockade
Portulaca *aff.* pilosa	L.	Ndara
TAMARISCINEÆ.		
Tamarix gallica	L.	Salt River, Nyika
HYPERICACEÆ.		
Hypericum intermedium	Steud.	Maka, Iveti Mts.
,, Schimperi	Hochst.	G. Nairotia, Laikipia
MALVACEÆ.		
Sida Schimperiana	Hochst.	Rift Valley, Kikuyu Scarp, and G. Malewa
,, rhombifolia	L.	Ndi, Taita Mts.
,, ,, *var.*	,,	Ndi, and steppes N. of Kiroruma
,, grewioides	G. and P.	Marungu, N. of Kiroruma
Abutilon asiaticum	G. Don.	Ngatana
,, indicum	L.	Njemps Mkuba
Pavonia arabica	Hochst.	Ngurunga Kifaniko, Tzavo; Njoro Larabwal, nr. Baringo
,, Schimperiana	,,	
var. tomentosa	,,	Kithu-Uri, Kikuyu; Maka, Iveti Mts.
Pavonia zeylanica	Cav.	Sabaki Valley
,, propinqua	Garcke	Lake Dumi, Tana Valley
Hibiscus *aff.* crassinervis	Hochst.	Kilungu, Iveti Mts.; Lanjoro on Kapte Plains
,, cannabinus	L.	Mwatchi; Ngurunga Kifaniko, nr. Tzavo
,, micranthus	,,	Tana Valley at Lake Dumi, Golbanti and Kau; Sabaki Valley at Tanganyika; Nyika at Maungu
,, gossypinus	Thunb.	G. Nairotia, Laikipia; Maka, Iveti Mts.
,, vitifolius	L.	N. of G. Thegu, Laikipia
TILIACEÆ.		
Grewia populifolia	Vahl	Witu
Triumfetta flavescens	Hochst.	Lake Kibibi, Rift Valley
,, *sp. nov.*	...	Ngomeni, nr. Tzavo

Species.	Author.	Locality.
LINACEÆ.		
Linum *sp.*	...	G. Laschau, Laikipia
GERANIACEÆ.		
Geranium simense	Hochst.	G. Mairi, Laikipia ; Lanjoro, S. of G. Thegu, Kikuyu
Oxalis *aff.* punctata	L.	Second Swamp, Kikuyu
Pelargonium glechomoides	Rich.	Mkuyuni, Foot-hills of Kamasia
Erodium *sp. nov.*	...	Kapte Plains, W. of Maka
Impatiens micrantha	Hochst.	Kenya, Bamboo Zone
RUTACEÆ.		
Clausena inæqualis	Benth.	
CELASTRINEÆ.		
Gymnosporia *n. sp.*	...	G. Nairotia, Laikipia
AMPELIDEÆ.		
Vitis Labrusca	L.	Witu.
SAPINDACEÆ.		
Cardiospermum Halicacabum	L.	Ngatana
LEGUMINOSÆ.		
Crotalaria glauca *var.*	Willd.	Ndi, Taita Mts.
,, senegalensis	Bacle	Witu
Parochetus major	Don.	Kenya, lower margin of Bamboo Zone
Trifolium Johnstoni	Oliv.	Kenya, Lower Alpine Zone
,, repens	L.	Kithungulu, Kikuyu
Indigofera *sp.*	...	N. end of Lake Losuguta
Tephrosia purpurea	Pers.	W. shore of Baringo
,, ,, *var.* β pubescens	Baker	Ndara, Taita Mts.
,, *aff.* Hildebrandtii	Vatke	Kilungu, Iveti Mts.
,, (*aff. sp.* from Shiré)	...	Mto Mkindu, Nyika
Æschynomene Telekii	Schwf.	Kenya
Stylosanthes mucronata	Willd.	S. of Kiroruma River
Clitoria ternatea	L.	Lava Plains, N. of Kiroruma River
Centrosema virginianum	Benth.	Ngatana
Glycine javanica	L.	Kikuyu Scarp of Rift Valley; Alng'aria, Laikipia ; Kithungulu, Kikuyu
Rhynchosia flavissima	,,	Ngatana
Derris uliginosa	Benth.	Witu

Species.	Author.	Locality.
Eriosema polystachyum	Baker	Alng'aria, Laikipia
Cæsalpinia pulcherrima	L.	Witu
Cassia *aff.* Hildebrandtii	Vatke	Maka, Iveti Mts.
Acacia arabica	Linn.	Sabaki Valley
Bauhinia Volkensii	Taubert	Kilungu, Iveti Mts.
ROSACEÆ.		
Alchemilla Johnstoni	Oliv.	Kenya, on terminal moraine in Lower Alpine Zone; Höhnel Valley in Upper Alpine Zone
,, argyrophylla	,,	Lower Alpine Zone, Kenya
,, pedata	Engl.	
var. gracilipes (?)	,,	Ndoro, at W. foot of Kenya
CRASSULACEÆ.		
Crassula *aff.* corymbulosa	Link.	Lower Alpine Zone, Kenya
Tillæa aquatica	L.	,, ,, ,,
LYTHRACEÆ.		
Ammania salicifolia	Mont.	Kenya, lower margin of Bamboo Zone
CUCURBITACEÆ.		
Cephalandra indica	Naud.	Ngatana
Momordica Charantia	L.	Ngatana and Witu
,, trifoliolata	Hook. fil.	Ngatana
UMBELLIFERÆ.		
Caucalis infesta	Curt.	Kikuyu Scarp in Rift Valley

PART II.—MONOPETALÆ

BY JAMES BRITTEN, F.L.S.

Species.	Author.	Locality.
RUBIACEÆ.		
Pentas carnea	Benth.	Magarini Hills, nr. Mambrui
Pentodon decumbens	Hochst.	Lake Dumi, Tana Valley
Diodia sarmentosa	Sw.	Marungu, steppes of the Kiroruma
Anthospermum lanceolatum	Thunb.	G. Laschau, Laikipia
Galium Aparine	L.	Mid. of Bamboo Zone, Kenya
,, simense	Hochst. (?)	Upper Alpine Zone, Kenya

Species.	Author.	Locality.
COMPOSITÆ.		
Ethulia conyzoides	L.	Ngatana
Vernonia cinerea	,,	,,
,, demulans	Vatke	Witu
,, pauciflora	Less.	Nzaoi, Iveti Mts.; Wakilomi, in Maka, Iveti Mts.; (?) Ndara, Taita Mts.
Ageratum conyzoides	L.	Mkonumbi; Ngatana; Kapte Plains; Foot-hills of Kamasia
Dicrocephala latifolia	DC.	Foot-hills of Kamasia; G. Thegu, Laikipia
Conyza Newii	Oliv. and Hiern.	Mto Ndangi, Ukamba; G. Narok, Laikipia
Tarchonanthus camphoratus	L.	G. Narok, Laikipia
Pluchea Dioscoridis	DC.	Ngatana; Makongeni, Sabaki
Achyrocline Hochstetteri	Sch. Bip.	G. Thegu and Lanjoro, S. of G. Thegu, N. Kikuyu; Steppes of Kiroruma; Wakilomi, in Maka, Iveti Mts.
,, glumacea	Oliv. and Hiern.	Foot-hills of Kamasia
,, Schimperi	Sch. Bip.	Maji Chumvi, nr. Mombasa; G. Thegu, N. Kikuyu
Helichrysum fœtidum	L.	G. Nairotia, Laikipia; Wakilomi, in Maka, Iveti Mts.
,, cymosum	Less.	G. Laschau, Laikipia
,, ,, *var.* compactum	...	Upper Alpine Zone, Kenya
Helichrysum Meyeri-Johannis	Engl.	Lower Alpine Zone, Kenya
,, nudatum	Less.	Gopo lal Mwaru, Laikipia
Polycline gracilis	Oliv.	Kapte Plains, W. of Maka
Astephania africana	,,	Wakilomi, in Maka, Iveti Mts.
Aspilia Holstii	O. Hoffm.	G. Thegu, N. Kikuyu
Melanthea Brownii	Sch. Bip.	Mkonumbi; Ngatana; Njemps Mkuba
Spilanthes Acinella	L.	Foot-hills of Kamasia; G. Mairi, W. foot of Kenya
Guizotia abyssinica	Cass.	G. Mairi, W. foot of Kenya
Chrysanthellum procumbens	Pers.	Mto Kiboko, Ukamba
Senecio discifolius	Oliv.	Kithungulu, N.E. Kikuyu; Steppes of Kiroruma
,, keniensis	Bak. fil.	Upper Alpine Zone, Kenya.
Tripteris Vaillantii	Decaisne	Ndi, Taita Mts.
Arctotis Ruppellii	O. Hoffm.	Upper Alpine Zone, Kenya
Lactuca capensis	Thunb.	Ngatana; Lake Dumi, Tana; G. Mairi, W. foot of Kenya; E. of Mangea
Sonchus Bipontini	Asch.	
var. pinnatifidus	...	G. Thegu, N. Kikuyu
CAMPANULACEÆ.		
Lobelia Gregoriana	Bak. fil.	Upper Alpine Zone, Kenya

Species.	Author.	Locality.
Lobelia Telekii	Schweinf.	Upper Alpine Zone, Kenya
,, Holstii	Engl.	Mt. Mbololo, Taita Mts.
,, fervens	Thunb.	Kilima Meza, E. of Lake Elmetaita; Kapte Plains, W. of Maka; Mt. Mangea, Giriama; Mkonumbi
Cyphia glandulifera	Hochst.	Kikuyu Scarp of Rift Valley
Lightfootia, *sp.*	...	Sabaki Valley
ERICACEÆ.		
Erica, *sp. n.*	...	Upper Alpine Zone, Kenya
Blæria, *sp.*	...	,, ,, ,,
PLUMBAGINEÆ.		
Plumbago zeylanica	L.	Nzaoi, Iveti Mts.; Kamasia
PRIMULACEÆ.		
Anagallis serpens *var.*	Hochst.	Nzaoi, Iveti Mts.; Kamasia
OLEACEÆ.		
Jasminum abyssinicum	Br.	Mambrui.
SALVADORACEÆ.		
Salvadora persica	L.	Lake Dumi, Tana Valley
ASCLEPIADEÆ.		
Gomphocarpus fruticosus	R. Br.	Marungu, Steppes of Kiroruma
,, ,,	,,	Alng'aria, Laikipia; Lokenya, Kapte Plains
var. angustissimus	Engl.	
Gomphocarpus *cfr.* rhinophyllum	K. Schum.	Alng'aria, Laikipia
Pentarrhinum abyssinicum	Decne.	Njemps Mkuba
Kanahia Delilei	,,	G. Nagut, N. of Elmetaita; Bondoni, Kapte Plains
Dæmia extensa	R. Br.	Kamasia
Baseonema Gregorii	Schl. & Rendle	Kinani, E. Ongalea Mt.
GENTIANEÆ.		
Sebæa trachyphylla	Griseb.	G. Mairi and Lanjoro, S. of G. Thegu, at W. foot of Kenya
Exacum quinquenervium	,,	Nzaoi, Iveti Mts.
Swertia pumila	Hochst.	Lower edge of Bamboo Zone, Kenya
,, Schimperi	Griseb.	Upper and Lower Alpine Zones, Kenya
Limnanthemum indicum	,,	Mkonumbi

Species.	Author.	Locality.
BORAGINEÆ.		
Trichodesma zeylanicum	Br.	Witu; Nzaoi, Iveti Mts.
Cynoglossum micranthum	Desf.	Kamasia; G. Narok, Laikipia; Wakilomi, Iveti Mts.
Heliotropium zeylanicum	Lam.	Mkonumbi; Lake Dumi, Tana Valley; G. Morendat and G. Nagut, Rift Valley; Steppes of Thika-thika; W. of Lugard's Falls, Sabaki River
,, strigosum	Willd.	Ndara, Taita Mts.; Steppes of Thika-thika
CONVOLVULACEÆ.		
Ipomœa tuberculata	Roem. & Sch.	Lanjoro, nr. Lokenya, Kapte Plains
,, ovalifolia	Chois.	Lake Dumi, Tana Valley
,, *cfr.* cordofana	,,	,, ,,
,, ophthalmantha	Hallier	Magarini, nr. Mambrui
,, tenuirostris	Chois.	Bank of Kiroruma
,, carnosa	Br.	Lake Dumi, Tana Valley
,, Britteniana	Rendle	Ngatana
,, blepharophylla	Hallier	
var. cordata	Rendle	Marungu, Steppes of the Kiroruma
,, *cfr.* tambalensis	Baker	Nzaoi, Iveti Mts.
Merremia Gregorii	Rendle	Golbanti, Tana Valley
Aniseia calycina	Chois.	Njemps Mkuba
Hewittia bicolor	,,	Banks of Kiroruma
Convolvulus sagittatus	Thunb.	
var. macroglottis	Baker	G. Mairi, W. foot of Kenya
Convolvulus malvaceus	Oliv.	Wakilomi, Maka, Iveti Mts.
Evolvulus nummularius	L.	Sabaki Valley
SCROPHULARIACEÆ.[1]		
Veronica, *sp. nov.*	...	Kenya
Craterostigma pumilum	Hochst.	Nr. Lokenya, Kapte Plains
,, *sp. nov.*	...	,, ,, ,,
Diclis ovata	Benth.	Lower edge of Bamboo Zone, Kenya
Ilysanthes gratioloides	,,	Gorge of Thika-thika
Striga senegalensis	,,	Lake Dumi, Tana Valley
,, *sp.*	...	G. Laschau, Laikipia
Alectra, *sp.*	...	G. Mairi, W. foot of Kenya;
Cycnium Herzfeldianum	Engl.	Kithungulu, Kikuyu
,, tubulosum	,,	Kikuyu Scarp of Rift Valley; G. Mairi, W. foot of Kenya
Sopubia ramosa	Hochst.	Gopo lal Mwaru, Laikipia
,, kituiensis	Vatke	Steppes of Thika-thika
,, trifida	Ham.	G. Mairi, W. foot of Kenya; Wakilomi, Maka, Iveti Mts.
ACANTHACEÆ.[2]		
Homilacanthus Gregorii	S. Moore	Lower edge of Bamboo Zone, Kenya

[1] The Solanaceæ, Selagineæ, and Verbenaceæ have not yet been identified.
[2] The Acanthaceæ are taken from the paper by Mr. Spencer Moore.

Species.	Author.	Locality.
Thunbergia Gregorii	S. Moore	Kilungu, Iveti Mts.
Dyschoriste Hildebrandtii	,,	
,, *var.* mollis	,,	Steppes of Thika-thika
Ruellia megachlamys	,,	Ngurunga Kifaniko, Taita Mts.
Crabbea velutina	,,	
Asystasia linearis	,,	Mwatchi, nr. Mombasa
Ecbolium amplexicaule	,,	Sabaki Valley
,, striatum	Balf. fil.	Ngatana
Justicia leikipiensis	S. Moore	Gopo lal Mwaru
,, Gregorii	,,	Mkuyuni, Foot-hills of Kamasia
LABIATÆ		
Moschosma polystachyum	Benth.	Kau; Lake Dumi, Tana Valley
Hoslundia verticillata	Vahl	Foot-hills of Kamasia and of Laikipia
Micromeria punctata	Br.	G. Mairi and Lanjoro, S. of G. Thegu, at W. foot of Kenya
Leucas Neuflizeana	Courb.	Ngurunga Kifaniko, S. of Tzavo; Bondoni, Kapte Plains
,, near grandis	Vatke	G. Thegu, S.W. of Kenya
,, oligocephala	Hook. fil.	Ndi, Taita Mts.
,, glabrata	Br. (?)	Kamasia
Tinnea æthiopica	K. and P.	Mkonumbi
Ajuga bracteosa	Benth.	G. Laschau, Laikipia; Marungu, Steppes of Kiroruma

PART III.—APETALÆ

By James Britten, F.L.S., and A. B. Rendle, M.A., F.L.S.

Species.	Author.	Locality.
AMARANTACEÆ.		
Ærua lanata	Juss.	Mt. Mangea, Giriama
,, ,, *var.* oblongata	Asch.	Kilima Meza, E. of Kariandusi
,, ,, *forma*	...	Maungu, Nyika
,, javanica	Juss.	Makongeni, Sabaki
Alternanthera nodiflora	R. Br.	E. shore of Lake Losuguta
Achyranthes aspera	L.	Kau, nr. Witu; Ndi, Taita Mts.; G. Nairotia, Laikipia
Celosia argentea	,,	Njemps Mkuba
,, anthelminthica	Asch.	G. Nairotia, Laikipia
Pupalia lappacea	Moq.	G. Tigirish, Njemps
Digera arvensis	Forsk.	
,, *var.* perennans	Moq.	Lake Dumi, Tana Valley
Amarantus chlorostachys	Willd.	Ngatana, Lower Tana; Banks of Kiroruma, or Upper Tana
Centema, *sp.*	...	Witu; Ngatana
,, ,,	...	Ngurunga Kifaniko, S. of Tzavo

Species.	Author.	Locality.
Cyathula cylindrica, *var.*	Moq.	Bamboo Zone, Kenya
Pupalia, *sp.*	...	Kikuyu Scarp of Rift Valley
CHENOPODIACEÆ.		
Chenopodium fœtidum	Schrad.	Kau, nr. Witu; Kilima Meza, E. of Kariandusi
,, album	L.	Kariandusi
Suæda monoica	Forsk.	Lugard's Falls, Sabaki
POLYGONEÆ.		
Oxygonum sinuatum	Benth. & Hook.	Kinani, Nyika; Malewa R., Rift Valley
Rumex Steudelii	Hochst.	Rangatan Ndari, Laikipia
,, nervosus	Vahl	Ndoli
Polygonum tomentosum	Willd.	Ngatana, on Tana
,, *aff.* ,,	,,	Golbanti, on Tana
LAURACEÆ.		
Cassytha, *sp.*	...	Lower edge of Bamboo Zone, Kenya
EUPHORBIACEÆ.		
Euphorbia pilulifera	L.	Nr. Mkuyuni, Foot-hills of Kamasia
,, hypericifolia	,,	,, ,, ,,
,, indica	Lam.	Ngatana
,, sarmentosa	Welw.	Marungu, E. Kikuyu
Phyllanthus maderaspatensis	L.	Ndara, Taita Mts.; Wakilomi, in Maka, Iveti Mts.
var. Thonningii	Müll. Arg.	
Phyllanthus, *sp.*	...	G. Thegu, N. Kikuyu
THYMELEACEÆ.		
Gnidia Vatkeana	Gilg. and Engl.	Nzaoi, Iveti Mts.
BALANOPHOREÆ.		
Sarcophyte sanguinea	Sparrm.	Kibwezi

PART IV.—MONOCOTYLEDONS, ETC.

BY A. B. RENDLE, M.A., F.L.S.

Species.	Author.	Locality.
HYDROCHARIDEÆ.		
Lagarosiphon tenuis	Rendle	Kinani
,, hydrilloides	,,	Kariandusi

Species.	Author.	Locality.
ORCHIDEÆ.		
Mystacidium, *sp.*	...	Thika-thika
Habenaria pedicellaris	Reichb.	,,
,, ndiana	Rendle	Ndi
Brachycorythis pubescens	Harv.	Kariandusi
var. minor	Rendle	
Disa Gregoriana	,,	Lower Alpine Zone, Kenya
HÆMODORACEÆ.		
Sanseviera cylindrica	Boj.	Nyika
,, guineensis	Willd.	,,
IRIDACEÆ.		
Acidanthera candida	Rendle	Kapte Plains
Gladiolus Quartiniano, *aff.*	A. Rich.	Gopo lal Mwaru, Laikipia; Kariandusi
,, watsonioides	Baker	Lower Alpine Zone, Kenya
AMARYLLIDEÆ.		
Hypoxis laikipiensis	Rendle	Alng'aria, Laikipia
,, Gregoriana	,,	Rift Valley, Kikuyu Scarp
Crinum, *sp.*	...	Rift Valley, Miviruni, and G. Nyuki
LILIACEÆ.		
Asparagus plumosus	Baker	Rift Valley, Kikuyu Scarp
,, medeoloides	Thunb.	,, ,,
Bulbine asphodeloides	Schult. fil.	Rift Valley, Kikuyu Scarp; Lake Dumi
Anthericum acuminatum	Rendle	Ngurunga Kifaniko, nr. Tzavo
,, purpuratum	,,	,, ,, ,,
,, Gregorianum	,,	Kapte Plains
Ornithogalum Eckloni	Schlechtd.	Kithungulu, Kikuyu
Gloriosa virescens	Lindl.	G. Laschau, Laikipia; Lanjoro, Kapte Plains
COMMELINACEÆ.		
Commelina africana	L.	Foot-hills of Laikipia
,, *cf.* purpurea	C. B. Clarke	Ndoro, Laikipia
,, subulata	Roth.	Kinani
,, albescens	Hassk.	Mkonumbi
Aneilema sinicum	Lindl.	G. Thegu and Kithungulu, Laikipia
,, Clarkei	Rendle	Lake Dumi
JUNCACEÆ.		
Juncus effusus	L.	Kenya, Lower Forest Zone
,, Fontanesii	J. Gay	Ndoro
Luzula spicata	DC.	
var. simensis	Hochst.	Kenya, Bamboo Zone
TYPHACEÆ.		
Typha angustifolia	L.	Cliffs above Lake Losuguta

Species.	Author.	Locality.
ALISMACEÆ.		
Alisma Plantago	L.	G. Nairotia, Laikipia
NAIADEÆ.		
Aponogeton abyssinicum	Hochst.	Second Swamp, Kikuyu
Naias graminea	Del.	Kinani
GRAMINEÆ.		
Eriochloa annulata	Kunth.	Banks of G. Tigirish, nr. Njemps
,, ,, *var. nov.*	...	Tanganyika, W. end of Sabaki
Panicum brizanthum	Hochst.	Ndoli, Iveti Mts.
,, fenestratum	,,	Ndara; Ngomeni
,, controversum	Steud.	G. Nagut and G. Tigirish, Rift Valley
,, scalarum	Schweinf.	Njemps Mkuba; Kikuyu Scarp of Rift Valley
,, Schimperianum	Hochst.	Mkonumbi
,, curvatum	L.	Second Stockade, Sabaki
,, maximum	Jacq.	Kau, nr. Witu
,, colonum	L.	G. Mairi, W. foot of Kenya
,, Crus-galli	,,	Mwatchi, nr. Mombasa
Tricholæna rosea	Nees	Njoro Larabwal and Gopo lal Mwaru, Laikipia
,, *cf.* ,,	,,	Hombe, Kikuyu
Paspalum scrobiculatum	L.	,, ,,
Setaria glauca	Beauv.	G. Nagut, Rift Valley; Foot-hills of Kamasia and Laikipia
,, *cf.* ,,	,,	Mwatchi, nr. Mombasa; Nzaoi, Iveti Mts.
Pennisetum ciliare	Link.	Njoro Larabwal, Laikipia; Steppes of Thika-thika
,, setosum	A. Rich.	Mkonumbi; Kau
Tragus racemosus	Desf.	Njemps; Foot-hills of Kamasia
Perotis latifolia	Ait.	Mkonumbi
Imperata arundinacea	Cyrill.	Ngatana
Andropogon Sorghum	Brot.	Shore of Baringo
,, Ischæmum	L.	Larabwal, nr. Njemps
,, polyatherus	Hochst.	Mkonumbi
,, finitimus	,,	Larabwal, nr. Njemps
Themeda Forskalii	Hack.	,, ,,
,, ,, *var.* punctata	,,	Kapte Plains
Sporobolus virginicus	Kth.	Witu
,, marginatus	Hochst.	Njemps
,, spicatus	Kth.	Kariandusi, Rift Valley
,, indicus	R. Br.	Larabwal, nr. Njemps
Aristida cœrulescens	Desf.	Ndara
Trichopterix *n. sp.*	...	Lake Losuguta, Laikipia
Cynodon Dactylon	Pers.	Mkonumbi; Kariandusi
Chloris abyssinica	Hochst.	G. Murentat, Rift Valley
,, punctulata	,,	Nzaoi, Iveti Mts.

Species.	Author.	Locality.
Chloris barbata	Swartz	G. Tigirish, Njemps
Eleusine ægyptiaca	Desf.	G. Tigirish and G. Nagut, Rift Valley; Witu
,, coracana	Gaertn.	Maka, Iveti Mts.
Leptochloa filiformis	R. and S.	Mwatchi
Phragmites communis	Trin.	Lugard's Falls, Sabaki Valley
Danthonia villosa (*vel var.*)	Nees	Lower Alpine Zone, Kenya
Eragrostis ciliaris	Link.	Kau ; Ngomeni
,, tenuifolia	Hochst.	Kithungulu, Kikuyu
,, *cf.* paniciformis	R. Br.	Lake Losuguta
,, Brownei	Nees	Alng'aria
,, megastachya	Link.	G. Nagut and G. Tigirish, Rift Valley
CONIFERÆ.		
Podocarpus, *aff.* Mannii	Hook. fil.	Bamboo Zone, Kenya
FILICES.[1]		
Davallia Lindeni	Hook.	East of Kariandusi
Adiantum caudatum	L.	Ndi
Cheilanthes multifida	Sw.	Alng'aria
,, Kirkii	Hook.	Alng'aria and G. Nairotia, Laikipia
Pellæa involuta	Baker	Kapte Plains ; Second Swamp, Kikuyu
,, hastata	Link.	Ndi, Taita Mts. ; G. el Narua, nr. Njemps ; G. Thegu, Laikipia
Pellæa calomelanos	Link.	Kapte Plains ; Larabwal, nr. Njemps
Pteris aquilina	L.	Kithungulu, Kikuyu ; Mbololo and Ndi, Iveti Mts.
Asplenium furcatum	Thunb.	Kariandusi, Rift Valley ; Alng'aria, Laikipia ; Lower Alpine Zone, Kenya
,, ceterach	L.	Ndi
Actiniopteris radiata	Link.	Ndara ; Kinani ; Kikuyu ; Larabwal, nr. Njemps
Nephrodium dilatatum	Desv.	Kilungu, Iveti Mts.
Polypodium lepidotum	Willd.	Kapte Plains ; Bamboo Zone, Kenya

PART V.—MOSSES, HEPATICS, AND LICHENS

BY A. GEPP, M.A., F.L.S.

Name.	Author.	Localities where collected.	Distribution.
MOSSES.			
Leucobryum cucullatum	Broth.	G. Nairotia, Laikipia	Usambara
Leptodontium repens	(C.Müll.)	Upper Alpine Zone, Kenya	Kilima Njaro

[1] By William Carruthers, Esq., F.R.S.

Name.	Author.	Localities where collected.	Distribution.
Grimmia calyculata	C. M.	Lower Alpine Zone, Kenya	Kilima Njaro
Macromitrium liliputianum	,,	G. Narok, Laikipia	
Funaria calvescens	Schwaegr.	Second Stockade, Sabaki Valley	Wide
Breutelia gigantea	Doz. and Molk.	E. of Mt. Mangea, Giriama; Kurawa, Khilifi River	Bourbon, Java
Orthodon borbonicus	Bory	Bamboo Zone, Kenya	Bourbon, Madagascar, Cameroons.
Bryum pseudotriquetrum	Hedw.	G. Nairotia, Laikipia	Cosmopolitan
Braunia Schimperi	Br. and Sch.	Upper Alpine Zone, Kenya	Abyssinia
Lasia producta	Jaeg.	Rangatan Ndari, Laikipia; G. Nairotia, Laikipia	S. Africa
Pterogonium gracile	Sw.		
var. capense	Rehm.	Lari lal Morjo, Laikipia; Marungu, Steppes of Kiroruma	S. Africa
Pilotrichella (Orthostichella) sericea	C. M.	Middle of Bamboo Zone, Kenya	Laikipia
Papillaria africana	Jaeg.	Lari lal Morjo and G. Laschau, Laikipia; Marungu, Steppes of the Kiroruma	E. and S. Africa
,, patentissima	C. M.	G. Thegana, Kikuyu	Usambara
Neckera (Rhystophyllum) Hoehneliana	,,	Lower Alpine Zone, Kenya	Laikipia
Calyptothecium africanum	Mitt.	Snowfields, Kenya	S. and E. Africa
Entodon rotundifolius	C. M.	G. Thegana, Kenya	Kilima Njaro
,, lacunosus	Broth.	,, ,,	Usambara
Hypnum cupressiforme	L.	Rangatan Ndari, Laikipia; G. Nairotia, Laikipia	Cosmopolitan
HEPATICS.			
Jungermannia (Chandonanthus) hirtella	Web.	Upper Alpine Zone, Kenya	Transvaal, Bourbon, India, Oceania
Fimbriaria Boryana	Mont.	Rangatan Ndari and G. Nairotia, Laikipia	Bourbon
LICHENS.			
Leptogium phyllocarpum	Nyl.	Upper Alpine Zone, Kenya	Wide
Stereocaulon ramulosum	Ach.	Upper Alpine Zone, Kenya; Kurawa, Khilifi River	E. Africa; Oceania; S. America
Cladonia pyxidata	Fr.	Snowfields, Kenya	Wide
Cladina rangiferina	Nyl.	Kurawa, Khilifi River	Cosmopolitan
Ramalina canaliculata	Tayl.	Gopo lal Mwaru, Laikipia	E. Africa, India, Oceania, Peru

Name.	Author.	Localities where collected.	Distribution.
Ramalina fraxinea *var.*	Ach.	G. Thegana and Karati, Kikuyu; Lower Falls of the Thika-thika; Maka, Iveti Hills; E. of Mt. Mangea, Giriama	Cosmopolitan
,, pusilla	Le Prèv.	Kibwezi; Rangatan Ndari and G. Nairotia, Laikipia; G. Thegana, Kikuyu; Upper Alpine Zone, Kenya; Falls of the Thika-thika; Ndoli, Iveti Mts.; E. of Mangea, Giriama	Africa, Java, Japan,
Roccella Montagnei	Bél.	Common all along Sabaki Valley	India, Java, Australia
Usnea florida, *var.* strigosa	Ach.	Lari lal Morjo and G. Narok, Laikipia; G. Thegana, Kikuyu; Upper Alpine Zone, Kenya; Karati, Kikuyu; Falls of Thika-thika; Sabaki Valley	Wide
Usnea articulata	Hoffm.	G. Thegana, N. Kikuyu; Lari lal Morjo, G. Narok and G. Nairotia, Laikipia; Upper Alpine Zone, Kenya	,,
,, ceratina	Ach.	G. Narok and Ndoro, Laikipia; G. Thegana, Kikuyu; Upper Alpine Zone, Kenya	Cosmopolitan
,, ,, *var.* aspera	Eschw.	Upper Alpine Zone, Kenya	Zanzibar, Abyssinia, Brazil
Usnea cornuta	Flot.	G. Thegana, Kikuyu; Voroni, Thika-thika Valley	E. Africa, India, Europe
,, longissima	Ach.	G. Thegana, Kikuyu	Wide
,, angulata	,,	G. Thegana, Kikuyu; Upper Alpine Zone, Kenya; Sabaki Valley	,,
Parmelia caperata	,,	G. Narok, Laikipia; G. Mwari, Lower Forest Zone, Kenya	,,
,, perforata	,,	Sabaki Valley; Kibwezi and Mto Ndangi, Ukamba; Maka and Ndoli, Iveti Mts.; Lari lal Morjo and G. Nairotia, Laikipia; G. Thegana and Karati, Kikuyu; Falls of Thika-Thika, Kapte Plains	,,
,, perlata	,,	Lower Forest Zone, Kenya, etc.	,,
,, latissima	Fée.	Karati, Kikuyu	S. Africa, China, Oceania, Tropical America
,, lævigata	Ach.	Kurawa, Khilifi River	Wide
,, physodes *var.*	,,	Upper Alpine Zone, Kenya	,,

Name.	Author.	Localities where collected.	Distribution.
Stictina limbata *var.* umbilicariæformis	Nyl. Hochst.	Karati, Kikuyu	Abyssinia, Kilima Njaro
Sticta aurata	Ach.	G. Thegana, Kikuyu	Tropical and Temperate Zones
Peltigera membranacea	Nyl.	Karati, Kikuyu	Wide
Physcia flavicans	D. C.	Karati and G. Thegana, Kikuyu; Upper Alpine Zone, Kenya; W. of Falls of Thikathika; Sabaki Valley	Cosmopolitan
,, parietina	de Not.	Maka, Iveti Mts.; Upper Alpine Zone, Kenya	Wide
,, ciliaris	D. C.	Upper Alpine Zone, Kenya	,,
,, leucomelas	Mich.	Terraces of Lake Suess, Kikuyu Scarp; Gopo lal Mwaru, Guaso Narok, and Guaso Nairotia, Laikipia; Guaso Thegana and Karati, Kikuyu; Upper Alpine Zone, Kenya	,,
,, speciosa	Fr.	Rangatan Ndari, Laikipia	
,, ,, *var.* hypoleuca	Nyl.	,, ,, ,,	Wide
Pyxine cocoes	,,	Gopo lal Mwaru and G. Nairotia, Laikipia	Tropical Zone

PART VI.—FUNGI

BY MISS A. L. SMITH AND JOHN B. CARRUTHERS, F.L.S.

Name.	Author.	Localities where collected.	Distribution.
POLYPOREÆ.			
Fomes lucidus	Fr.	Ngatana; E. of Mangea, Giriama	Wide
Schizophyllum commune	,,	Gopo lal Mwaru, Laikipia	,,
Polystictus elongatus	Berk.	Kibwezi	,,
,, Gregorii	A. L. Smith	E. of Mangea, Giriama	
,, chrysites	Berk.	G. Narok, Laikipia	South America and Cape Town
,, sanguineus	Meyer	Nr. Makongeni, Sabaki Valley	Wide
,, versicolor	Fr.	Lugard's Falls, Sabaki Valley	,,
,, velutinus	,,	Ngomeni	,,

Name.	Author.	Localities where collected.	Distribution.
Polystictus versatilis	Berk.	E. of Mangea, Giriama	Tropics, E. and W. Indies, Brazil
Trametes Sprucei	,,	Ngatana	Cuba, Brazil
Stereum hirsutum	Fr.	Rangatan Ndari, Laikipia; Ngatana	Wide
,, lobatum	,,	G. Narok, Laikipia; Rangatan Ndari, Laikipia; Upper Alpine Zone, Kenya; Makongeni, Sabaki Valley	,,
Favolus Rhipidium	Berk.	Upper Forest Zone, Kenya	Asia, N. and S. America
TREMELLINEÆ.			
Hirneola Auricula-Judae	Berk.	Upper Forest Zone, Kenya; Kibwezi	Wide
GASTEROMYCETES.			
Lycoperdon gemmatum	Batsch.	Kibwezi	Wide
PYRENOMYCETES.			
Xylaria Hypoxylon	Grev.	Upper Forest Zone, Kenya	Wide
Phyllachora graminis	Fuck.	Kapte Plains, W. of Maka	,,

PART VII.—CYPERACEÆ[1]

BY C. B. CLARKE, F.R.S., Pres. Linn. Soc., etc.

Name.	Author.	Name.	Author.
Kyllinga oblonga	C. B. Clarke	Cyperus difformis	L.
,, cylindrica	Nees	,, dichrostachyus	Hochst.
var. β major	C. B. Clarke	,, aristatus	Rottb.
Pycreus flavescens	Nees	,, distans	Linn. f.
,, *forma* Abyssinica		,, bulbosus	Vahl
,, nitens	Nees	,, fissus	Steud.
,, Hildebrandtii	C. B. Clarke	,, Fenzelianus	,,
,, debilissimus, *sp. nov.*		,, rotundus	L.
Juncellus lævigatus	C. B. Clarke	,, ,, *var.* spadicea	Boeck.
,, minutus, *sp. nov.*		,, rigidifolius	Steud.
Cyperus compactus	Lam.	,, exaltatus	Retz.
,, Teneriffæ	Poiret	,, ,, *var.* dives	C. B. Clarke
,, amabilis	Vahl	Mariscus leptophyllus	,,
,, denudatus	Hook. fil.	,, concinnus, *sp. nov.*	

[1] The localities have not been inserted owing to late date of receipt of the list. Many of the species were found in several localities.

Name.	Author.	Name.	Author.
Mariscus coloratus	Nees	Scirpus costatus	Boeck.
,, psilostachys, *sp. nov.*		,, quinquefarius	Buch. Ham.
,, globifer, *sp. nov.*		,, corymbosus	Roth
,, maritimus, *sp. nov.*		Bulbostylis capillaris	Kunth
,, hemisphæricus	C. B. Clarke	*var.* trifida	C. B. Clarke
,, Gregorii, *sp. nov.*		Bulbostylis Schimperiana	,,
Courtoisia cyperoides	Nees	Fuirena glomerata	Lam.
var. β Africana	C. B. Clarke	,, ,, *var.* β Andongensis	C. B. Clarke
Eleocharis marginulata	Steud.	Lipocarpha pulcherrima	Ridley
Fimbristylis diphylla	Vahl	Scleria bulbifera	A. Rich.
,, monostachya	Hassk.	Carex monostachya	Hochst.
,, exilis	Roem. & Sch.	,, Koestlini	,,
Scirpus setaceus	L.	,, Æthiopica	Schkuhr

APPENDIX C[1]

MAMMALIA

By Oldfield Thomas, F.L.S.

Name.	Author.	Localities where collected.	Distribution.
Colobus occidentalis	Rochebrune	Forests on Laikipia, and Kenya	
Herpestes Kaffir	Gmel.		
,, albicaudatus	G. Cuv.	Kibwezi	
Galago *sp.*	...	,,	
Nyctimonus angolensis	Pet.	Golbanti, Tana Valley	
,, pumilus	Cretz.	Ngatana, Tana Valley	
Erinaceus albiventris	Wagn.	Wakilomi, in Maka, Iveti Mts.	
Macroscelides rufesceus	Pet.	Kibwezi	Melanic *var.*
Crocidura 2 *spp.*	...	Lake Kibibi, Rift Valley; Telphusa Swamp, Laikipia; Karati and Thiriati, N. Kikuyu	
,, gracilicauda	Pet.		
Sciurus undulatus	True	Witu	
Gerbilus leucogaster	Pet.	Ndaro, Taita Mts.	
,, pusillus	,,	,, ,,	
Otomys irroratus	Bts.	Upper Alpine Zone, Kenya	
Arvicanthis barbarus	L.	Thiriati, Kikuyu	
Mus *spp. var.*	...	Fort Smith; Njemps Ndogo; Ngatana; Karati and Kithu-Uri, N. Kikuyu	
Homo sapiens	L.	...	Skulls of Wakamba dug up at Tzavo
Aulacodus swinderenianus	Temm.	Ngatana	
Aulacodus gregorianus	Thos.	Guaso Uini, N. Kikuyu	
Hystrix galeata	,,	Various localities	
Procavia shoana	Gigl.	Upper Alpine Zone, Kenya	(Skull)
Equus boehmi	Matschie	Thika-thika; Njemps	
Kobus defassa	Rüpp.	Njemps	
,, ellipsyprymnus	Ogilby	Ngatana	
Æpyceros melampus	Licht.	Njemps	
Cervicapra arundineum	Bodd.	,,	

[1] Most of the remaining zoological collections have been described in separate papers. (See App. A, pp. 387-388.)

GLOSSARY AND INDICES

GLOSSARY OF NATIVE WORDS AND TECHNICAL TERMS

[Those which are explained in the context are excluded. The following abbreviations are used: Bot. = Botany; Geol. = Geology; Hind. = Hindustani; Mt. = Mountaineering; Suah. = Kisuahili]

Agglomerate (Geol.), a volcanic rock composed of coarse blocks of lava and tuff

Alluvium (Geol.), a muddy deposit of recent age, such as that which forms river deltas and plains of rivers

Andesite (Geol.), a lava, intermediate in character between basalt and trachyte. There are many varieties. The name was given to them owing to their abundance in the Andes

Arête (Mt.), the ridge leading to the summit of a peak

Askari (Suah.), a sergeant in a caravan of Zanzibari

Barra (Suah.), open grass-covered country, in contradistinction to cultivated areas and forests

Basalt (Geol.), a lava, dark in colour (generally black), heavy in weight, and containing a small percentage of silica

Bergschrund (Mt.), a great crevasse across a glacier, generally occurring between the unconsolidated snow (or névé) of the collecting fields, and the solid ice of the lower course

Boma (Suah.), a stockade or zeriba, generally made of thorn bush

Bwana (Suah.), master

Chit (Hind.), a paper, generally meaning a cheque or promissory note

Col (Mt.), a high mountain pass

Corrie glacier (Geol.), a small glacier resting on a ledge on a mountain face

Cwm (Mt.), a type of valley

Domo (Suah.), a door, used also by Zanzibari for a mountain pass

Exosmosis (Bot.), the escape of water from a cell by gradual passage through its walls

Fault (Geol.), a break in the continuity of a bed of rock by downward movement on one side of a fracture

Ghee (Hind.), a native butter

Gneiss (Geol.), a rock composed of the same minerals as granite (viz. quartz, felspar, and mica), but in which these are arranged in regular layers

Gopo (Masai), open grass steppes

Hongo (Suah.), tax levied by a tribe for right of passage through their country

Hornblende (Geol.), a dark-greenish or brown mineral, the commonest of the amphibole group. It is a constituent of syenite, hornblende-schist, etc.

Igneous (Geol.), a class of rocks formed at high temperature, whether underground, such as granite, or on the surface, such as lavas

Jemadar (Hind.), native officer acting as adjutant or sergeant-major to Indian troops

Kanzu (Suah.), the long robe worn by Arabs and better-class Suahili

Kiringozi (Suah.), the guide of a caravan

Kiroboto (Suah.), the name of the Arab and Beluchi soldiers in the army of the Sultan of Zanzibar

Masika (Suah.), a rainy season

Mau (Suah.), a canoe
Mhogo (Suah.), cassava, the root of native arrowroot
Munipara (Suah.), a native headman
Mvita (Suah.), one of the native names of Mombasa. The term means "battle"

Obsidian (Geol.), a volcanic glass

Pegmatite (Geol.), an igneous rock which occurs as dykes
Permo-Carboniferous (Geol.), the period in the earth's history in which were deposited the Mountain Limestone, the Coal Measures, and the Magnesian Limestones
Phonolite (Geol.), or "clinkstone," a lava the thin slabs of which are resonant when struck
Porphyritic (Geol.), a term applied to those igneous rocks in which one of the constituents occurs in much larger crystals than the rest
Posho (Suah.), a day's food allowance
Potiss (Suah.), a native food made of boiled flour

Quartz (Geol.), rock crystal, the crystalline form of silica, and one of the commonest rock-forming minerals
Quartzite (Geol.), a rock formed of granular quartz, generally formed by the alteration of sandstones

Rhizome (Bot.), the creeping underground stem of a plant
Roches Moutonnées (Geol.), rock surfaces which have been rounded by the flow of a glacier across them

Safari (Suah.), a journey or a caravan
Sanidine (Geol.), a glassy mineral belonging to the group of felspars found in some lavas
Schist (Geol.), a rock composed of alternate layers of two different minerals. The schists are produced by alteration of other rocks; they are of considerable geological antiquity, and are mostly (according to some geologists, always) of Archean age.
Sedimentary (Geol.), a group of rocks, including those formed by the deposition of sediments, such as limestones, clays, sandstones, etc.
Shadda (Suah.), ten strings of beads, a Suahili measure of value
Shale (Geol.), a rock composed of clay, which splits readily into flakes
Shamba (Suah.), a plantation or cultivated field
Shauri (Suah.), a conference
Silica (Geol.), a compound of silicon and oxygen, which occurs as quartz, flint, etc., and in combination with other materials in rocks

Taenda (Suah.), the order to "start"
Talus (Geol.), loose, fallen material, that collects at the foot of cliffs and in gorges
Trachyte (Geol.), a lava rich in silica
Transpiration (Bot.), the evaporation of water from the surfaces of leaves
Traverse (Mt.), a term used in mountaineering for a climb across the face of a slope from one ridge to another
Tuff (Geol.), a rock formed of small fragments of volcanic ashes and lava

Waschenzi (Suah.), a term applied by Zanzibari to up-country natives. It means "savages." It is sometimes used in a more limited sense for a tribe near Bagamoyo

Yambo (Suah)., the common Suahili greeting

INDEX OF AUTHORS AND PERSONS

INDEX OF LOCALITIES[1]

[1] Owing to the confusion existing in respect to East African geographical names, the longitude and latitude of localities in British East Africa is given. It is hoped that the list will thus be of use as a contribution to a complete index of British East African place-names. In the case of rivers the latitude and longitude are given either for the mouth (m.) or for the principal ford (f.)

SUBJECT INDEX